UNNATURAL PHENOMENA

A Guide to the Bizarre Wonders of North America

He got half a mile start, but looking back at a bend of the road he saw the snake grasp its tail in its mouth, forming a hoop, that swiftly rolled after him. The hoop-snake came on with electric speed. (See "1891: Hoop-Snake," page 205.)

UNNATURAL PHENOMENA
A Guide to the Bizarre Wonders of North America

Jerome Clark

Illustrations by John Clark

A B C CLIO

Santa Barbara, California   Denver, Colorado   Oxford, England

Library of Congress Cataloging-in-Publication Data

Clark, Jerome.

    Unnatural phenomena : a guide to the bizarre wonders of North America / Jerome Clark
       p. cm.
    Includes bibliographical references and index.
    ISBN 1-57607-430-7 (alk. paper) — ISNBN 1-57607-431-5 (ebook)
    1. Ghosts—United States.   2. Curiosities and wonders—United States.   I. Title.
    Bf1472.U6U55   2005
    001.94'0973'09034—dc22   2005011206

09   08   07   06   05   10 9 8 7 6 5 4 3 2 1

This book is also available on the World Wide Web as an eBook. Visit abc-clio.com for details.

ABC-CLIO, Inc.
130 Cremona Drive, P.O. Box 1911
Santa Barbara, California 93116-1911

This book is printed on acid-free paper.

Manufactured in the United States of America

To Lucius Farish,
old friend and fellow traveler on the unnatural landscape

Contents

Acknowledgments

Though the vast majority of accounts that appear in these pages are the product of my own researches, I have borrowed a small number of reports from several esteemed colleagues whom I wish to credit here, with gratitude: Chris Aubeck, Rod Brock, Thomas E. Bullard, Joel Carpenter, John Hudson, and Theo Paijmans.

"WHAT IS THE STYLE OF SPRING LIE THIS SEASON?" asked a one-eyed man, as he dropped down by the exchange editor's side. "Have they retrimmed the interconvertible snake, that breaks up into precincts and comes together at the call of the presiding link?"

"Haven't seen anything of it," replied the exchange editor, plunging his shears into the account of a tornado that lifted a State line a hundred feet and landed a river in a cottonwood grove without disturbing a ripple.

"I'm glad of that," said the one-eyed man, rubbing his hands slowly. "I never liked that lie. It always seemed far-fetched and unwholesome; besides, you couldn't help thinking a frog might swallow a quorum of the snake and not leave enough to transact business. Do you see anything of that lie about the dragon out in Illinois, with wings like a corn-patch and a smell of brimstone? Is the dragon current this season?"

"Haven't struck him yet," answered the editor, eliminating the profanity from a far Western account of a mine explosion, in which forty men were blown through the side of a mountain seven miles thick. "If he's around he's flying very low."

"That pleases me, too," smiled the one-eyed man, stroking his chin. "The only objection I ever had to that dragon was the smell. It never sounded reasonable. If they had said he smelled of brimstone and molasses it might have worked in a girl's boarding-house as a fair spring lie, but they couldn't fit it to do for men unless they perfumed him with blue pills. It was a great mistake. Have you stuck any fresh particulars about the sea-serpent since the thaw?"

"Too early for him," returned the editor, writing a new head-line to an account of a baby who fell out of a window sixty feet and bounded back without split, bruise or chip off. "He'll be around by the 14th of July."

"The 14th of July," murmured the one-eyed man. "I reckon you're right. Yes, you're right; 'with a head like a barrel and eyes like coffee cups; estimated by Mr. William Jennings, of Tobago township, to be at least two-thirds of a mile long.' Yes, that's his date, July 14. I like to read about him. There's always something breezy and fresh about that serpent, though he must be getting on in years now. What's become of the two old people that were born at the same moment, and died within ten minutes of each other, at the advanced age of 104? Ain't they dying this season?"

"Not that I have heard of," rejoined the exchange editor, pruning down a long article on a boy who was cut in half lengthwise by a steam grindstone, and whose recovery was confidently anticipated by the eminent local physicians.

"I hope they haven't quit the business," observed the one-eyed man with anxiety. "I've kind o' warmed up to those two old chums. There was something unanimous about 'em that caught me, and I count on getting around to 'em regularly if I am going to keep my health. May be the backward season has been against 'em. What's the news about the skeleton in the tree with a bag of money tied to its spine? Let's see, he's a spring product, isn't he?"

"No, fall," replied the editor, glancing over a report of a man who had just been relieved of a live lizard that had fed on his kidneys for forty years. "He'll come around about the 3rd of October."

"Just so, just so. I was misled on him. He's an old friend of mine, seems like one of the family, and if they should go over this season without finding him, I should commence to pine. Is there anything new this season, any servant-girls making Greek poetry in their sleep, any live frogs found concealed in a Philadelphia brick and springing eight feet in the air after an imprisonment of eighteen thousand years? Anything of that sort?"

"Nothing," sighed the exchange editor, putting sub-heads into an account of a whale climbing to the top of Absecon lighthouse to borrow a match. "Nothing fresh, except this one about the payment of a church mortgage out in Wisconsin, but that won't be popular among the Christians."

"I suppose not. I suppose not," murmured the old man. "Well, I'm much obliged. So long! It warms me up to see the old ones come around. A man of my age would miss 'em if they let up, and I began to be a little ticklish about the serpent and the skeleton until you explained the dates."

And as he went out the exchange editor turned over an article on an old woman of 90 who was cutting her eighth set of false teeth and fourth head of hair.

—*Butte [Montana] Daily Miner*, July 9, 1882, reprinted from the *Brooklyn Eagle*

Introduction

When I was a sixth-grade kid in the late 1950s, I discovered the science-fiction magazine *Fantastic Universe*—now defunct and long forgotten—in a drugstore in my small Midwestern town. Enthralled and enchanted, I never missed an issue after that. Through an advertisement on its back cover, I learned of the Science Fiction Book Club and hastened to join. As an incentive to potential members, the club offered up three (or maybe it was four) titles for a dollar. One book—a nonfiction work with the intriguing title *The Report on Unidentified Flying Objects*—caught my attention, and I put it on the list. That decision proved to be one of those life-altering moments into which we sometimes stumble, none the wiser at the moment it is happening.

The author, Edward J. Ruppelt, who led the U.S. Air Force UFO project in the earlier part of the decade, treated this already contentious subject in an evenhanded fashion that I found appealing, and still do. Neither credulous nor dismissive, the book acknowledged that some sighting reports were indeed puzzling but that their significance remained to be determined. At the tender age of eleven, I set out to determine their significance for myself by reading everything I could get my hands on about the subject. That turned out to be everything from debunking books to others that detailed meetings with angelic Venusians.

Very early on, I started paying attention to an ongoing *Fantastic Universe* series running under the enigmatic byline "Civilian Saucer Intelligence" (in prosaic reality three New Yorkers, one of whom, Ted Bloecher, became a good friend in my later adult life). "Shapes in the Sky" was an intellectually restrained survey of the UFO phenomenon as it appeared a decade into a controversy that had begun with some much-publicized sightings in the Pacific Northwest in the early summer of 1947, when the concept of "flying saucers" entered popular culture. One piece, in two parts, reviewed the work of somebody I had never heard of, a strange man named Charles Fort (b. August 9, 1874, Albany, New York; d. May 3, 1932, New York City).

Fort, I read, was a failed novelist who devoted the latter half of his life to collecting reports, mostly though not exclusively from newspapers, of what he called the "damned"—alleged occurrences and phenomena that mainstream science tried its best to ignore. When forced by insistent inquirers to respond to them, representatives of the institution would typically give vent to bristling hostility. If the response went beyond mere blistering ridicule, it found expression in "explanations" that, often enough, failed to explain anything at all.

Thus, the opening lines of Fort's first major work, *The Book of the Damned* (1919), said what the book would describe: "A procession of the damned. I mean the excluded. We shall have a procession of data that Science has excluded."[1] The "data" were less the stuff of the paranormal (ghosts, clairvoyance) than of the ostensibly natural and material: things, organic and inorganic, raining from the sky; colored precipitation; mysterious manifestations and presences in space; unknown animals and creatures; puzzling luminosities; freak weather; archaeological oddities; and unidentified lights and structures in the earth's atmosphere. These last made Fort the first ufologist—not the first observer to take note of reports of aerial oddities as such, but undeniably the first to recognize them as a phenomenon—in other words, something reported over decades and geographical boundaries.

But Fort's range of interests was wider than that. His contemporaries and others long before him had studied the supernatural; few, however, had ever focused so specifically on otherworldly phenomena as—at least ostensibly—nuts-and-bolts, flesh-and-blood evidence of visitation in the physical (one might say this-worldly) sense. In three books—the already-mentioned *Damned, New Lands* (1923), and *Lo!* (1931)[2]—he considered the possibility that beings and creatures from other planets (or, more fantastically and less seriously, giant floating land masses in the earth's atmosphere) were interacting with us, notwithstanding our furious denials of what, metaphorically and sometimes literally (in the experiences of some of our fellows), stood in front of our faces. Someone had to coin a phrase to characterize these sorts of manifestations. Like other maverick writers, intellectuals, and radicals of the time (including Theodore Dreiser, Oliver Wendell Holmes, Lincoln Steffens, and Clarence Darrow) an admirer of Fort's humor and iconoclasm, the journalist Ben Hecht proposed "Fortean."

It stuck. The Fortean Society—with which Fort, suspicious of all authority including his own, refused to associate himself—was formed in 1931, a year before its namesake's death in a New York hospital. It folded in 1959, but an International Fortean Organization survived into the 1990s. A popular magazine based in London, *Fortean Times,* has an international readership. The expression "Fortean phenomena" is not exactly in household usage, but it is in the collective vocabulary of a small army of anomaly enthusiasts (as well as anomaly detractors) around the world. The well-regarded science-fiction writer and critic Damon Knight published an admiring biography of Fort.[3] Books on Fortean subjects have been published in the thousands and, if anything, are growing in number as we enter a new century. Fort lives.

What Fort did specifically for me was to inform me that the power of books—especially on young, impressionable minds—is no sentimental fiction. That certain knowledge looks over my shoulder like a ghost as I write my own books and—very carefully—respond to letters from youthful readers. Reading *The Books of Charles Fort,* the omnibus volume Henry Holt and Company released in 1941, changed my life. In many senses, I have never recovered from it.[4]

Though my interest in Fortean phenomena waxes and wanes, Fortean weirdness is never, even in the ebbs, entirely absent from my thoughts. It has accompanied me all through the intellectual journey of my existence, making the world a vaster place, full of shadows and mysteries, and absent the certainties by which so many of my fellow humans—foolishly, in my opinion—guide their lives. Fort insisted that he was a skeptic, if not in the ordinarily understood sense, certainly in a more profound and, if less easy, ultimately more rewarding one. He was not a "skeptic" who declares his faith in the prevailing orthodoxy, but one who held that it is such received wisdom that we ought to be skeptical of because the world and the universe and our very consciousness resist anything more than ephemeral definition by us. "I cannot accept that the products of minds are subject-matter for beliefs," he said, claiming as his mission "the overthrow of old delusions by new delusions."

I didn't understand all of that at the time, of course. I read Fort for the stories and the rich, peculiar atmosphere of his prose: jokey, good-natured, finely tuned to absurdity in the most sober-sided quarters (especially universities, the scientific press, and astronomical observatories), and infused with a sense of our inherent haplessness. One of my favorite of Fort's stories concerns an observation of a city in the sky over Ashland, Ohio, on March 12, 1890. Popular opinion held that it was, Fort wrote, a "vision of the New Jerusalem." Through what Fort judged a dubious application of physics, a scientist explained the apparition as a mirage of Sandusky, some 60 miles away. "May have been a vision of heaven," Fort opined, "and for all I know heaven may resemble Sandusky, and those of us who have no desire to go to Sandusky may ponder that point."[5]

Though rarely, Fort had his explicitly serious moments, too. He wondered if so-called scientific explanations were like "a vampire that lulls consciousness that might otherwise foresee catastrophe. . . . Extramundane vandals may often have swooped down upon this earth, and they may swoop again; and it may be a comfort to us, some day, to mention in our last gasp that we told about this." Through all the wisecracks and calculated outrageousness, Fort linked all Fortean phenomena, even ones that most would think to be a very rough fit, in an overarching theory of interplanetary visitation, the outlines of which can be discerned, if at times only dimly, by an attentive reader. He anticipated pop-culture fads and speculations that would emerge only years and decades later: ufology's extraterrestrial hypothesis, of course, but also ancient astronauts, claims of personal contact with space people (exemplified in the contactee movement of George Adamski, Billy Meier, Heaven's Gate, and many others), cryptozoology (the study of unknown-animal reports), the Bermuda triangle, and more.

In Fort's view, entities from many planets, with different missions and varying degrees of interest in the planet and its inhabitants, have come to the earth over the centuries, and maybe—certain enigmatic archaeological artifacts may hint—far longer. Or possibly, extraterrestrials shaped human evolution and in later prehistory were taken to be ghosts, demons, and gods that survive in folk memory. Over the centuries, secret societies of earthlings have maintained contact with some of the visitors, while other aliens seek open interaction with us. Alien ships in the oceans have been mistaken for sea serpents, and the otherworldly spacecraft have kidnapped persons and ships, thus accounting for mysterious disappearances around the world. Other entities—in the early UFO age some theorists, including at least one employed by the U.S. Air Force's UFO-investigative agency, Project Sign, would call them "space animals"—are living organisms that dwell in space or in the atmosphere. Extraterrestrial forces are responsible for rains of stones, frogs, and other materials.

In later decades, some critics would call this "mixology" and decry efforts—by Fort and many of his rhetorical descendants—to cram all allegedly mysterious occurrences into a single box, when in fact most are probably unrelated, some even probably nonexistent, and a bad fit all in all. One doesn't always know how seriously to take Fort, and one suspects that if we could have seen his face as he wrote, we would have noticed the protruded cheek from the extended tongue, not to mention the merry glint in the eye. Of the notion of esoteric societies and secret meetings with extraterrestrials, no sooner did he propose it than he added, "Of course it sounds—," and indeed it does.

More conservatively, relatively speaking, he conjures up "super-things that have passed close to this earth with no more interest in this earth than have passengers upon a steamship in the bottom of the sea—or passengers may have a keen interest, but circumstances of schedules and commercial requirements forbid investigation of the bottom of the sea." In a letter published in the *New York Times* for September 5, 1926, Fort outlined possible evidence of interplanetary visitation, then remarked, perhaps presciently, "If it is not the conventional or respectable thing upon this earth to believe in visitors from other worlds, most of us could watch them a week and declare that they were something else, and likely enough make things disagreeable for anybody who thought otherwise."

WEIRD STUFF HAPPENING

Anomalies persist, of course, as do ridicule and wonder and speculation. Not a day goes by that UFOs, entity encounters, creatures, ghosts, poltergeists, sky falls, and even weirder things aren't reported somewhere.[6] According to one often-stated but so far as I know undocumented assertion, web searches for UFOs are exceeded only by those for pornography. Every bookstore has a section on anomalies and the paranormal (usually, unfortunately, under the misleading heading "New Age," as if anomalistics were just horoscope-reading and crystal-gazing by another name), and just about any day or night a cable network is running a documentary—an almost predictably cheesy one— on a Fortean subject. None of this has made anomalous claims respectable; if anything, it's pushed them so far out on the margins that any moment now you expect them to fall off the cliff and into a bottomless pit, never to be spoken of again.

That will never happen, it is no doubt needless to say. Though official culture spurns them, at life's ground level, where the unofficial culture and most human beings can be found, the fascination continues, and so do direct experiences of the anomalous. And it is experience, more than anything else, that keeps interest and conviction from fading away as surely as belief in unicorns or nature sprites. Persons who are not liars or lunatics have strange experiences—which have, they insist, the resonance of reality, not of dreams—in no inconsequential numbers, even if most are reluctant to speak of them for the oldest and saddest reason of all: fear of being made fun of. Still, as the folklorist Bill Ellis succinctly puts it, "Weird stuff happens."

The problem is the definition of "happens." Anomalous occurrences may be *experientially* real, but it does not follow that all of them are "real" on an event level. When something intrudes into our line of sight, we naturally assume that the something that seems to be registering in our eyes is a part of this world. We do not consider, in other words, that "experience" and "event" may not be synonymous in all circumstances. An event occurs in the world and leaves its mark there; an experience of the anomalous may register only in consciousness, and however vividly it may affect us, in astonishment or fright, it is likely to prove vexingly difficult to prove to a doubting world that you really did see something that all authoritative opinion declares impossible.

Even the most unlikely beliefs about strange phenomena are based in experiential claims, as I learned years ago when, out of idle curiosity, I began reading about merbeings. (I call them "merbeings" because the word "mermaids" is, one, corny, and two, inaccurate, inasmuch as mermen are nearly as prominent a feature of the tradition.) Like any sensible person, I assumed implicitly that nobody actually *saw* such things; they were extant only in legends, myths, rumors, and tall tales. I was wrong. People believed in merbeings because people reported seeing them, and some of the reports are indeed impressive.

A particularly well-attested series of sightings occurred in the late eighteen and early nineteenth centuries off the coast of remote northeastern Scotland. In 1797, William Munro, a parochial schoolmaster, claimed to have seen, on a rock extending from a beach into the sea, "an unclothed human female" combing its long, light brown hair. The figure, which he observed for three or four minutes in bright sunlight, was fishlike from the waist down. After a time it dropped into the water and looked toward him before disappearing permanently beneath the waves.[7]

Between 1811 and 1814, a spate of similar sightings in the area led to sheriff's investigations and the taking of sworn statements. Reports came from individuals whom the authorities deemed credible, and some were made in daylight from a distance of no more than a few feet. In most cases, there would be what in the next century would be called an "escalation of hypotheses." The witnesses, who were not believers in merbeings, first interpreted what they were seeing in conventional terms—as, for example, a drowning swimmer—until forced to credit what appeared to be before their eyes. They were reluctantly led to conclude that they were observing something far, far out of the ordinary. Some of the incidents involved multiple witnesses.[8]

Some reports of merbeings saw print in such reputable periodicals as the *American Journal of Science,* which in an 1820 issue published an extract from the log book of a ship that had been sailing between New York and France. Over a period of six hours, beginning at two in the afternoon, a strange creature was observed intermittently swimming and diving in the ocean half a ship's length away. It was described as having lower parts "like a fish; its belly was all white; the top of the back brown, and there was the appearance of short hair as far as the top of its head. From the breast upwards, it had a near resemblance to a human being and looked upon the observers very earnestly. . . . No one on board ever saw the like fish, before; all believe it to be a Mermaid. . . . Its hair was black on the head and exactly resembled a man's; that below the arms, it was a perfect fish in form, and that the whole length from the head to the tail [was] about five feet."[9]

Reports like these—there are many more—have been taken seriously enough that scientists and scholars have acknowledged that a solution, not simple dismissal, is called for. The most favored explanation has been that such sightings result from misidentifications of manatees ("sea cows"). Some may well be so explainable. In the more interesting cases, however, manatees are so at variance with what the witnesses claim to have seen that it would be easier to believe they were lying outright.

Furthermore, British folklorists Gwen Benwell and Arthur Waugh, authors of what is generally considered the standard text on mer-being legends and beliefs,[10] determined that three-quarters of the sightings have occurred in areas where manatees do not exist. "It is asking a lot of a maritime race," they remarked, "to believe that sailors, with the trained powers of observation which their own safety, and that of their ship, so often depend, could commit such a blunder" as to confuse a sea cow with a mer-being.[11]

And yet there are excellent reasons underlying conventional skepticism. A half-fish/half-human hybrid is a zoological absurdity. Even if such creatures could exist in any biological sense, more than sighting reports would be needed to attest to their presence. Because so many of the sightings occur in shallow coastal waters, bodies of dead or dying animals of this sort would have washed up on shore, and we would have conclusive physical—and massive—proof of their reality.

On its most immediate and obvious level, the paradox in evidence—where extraordinarily fantastic claim collides with manifestly unsatisfactory prosaic explanation—underscores the futility of the believer/debunker conflict that has raged for centuries. It will not do to retort lamely that yes, manatees may seem an unconvincing identification, but it will have to suffice because, while—to put it mildly—we do not know that merbeings exist, we do know that human beings are sometimes mistaken about what they think they are seeing. Two weak hypotheses, in other words, but one is based on the

conceivable and demonstrable, and sanity demands that we embrace it.

This is in no sense an unreasonable position on its face. Its only problem is its inadequacy, its presumption that any and all anomalous testimony is based on error, at least when it does not come from deceit or delusion. The folklorist David J. Hufford, turning the phrase "traditions of belief" on its head, complained that this sort of approach is based in "traditions of disbelief": the complacent conviction that some things just aren't possible, and therefore that the ordinary care with which scientists and academics address mainstream questions is unnecessary when they turn to fringe claims. Or put another way, any old guess will do.

That leaves those claiming the experiences left to feel foolish, frustrated, or furious, and finally (surely the real point) to keep their mouths shut. Never mind the witness's foolish insistence on his or her good sense and powers of observation; as "a highly respected small landholder" told the Scottish folklorist R. Macdonald Robertson of his own 1900 encounter: "What I saw was *real.* I actually encountered a mermaid."[12]

I am as certain as I am of anything that merbeings do not live and breathe as zoological creatures in our waters. Yet this remains a curious and puzzling business. The traditional skeptical theory—that witnesses were in the grip of popular superstitions that affected their sense of reality—does not hold up for the simple reason that what witnesses report about merbeings and what legends and folklore tell us about such creatures do not match. In traditional lore, merbeings are intelligent entities with supernatural powers. Besides speaking like normal human beings, they even shed their fishy bottoms to live on land and romance or wed dwellers on the land. Yet the merbeings of sightings neither speak nor communicate anything but animal-like sounds, if that. For that matter, they give no particular indication of possessing more than an animal's level of intelligence. Witnesses nearly always refer to what they have observed as an "animal" and assign the pronoun "it" rather than "him" or "her."

Though this sort of testimony raises questions for which there are no immediate answers, it does tell us something we need to know, which is that phenomena that absolutely do not exist in this world can yet be vividly experienced. That understanding relieves us of the responsibility of conjuring up "explanations" that do not explain, and it also warns us to be cautious when we consider comparably extraordinary testimony—say, of visiting extraterrestrial aliens—and we are less certain that the anomalous entities in question are not present in the world. It enforces a healthy skepticism that addresses equally what we know and what we do not know. It does not push us into unsustainable claims of certainty when in truth, as all of human history testifies, this world is full of unfathomable surprises and vexing ambiguities.

I want to stress that I am *not* talking about hallucinations, certainly not in any conventionally understood sense of the term. In my judgment, there is no avoiding the defiance of current knowledge that the most compelling anomalous experiences suggest. These really *are* strange and unaccountable. But much of anomalous experience—the rest is discussed below— "exists" only in memory and testimony. Strictly speaking, we do not know that the monster or entity or apparition has any independent existence outside our experience of it. When we are not "seeing" it, where is it? Is it anywhere? Of the most highly strange anomalous experience, all we can state with a reasonable degree of sureness is that it can be experienced. Just what is being experienced—and how—*that* is the question. Meantime, don't laugh. It can happen to anyone. It could happen to you.

ANOMALY AS QUASI-EVENT

Some strange things, however, leave strange traces, thereby indicating that whatever they are, they are also here.

On the late afternoon of January 8, 1981, at the small French village of Trans-en-Provence, an elderly man heard a whistling sound and saw something "in the form of two saucers upside down" land in an alfalfa field on his property. After resting there for a short time, it ascended to treetop height, then shot off toward the northeast.

This would have been only one more unverified UFO report if not for the fact that the gendarmerie investigated, noted the presence of two concentric circles, and took soil samples both from within the imprints and outside them (as control samples). France's official UFO-investigative agency, associated with that nation's NASA equivalent, took over the probe. Besides reinterviewing the witness, the investigators took the samples into three university laboratories for analysis. At the conclusion, the agency prepared a lengthy technical monograph that documented "deformations of the terrain caused by mass, mechanics, a heating effect and perhaps certain transformations of trace minerals. . . . [A] very significant event . . . happened on this spot."[13]

The Trans-en-Provence event—and event it is, to every available appearance—is the sort of very rare instance in which independent evidence can be brought to bear on an extraordinary claim. Such instances are rare because, for one thing, circumstances do not often collude to render a real scientific investigation possible. Scientific resources are seldom employed to study evidence in cases of anomalies, for various reasons, only one of which is that anomalies are not respectable and scientists sensitive to their reputations prefer to give them wide berth. But a larger, or at least equally important, reason is that physical phenomena hardly ever accompany anomalous experiences. Worse, the content of the latter makes such huge demands on acceptance and (yes) common sense that no sane person will demand a reinvention of the world on no more than testimony, even manifestly sincere and otherwise puzzling testimony, about something that every instinct says cannot be.

There are, in other words, experience anomalies, and then there are, much less often, anomalous events. The two, however, are not always discrete entities. There is a phenomenon— which one morning in October 2003, driving on a rural Minnesota highway as an electric storm loomed, I witnessed personally—called ball lightning. For a long time many physicists and meteorologists denied that it existed, but it is now almost universally accepted as a little-understood, mysterious, but natural phenomenon. Sociologist James McClenon has remarked, "Most scientists would grant higher status to ball lightning than they would to UFOs or psi [ESP and psychokinesis] merely because ball lightning seems to have a greater probability of being explained within the realm of present science." Yet he went on to write that how balls of light are interpreted depends on the context in which they are observed. "The line of demarcation between ball lightning and 'nocturnal lights' is not at all clearly defined."[14]

For example, an "effect that occurred during an electrical storm would be termed 'ball lightning.' . . . Other cases with the exact same appearance but occurring in other circumstances would be called UFOs, psychic lights, or will-o'-the-wisps. . . . Ball lightning is merely the category of experiences that have been observed under situations which point to an explanation, rather than a 'nonmechanistic' explanation."[15] As an example of the latter, McClenon cited an interview with someone who, "as a young child and while alone, observed a ball of light that was approximately one foot in diameter. As it approached, a window magically opened and the light entered the room. It then proceeded through the house and exited through the front door, which also magically opened and closed."[16] Later, when the parents returned, they refused to believe his story but were baffled that the windows, which had been closed and could not have been opened by a child, were open. McClenon also relates a story in which globular balls appeared in the midst of a classic haunting.[17]

Perhaps real, physical, this-world anomalies are a small signal barely audible under the noise of experience anomalies. Perhaps ghostly, intelligent-seeming lights are to ball lightning what UFO abductions are to the Trans-en-Provence case. If experience anomalies are shaped by images and motifs at least familiar to (if not necessarily believed by) those who live within the culture in which such experiences are perceived, they could borrow from authentic physical anomalies at least as easily as from fantasies. Just as both ball lightning and pseudo–ball lightning may coexist, so may unidentified flying objects that leave ground traces and show up on radar screens alongside fantastic pseudo-entities that can be experienced but never demonstrated. Or flesh-and-blood, albeit so far undocumented, animals alongside fantastic beasts whose only reality is an experiential one. A useful rule of thumb may be that the higher the strangeness of the reported encounter, the more likely it is to be unprovable outside memory and testimony.

OZ CONSCIOUSNESS

The British ufologist Jenny Randles has written on what she calls the "Oz Factor," the curiously altered state of consciousness in which many anomalous experiences seem to occur. It is, she said, the "sensation of being isolated, or transported from the real world into a different environmental framework." Cases of it are not hard to find. She rightly observes that is "very common."[18]

Randles cites instances from her own investigative experience. In one case, at 10:15 on a warm summer's night in July 1978, a couple in Manchester, England, observed a dark, disc-shaped object hovering in the sky and emitting brilliant purple rays like spokes from a wheel. After the display, which lasted a minute and a half, "the 'rays' collapsed inwards in sequence, and the object slowly extinguished itself. It was massive, compared with the size of the rooftop opposite." Weirdly, during the sighting the normally busy city streets were empty of both cars and pedestrians. The witnesses told Randles that they felt "singled out" and "alone."[19]

While working on another project a few years ago, I found a never-published report in the files of a UFO-research group. At 5:45 one morning in December 1959, as a man was driving to work on U.S. Highway 99 half a mile south of Proberta, California, his radio started acting up and his car lights dimmed. He pulled over to the side of the road to see what the problem was. While doing so, he spotted a huge, bluish-green, crescent-shaped object—perhaps 90 feet across and 15 to 20 feet thick—hovering about 60 feet above the road a quarter of a mile away. The witness experienced odd, frightening physiological sensations, including the feeling that he "was being sucked up in space toward the object, as by a magnet." Soon afterward, the object flew off to the northeast and disappeared in the Sierra foothills.[20]

Investigators who interviewed the witness in 1968 noted, "Something else that stood out in his mind was that he did not encounter a single car on this highway from shortly before this episode until he was about 1/2 mile north of Proberta. He believes this is the only time in his life that has happened. U.S. 99 W is the main highway from San Francisco to Portland and Seattle. It carries heavy traffic."[21]

On June 8, 2004—which as I write these words is two days ago—two individuals delivering early-morning newspapers at around 3:30 in south-central Winnipeg, Manitoba, reportedly observed a light approaching them, then hovering for a brief period at an altitude "three or four stories high." It then shot off toward the west. A few minutes later, less than a mile from the original location, the same two persons were frightened when they saw an oblong-shaped object, with white lights moving along in its mid-section, ascend from behind a wooded area and move silently toward them. "It was definitely not a plane or anything else," one would insist. "I saw it very clearly, and it was something very strange."[22]

As it swooped over their vehicle, the two noticed something that struck them as equally eerie: The streets were entirely deserted. "Usually, we see some sporadic traffic, even at this hour," one said to investigator Chris Rutkowski. "But there was absolutely no one around at all. It was like a ghost town in the middle of the city."[23]

An hour later, as they continued on their route, what they called a "strange plane," windowless and emitting a peculiar whirring sound, flew overhead. Two more unusual aircraft appeared shortly thereafter. Then, one witness reported, "Suddenly, everything changed and there were other cars around, people walking, and ordinary planes were flying over us."[24]

In this anomalous state of consciousness, the world takes on a kind of magical quality. What seems a familiar landscape proves to be only a psychic replication of that landscape—an, in truth, unnatural landscape—in which the "ob-servers" (if that is the word) stand eerily alone, removed from the real location their physical selves share with their fellow humans and associated activity. In this deeply anomalous state of consciousness, they encounter the profoundly strange. When the state passes, the fantastic object or entity or creature passes, too, surviving only in memory and testimony, leaving no other trace in the world.

EVENT ANOMALIES

One variety of well-attested extraordinary anomaly—many examples of which are recounted in the pages that follow—is the fall from the sky. Things come down to earth, with the rain or out of the blue, everything from fish, frogs, other reptiles, and amphibians to grain, leaves, straw, nuts, stones, and other materials. Not to mention flesh and blood.

This sort of thing has been going on for so long that the first known "scientific" explanation dates to A.D. 77 and Pliny's *Natural History.* Pliny ascribed the belief that living creatures were descending with the rain to popular ignorance of natural processes. The natural process in question was spontaneous generation. Laypeople, he said, just did not understand that when water interacts with mud, slime, and dust it creates frogs and other small creatures. In other words, the frogs (for example) did not come down; the elements out of which they were created were on the ground, and that is where they were generated. Later generations recognized the fallacy of spontaneous generation, but they kept the notion that the animals were there in the first place. The rain just brings or drives them out of their hiding places, it was held, and only the foolish or near-sighted thought they were sailing out of the sky.

No doubt this is true of some claimed falls, but on the whole it is a strained explanation and will not do for the phenomenon generally. For example, there is the fall that took place at Mountain Ash, Glamorganshire, Wales, late on the morning of February 9, 1859, witnessed by numerous residents, including members of the clergy. One observer, John Lewis, provided this account:

"I was getting out a piece of timber, for the purpose of setting it for the saw, when I was startled by something falling over me—down my neck, on my head, on my back. On putting my hand down my neck I was surprised to find they were little fish. By this time I saw the whole ground covered with them. I took off my hat, the brim of which was full of them. They were jumping all about. They covered the ground in a long strip of about 80 yards by 12, as we measured afterwards. [The] shed was covered with them, and the shoots were quite full of them, scraping with our hands. We did gather a great many, about a bucketful, and threw them into the rain pool, where some of them now are. There were two showers, with an interval of about 10 minutes, and each shower lasted about two minutes or thereabouts. The time was 11 a.m. The morning up-train to Aberdare was just then passing. It was not blowing very hard, but uncommon wet. . . . They came down with the rain [as if] in a body."[25]

Over the past century or more, as the reality of falls (some of them witnessed by scientists, in some cases not during storms) became unavoidable, the favored explanation became waterspouts. It has been supposed that fierce windstorms passing over bodies of water have sucked their inhabitants into the air and dropped them on land. Again, there can be no doubt that such occurrences happen and that some falls can be so explained. But once more, the "scientific" explanation does not explain some of the most curious features of the phenomenon, such as its weird selectivity. Almost always, *only one thing falls,* whereas a tornado or cyclone picks up everything and drops it without discrimination. As Fort pointed out, "When a whirlwind strikes a town, away go detachables in a monstrous mixture, and there's no findable record of washtubs coming down in one place, and all the kittens coming down together somewhere else."[26]

More recently, William R. Corliss, the respected, relatively conservative compiler of scientific anomalies, has written:

"The stranger aspects of these falls appear only after reviewing many reports. First, the transporting mechanism (whatever it may be) prefers to select only a single species of fish or frog or whatever animal is on the menu for that day. Second, size selection is also carefully controlled in many instances. Third, no debris, such as sand or plant material[,] is dropped along with the animals. Fourth, even though saltwater species are dropped, there are no records of the accompanying rainfall being salty. All in all, the mechanism involved is rather fastidious in what it transports. The waterspout or whirlwind theory is easiest to swallow when the fish that fall commonly shoal on the surface in large numbers in nearby waters. It is much harder to fit the facts when the fish are from deep waters, when the fish are dead and dry (sometimes headless), and when animals fall in immense numbers."[27]

Moreover, the presence of a waterspout is more often asserted than demonstrated—in other words, simply inferred from the fact of the fall, a tautology that can be expressed as follows: How do we know there was a waterspout? Because fish (or whatever) fell. Why did fish (or whatever) fall? Because there was a waterspout.

A final, seemingly insurmountable difficulty with conventional hypotheses is the sheer amount of material or number of animals that figure in some accounts. Tens of thousands of toads—and only *young* toads—fell on Brignoles, France, on September 23, 1973, in a sudden storm. In September 1922, another French village, Chalon-sur-Saone, reported a young-toad fall that lasted *two days.* Sometimes, as Corliss remarks, the species are not recognizable as anything local.

Another phenomenon, albeit a quite different one, that passes more or less unchanged through the ages is the poltergeist (a German word meaning "noisy ghost"). As early as 1599, the Jesuit writer Martin Del Rio referred to "those specters which in certain times and places or homes are wont to occasion various commotions and annoyances. I shall pass over examples, since the thing is exceedingly well known, and instances can be read in older and more recent authors."[28] Poltergeists are assorted with unexplained raps and blows, movement of objects, destruction of property, and—more rarely—fire-lighting, apparitions (usually glimpsed only briefly; ordinarily, poltergeists are invisible), and biting. The phenomenon, with some variation, has persisted relatively un-changed through recorded history, shaped somewhat by cultural expectation but not so radically that the core phenomenon is not rec-ognizable to a modern inquirer. One authorita-tive source[29] identifies the first recorded case as having occurred in Ravenna, Italy, in A.D. 530, another[30] in Germany in A.D. 355. The noted folklorist and scholar of the occult Leslie A. Shepard has observed, "The poltergeist is by no means indigenous to any one country, nor any particular period."[31]

Attempts to explain the poltergeist have changed over time, however. For many cen-turies, two explanations predominated: sinister supernatural agencies (demons and witches) and fraud. In the twentieth century, as the phenom-enon attracted the attention of scientifically ed-ucated investigators, a new view emerged that poltergeist phenomena come from unconscious psychological processes, typically from a dis-turbed adolescent, who employs mind-over-matter—psychokinetic—powers to wreak havoc. Skeptics disinclined to accept the exis-tence of so speculative a concept have continued to argue that the entire phenomenon, wherever and whenever it occurs or has occurred, is no more than a series of hoaxes. Which side one comes down on in this debate is wholly de-pendent on one's personal vision of the possible. Each is in its own way an extraordinary claim.

FORT'S LANDSCAPE

The book you are holding in your hands owes its inspiration to a longtime ambition of mine: to go back to the nineteenth and early twentieth centuries and conduct a comprehensive exami-nation of Fort's sources, or what could have been his sources. Fort's treatment of the phe-nomena that engaged his attention is often un-satisfactory—sometimes amounting to no more than a sentence, and sometimes less than that. For years, when time, money, and circumstances permitted, I would visit newspaper archives in various places and dig up what I could find.

If nothing else, that experience forced me to recognize the magnitude of Fort's accomplish-ment. He had some help from journal indexes, but for the most part he was engaging in a hit-and-miss—and hugely time-consuming—enter-prise. He perused many hundreds of newspapers one issue at a time. It should be noted, too, that newspapers of earlier ages often carried print so tiny that it almost defies reading; no wonder he lost his sight for a period of time.

The Internet age has made concentrated re-search into old newspapers doable. I didn't fully understand this until the ufologist Barry Green-wood, who himself has a long-standing interest in historical UFOs and UFO-like anomalies, alerted me to the possibilities of Internet re-search and told me how it could be done. What followed was a solid year of long days, usually seven of them a week, doing searches for a range of anomalous claims and reports that would have caught Fort's attention if he had come across them. (Just trying to think up productive search words was a headache in itself.) Over-whelmingly, the reports that follow are not in Fort's books, and 90 percent or more have never been between book covers before.

I joined a small, private e-mail list, Magonia Exchange, ably directed by Chris Aubeck of Spain and Rod Brock of Washington state. Magonia Exchange exists for the benefit of re-searchers doing archival research into historical anomalies. (Magonia is Latin for "magic land.") I confined my own searches to the United States—a wider search would have immeasur-ably complicated an already exhausting en-deavor—and found material in forty-nine of the fifty states; only Hawaii is unrepresented here. It does not follow that some states are more mys-terious (or imaginative) than others. In that re-gard I am pretty certain they are all just about the same. The disparity between, say, Ohio and Delaware is that Ohio newspapers so far are on line in greater numbers than their Delaware counterparts.

There were long, unproductive, tedious stretches that taught me, for example, all the ways, often inane and hackneyed, in which nineteenth-century Americans used words such as "creature," "beast," "supernatural," "un-canny," "weird," "ghost," and the like. I also learned, perhaps more usefully, an appreciation for the relatively enlightened discourse of our time, where open expressions—much less casual ones—of racism and white supremacy are happily unacceptable. In that regard, the nineteenth-century American press is sufficient to generate an unexpected appreciation for political correctness.

By the time the project was nearing completion (augmented on infrequent occasion by non-newspaper, secondary, and more recent sources when they seemed useful for one reason or another), I came to the conclusion that the material on anomalies and fantastic claims would be best presented exactly as written, and I have done that for the most part. The wording gives the reader a clear sense of what exactly was being written about, as paraphrase does not. For the most part, I have written entries in my own words only when there was no other way to relate the story, as when it had a long, complicated history that needed my intervention for it to make sense. Two examples are my treatments of two notorious episodes, the Cardiff Giant (New York) and the Kensington rune stone (Minnesota). The former was recognized as a hoax in its own time, but the latter still has its adherents—though, for reasons I explain, I think they are wrong.

Mostly, almost entirely, this book recounts discrete (alleged) experiences and events. The reporting usually leaves much to be desired, though that sort of journalistic failure was not confined to extraordinary phenomena. Professional journalism as we understand it today from our consumption of print and electronic media is very much a product of the twentieth century, and not even the very early twentieth century into which this book ventures. Sometimes it is not even entirely clear what sort of phenomenon is being claimed. Sometimes it is fairly or blatantly apparent that somebody—the "witness" or the reporter or an editor—is just making something up to fill space, amuse readers, or both. There are not a few stories here that I flat-out disbelieve and regard as purely imaginary in the quotidian definition of the adjective. And so should you. Though I am convinced that sincere persons believe they have glimpsed fantastic creatures, I am just as convinced that they do not experience them as, say, Indiana farmer Jacob Rishel is said to have experienced

his encounter with a dangerous dragonlike beast in 1879.

Of course I can't *prove* that Rishel, if he indeed existed (I was unable to find any record of him), was, shall we say, stretching the truth a millimeter or two. After all, as with comparably unlikely entities as merbeings and fairies, people once believed that it was possible to encounter dragons and that they were best avoided. A pamphlet from 1669, *The Flying Serpent or Strange News out of Essex,* recounts many sightings, mostly from the previous spring and summer, of a serpent "8 or 9 feet long, the smallest part of him about the bigness of a Mans leg, on the middle as big as a Mans thigh[;] his eyes were very large and piercing, about the bigness of a Sheeps eye[;] in his mouth he had two rows of Teeth which appeared to their sight very white and sharp, and on his back he had two wings indifferent large but not proportionable to the rest of his body, they [two witnesses] judging them not to be above two handfuls long, and when spreaded, not to extend from the top of one wing to the utmost end of the other above two feet at the most, and therefore all too weak to carry such an unwieldy beast."[32]

Keep that description in mind, because you're going to run into it on the American landscape several hundred years later. And smaller dragons, too, such as the "small winged snakes" mentioned by Herodotus in the fifth century B.C.

By the nineteenth century, though, dragons—seldom called by the name by then—could still be "seen" and experienced, but in lesser form: as big snakes,[33] with or without wings; monstrous reptilian forms in oceans, lakes, and rivers; or sky serpents. But as befits their shadowy presence in this world, as anomalies of the imagination rather than of nature, they may frighten, but they are not truly menacing. They are, one might say, not real enough to do any harm except to the observer's peace of mind. I am prepared to accept testimony to unusual experiences and perceptions, but not to the this-world quality of fantastic creatures and beasts. You can "see" a merbeing; you can't capture it and put it in a fish tank for everybody who wanders by to take in. You can "see" a dragon, but it can no more kill or injure you than you can shoot it, stuff it, and deliver it to a natural-history museum.

The wonders and marvels of the America of the nineteenth century and the early twentieth are caught between the medieval and the modern. Armies in the sky compete for space in the atmosphere with miraculous airships piloted by mysterious mechanical geniuses—or by Martians or by divine beings—on their way to becoming UFOs. Apparitions of machines—ghost trains—speed down phantom tracks, eerie reminders of the many employees and passengers who were killed as America rapidly and recklessly industrialized, and a link between supernatural traditions of the past and an ever more sophisticated technology that would define the future. The "wild men" of past centuries wander through, somewhere between their original home in the wilderness of the psyche to rebirth in the next century as Bigfoot and Sasquatch. Mediums find themselves communicating not just with the dead but with newly arrived extraterrestrial visitors. Men in black menace people, but not, as they would in the next century, UFO researchers and witnesses.

Meantime, familiar folkloric figures show up on the New World landscape: black dogs (of the sort memorialized in Robert Johnson's classic 1938 blues "Hell Hound on My Trail"); vampires and bedroom invaders; giants (in the form both of ghostly entities and of skeletons dubiously claimed to be of a vanished race); centaurs and minotaurs; monsters of all kinds; angels and other divine entities; omens in nature; and more.

In the pages to come, expect to encounter the full range of phenomena defined as Fortean, from out-of-the-ordinary but natural occurrences to puzzling event and experience phenomena to outright yarns and fabrications on fantastic themes. They include:

Animals and birds of unknown kind
Anomalous airships
Archaeological oddities
Armies in the sky
Astronomical oddities
Ball lightning

Big snakes
Creatures of all kinds
Eccentric personalities
Entombed animals
Falls from the sky
Figures in glass
Flying entities
Ghost ships
Ghost trains
Ghosts, human and animal
Giants
Hairy bipeds
Hoaxes of the fantastic
Lake and river monsters
Landscape anomalies
Men in black
Merbeings
Meteorological anomalies
Meteors and meteorites of unusual character
Mirages of extraordinary character
Misperceptions
Nocturnal lights and mysterious luminosity
Omens in nature
Phantom troops
Poltergeists
Psychic communications from otherworldly entities
Psychological anomalies
Religious visions and miracles
Reptiles encased in human stomachs
Sea serpents
Sky serpents
Sounds of uncertain nature and origin
Vampires and other bedroom intruders

Though this book covers the bulk of the time period documented in Fort's works, there is, as already stated, little overlap. There is also no question that an enormous amount of material on wonders and marvels still lies buried and unrecovered in the thousands of newspapers from the big cities and small towns of a vanished America. Fort only started the search. This book should be seen as a continuation, but by no means the completion, of Fort's pioneering work.

My own inquiries confirm, of course, that weird stories were out there in no small number. That was no surprise. After all, no one, at least to my knowledge, has ever accused Fort of inventing reports and sources. What is absent, however, is just about any evidence of something that we would now call the UFO phenomenon. Certainly there are hints of it, most prominently in a handful of puzzling nocturnal-light sightings, but the more dramatic stories have a strange way of starting in a familiar way, then abruptly heading off in a startling direction that radically redefines their character. For a couple of especially striking examples, see the 1893 "colored discs" story from Kentucky and the 1907 Tennessee landing report.

I have concluded, to my satisfaction at least, that Fort was almost certainly wrong to think that he had uncovered evidence of interplanetary visitation in these and comparably ambiguous, inscrutable allegations. I also think that subsequent generations of ufologists have mistakenly read his books as if they set forth a catalog of UFO reports essentially indistinguishable from the "flying saucers" of 1947 and after. But even if his interpretations are suspect, Fort surely succeeded in showing how oddly vivid our dreams are, and how peculiar our experiences. The largest lesson may be that one doesn't have to conjure up other worlds to find that ours alone harbors yet enough mysteries to engage scientists, scholars, and curiosity-seekers—anyone, in fact, open-minded enough to see them all around us—for decades and centuries to come. That landscape only looks unnatural to those of us who are casual and ignorant observers. To those who are willing to pay attention, a new age of exploration awaits.

Jerome Clark

Notes

1. Charles Fort, *The Book of the Damned* (New York: Horace Liveright, 1919), 7.

2. Charles Fort, *New Lands* (New York: Boni and Liveright, 1923); Charles Fort, *Lo!* (New York: Claude Kendall, 1931). Fort's fourth and final book, *Wild Talents* (New York: Claude Kendall, 1932), has a different subject: extraordinary alleged abilities associated with individual human beings.

3. Damon Knight, *Charles Fort: Prophet of the Unexplained* (Garden City, NY: Doubleday, 1970).

4. Charles Fort, with an introduction by Tiffany Thayer, *The Books of Charles Fort* (New York: Henry Holt, 1941).

5. Ibid., 459.

6. As I write, in early June 2004, the Chilean press has printed a photograph, taken in a Santiago park, of a fuzzily focused but nonetheless discernible humanoid shape, and its Syrian counterpart reports "eyewitness accounts" of a red-colored entity, with a "peasantish" appearance, that ran up a tree and then "rushed skyward with incredible swiftness after leaving a wake of smoky white bubbles."

7. Gwen Benwell and Arthur Waugh, *Sea Enchantress: The Tale of the Mermaid and Her Kin* (New York: Citadel, 1965), 111–113.

8. Ibid., 113–117.

9. "Mermaid," *American Journal of Science* 1, no. 2 (1820): 178–179. Cited in William R. Corliss, ed. *Incredible Life: A Handbook of Biological Mysteries* (Glen Arm, MD: Sourcebook Project, 1981), 329–330.

10. Benwell and Waugh, *Sea Enchantress*.

11. Ibid., 14. Those inclined to relegate such sightings to an earlier, presumably more ignorant, superstitious age are referred to Chapter 15 of Benwell and Waugh's *Sea Enchantress* for some intriguing twentieth-century reports.

12. The man, identified as Alexander Gunn of Sutherland, told Robertson that one day in 1900 he and his dog had gone to the seaside looking for a lost sheep. The dog suddenly started howling, and Gunn glanced up to see—on a ledge only six or seven feet away—a reclining figure, human-sized, with reddish-yellow curly hair and blue-green eyes. It was, of course, a mermaid. Both Gunn and the creature were frightened, and the man fled. He thought it was waiting for the next tide to come in and float it out to sea. Gunn swore to the truth of the encounter till his death in 1944. See Benwell and Waugh, *Sea Enchantress,* 261.

13. For a fuller discussion of this remarkable incident, see *Enquete 81/01: Analyse d'un Trace* (Toulouse, France: Groupe d'Etudes des Phénomènes Aérospatiaux Non Identifiés (GEPAN), March 1, 1983). A partial translation appears as "GEPAN's Most Significant Case," *MUFON UFO Journal* 193 (March 1984):

3–16. See also Michel C. L. Bounias, "Biochemical Traumatology as a Potent Tool for Identifying Actual Stresses Elicited by Unidentified Sources: Evidence for Plant Metabolic Disorders in Correlation with a UFO Landing," *Journal of Scientific Exploration* 4, no. 1 (1990): 1–18.

14. James McClenon, *Deviant Science: The Case of Parapsychology* (Philadelphia: University of Pennsylvania Press, 1984), 60–63.

15. Ibid.

16. Ibid.

17. Ibid., 61–62.

18. Jenny Randles, *UFO Reality: A Critical Look at the Physical Evidence* (London: Robert Hale, 1983), 100.

19. Ibid., 98–100.

20. Paul Cerny, Kenneth Tice, and Robert Staver, *UFO Sighting Report* (1968). In files of National Investigations Committee on Aerial Phenomena, archived by J. Allen Hynek Center for UFO Studies, Chicago.

21. Ibid.

22. Posting by Chris Rutkowski on UFO Updates e-mail list, June 9, 2004.

23. Ibid.

24. Ibid.

25. From a March 8, 1859, letter written by John Griffith and published in *London Times,* March 10. Cited by Robert Schadewald, "The Great Fish Fall of 1859," *Fortean Times* 30 (Fall 1979): 39.

26. Fort, *The Books of Charles Fort,* 545.

27. William R. Corliss, ed., *Handbook of Unusual Natural Phenomena* (Glen Arm, MD: Sourcebook Project, 1977), 477.

28. From *Disquisitionum Magicarum.* Cited in Alan Gauld and A. D. Cornell, *Poltergeists* (London: Routledge and Kegan Paul, 1979), 3.

29. Gauld and Cornell, *Poltergeists,* 366.

30. D. Scott Rogo, *The Poltergeist Experience* (New York: Penguin, 1979), 41.

31. "Poltergeist," in *Encyclopedia of Occultism and Parapsychology,* 3d ed. (Detroit: Gale Research, 1991), 1305.

32. Cited in George Monger, "Dragons and Big Cats," *Folklore* 103, no. 2 (1992), 203.

33. When you encounter big-snake stories in the text, keep in mind that according to conventional herpetology, the biggest snake on earth is the reticulated python, a tropical creature that can grow to about 30 feet. Herpetologists have yet to document snakes of larger size.

Alabama

1855 and Before: Tracks of Giant Snakes

THE LONG DROUGHT, it would appear, is sorely felt by the snake tribe, forcing their snakeships to travel over the country in search of water. A short time back a rattlesnake was killed near the head of Evans' mill pond, seven feet in length, and eighteen inches in girth. The track of another one has been seen in the same neighborhood, supposed to be much larger, his track measuring eight inches across. The track of a still larger one has been seen regularly in August of every year for the past 37 years crossing the Dehaba road, about four miles west of this [Greenville], in the direction of a swamp, which is supposed to be his regular hiding place. The track of this monster is thirteen inches across. Repeated attempts have been made to capture this patriarch of the snake tribe, but without success. He crossed on his usual track last August."

Source:

"Monster Snake," 1855. *Janesville* [Wisconsin] *Gazette* (July 7). Reprinted from the *Greenville Alabamian.*

1870: Rain Mystery

FOR SEVERAL DAYS THERE has [*sic*] been mysterious and vague rumors of a most remarkable meteorological phenomenon out at the Catholic graveyard on Stone street [in Mobile], above Three Mile Creek. It is asserted by those who say they have seen it, that for the last five days a gentle shower has fallen continuously on the lot of the Lemoine family in which are buried Mr. Victor Lemoine and many others of his family. With a view of getting at the facts of this most extraordinary affair, we had last night

an interview with Mr. Louis B. Lemoine, employed at Asa Holt's, a son of the deceased Victor Lemoine who died in 1851, who related the following startling particulars:

"'Having heard that it was reported that it had been raining for several days on the enclosed ground which forms my family burying ground in the Catholic burying ground on Stone street, above the Three-mile Creek, I drove out there last evening to satisfy myself, and to my intense astonishment, I saw that a column of rain was coming down without ceasing, which, although hardly powerful enough to lay the dust, was enough to wet the hands or any article, and at times it rained quite hard. The volume of rain fell inside of the enclosure, and nowhere else, as the weather was and has been bright and clear, all the time, during the five days the rain has been falling on those graves. There are thirteen of my family buried in the lot of ground upon which it has been raining. My mother, brother and sisters visited the spot yesterday and the day before to satisfy themselves about the truth of the matter, and declare that they too saw this wonderful phenomenon. It has been seen by over two hundred persons. I took a friend with me when I visited the spot, who also saw the rain falling as described. Mr. John Rosset, keeper of the cemetery, told me that the rain had commenced falling in heavy drops about five days ago. I am willing to take my oath as to the truth of this statement.'

"So incredible did this extraordinary affair seem, that those who saw it several days ago refrained from stating or asserting what they had seen, for fear that not only their veracity but sanity would be questioned, and it was not until a number of gentlemen of the first respectability had seen and reported the result of their personal observations, any credence was attached to the truth of the matter. Take it altogether, it is certainly the most astounding and miraculous atmospheric wonder that has ever been witnessed in this part of the world, and will doubtless afford abundant food for thought, research and observations not only among scientific men, but among all classes. There are so many who vouch for the truth of Mr. Lemoine's statement, and his character for veracity is such, that there can be no longer any doubt of the fact that it has been raining for the past five days on the graves of his kindred."

Source:

"Miraculous Meteorological Phenomenon," 1870. *Titusville* [Pennsylvania] *Morning Herald* (November 12). Reprinted from the *Mobile* [Alabama] *Register,* November 3.

1882: Falling Perch

WE HAVE OFTEN HEARD of fish being rained from the clouds, but have never seen an eyewitness to the phenomenon until last week. S. P. Thomson, one of our leading prairie planters [in Kansas], assures us that on the 12th of October last he saw three fish of the perch variety fall into his front yard during a shower of rain. Mr. T. gathered up the fish, which were still alive, notwithstanding their long and rapid journey through space toward the center of gravity, and, after satisfying himself that they were only ordinary perch, placed them in an adjacent stream, where they swam off as lively as if terranean streams and not celestial vapors were their natural element. Mr. Thomson says he is not natural philosopher enough to explain this strange occurrence, but he is absolutely certain that the fish did fall from the clouds."

Source:

"A Strange Sight in Alabama," 1882. *Atchison* [Kansas] *Globe* (November 16). Reprinted from the *Union Springs* [Alabama] *Herald.*

1888–1889: Ghost Breaks Up Marriage

A GHOST HAS CAUSED a divorce suit and broken up a once happy family in Blount county, Alabama. About five years ago James Martin married a Miss Noel, one of the belles of the county. The young couple went to live at the old Martin home and all went well until about a year ago. Mrs. Martin, naturally very timid, heard a ghost rambling through the old house one night and was badly frightened. She told her husband about it, but he could hear nothing, he said.

"From that time it became a nightly visitor at the Martin home. Mrs. Martin wanted to leave the old house at once, but her husband objected, declaring the strange noises were made by rats. Several times Mrs. Martin, so she says, saw a white robed figure wandering through the old house, and soon her nerves and health began to give way under the strain. She was finally prostrated by her fear of her ghost, and went to the home of her parents to recover her health and strength. Fear of the ghost overcame love of husband, and Mrs. Martin refused to live with him again. Martin tried in vain to induce his wife to return to the haunted house to live, but she refused, and he filed a suit for divorce on the ground of abandonment."

Source:

"An Alabama Ghost," 1889. *Marion* [Ohio] *Daily Star* (March 5).

1889: Scaring the Horse

A GHOST HAS APPEARED near Akron, Ala., at a point on the railroad where a man was run over and killed last summer. A ghostly figure in white, with arms extended, was seen there by a young couple who were out riding the other evening. Both were badly scared and so was the horse."

Source:

"Curious Things of Life," 1889. *Marion* [Ohio] *Daily Star* (May 22).

1897: Snake from Toe

A YOUNG NEGRO GIRL living on Birmingham street [in Gadsden] became strangely afflicted some time ago, complaining of acute pains all over her body. Yesterday, while her mother was rubbing her lower limbs her great toe burst and something resembling a small snake about three inches long popped out. A physician procured a magnifying glass and examined it, pronouncing it an embryo snake. The girl seems to be in terrible agony and claims that she can feel the things crawling about in her body. Her flesh is wasting away and the doctors say she will die.

"No one seems to understand her case or to be able to render her any assistance. Hundreds of people have visited the house and the statements can be thoroughly substantiated by reputable people."

Source:
"Negro Girl Full of Snakes," 1897. *Atlanta Constitution* (January 4).

1900: Flying Cross

RESIDENTS OF OZARK, a village in the southeast part of the state, reported that at 11:30 on Saturday night, May 5, a cross of fire appeared in the western sky and sailed eastward. Even in the late-evening darkness it was brightly colored, like an illuminated cloud. The long piece seemed to be 10 feet long, the shorter one 3 feet long.

At an estimated altitude of 1,000 feet, it moved rapidly and emitted a trail of sparks in its flight. Onlookers predictably took it to be a sign of the end of the world.

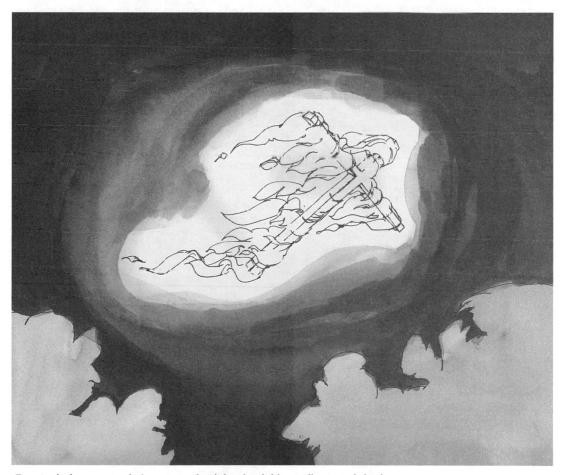

Even in the late-evening darkness it was brightly colored, like an illuminated cloud.

Source:
"Cross of Fire in the Heavens," 1900. *Fort Wayne* [Indiana] *News* (May 7).

1900: Shot by a Light

On the hot and sticky evening of September 3, T. T. Roche and his sister sought relief in their backyard in Mobile. Roche lay in a hammock under a big walnut tree, and his sister seated herself on an armchair in the middle of the yard. They were positioned so that they faced each other from a distance of only a few yards.

As the two chatted, Roche noticed something coming out of the sky and descending directly over his sister's head. At first the thing was "quite brilliant," he would report, but it grew fainter as it fell. Now it had the appearance of a very large firefly. Just as he opened his mouth to warn his sister, she jumped to her feet and screamed that something had shot her in the arm.

"I sprang up and ran to her," Roche wrote to a science magazine, "and on examining the arm we found a small red spot just above the elbow, which rapidly grew larger, and rose, a blister forming almost at once. It was exceedingly painful and had all the appearance of a fresh burn. The arm swelled, and the place made quite a painful sore. At present (four days later) it looks exactly like an old vaccination, being quite deeply pitted. On examining the dress, we found the sleeve scorched and burned through, in a tiny hole, just above the spot where the wound is."

Source:
"Struck by a Fragment of a Meteor," 1900. *Scientific American Supplement* 49. Cited in William R. Corliss, ed., 2001. *Remarkable Luminous Phenomena in Nature: A Catalog of Geophysical Anomalies.* Glen Arm, MD: Sourcebook Project, 146.

1910: Rain without Explanation

Reports were brought here today [October 27] of a singular phenomenon of nature at Heflin, Cleburne county, this state, which has attracted the attention of the population of the entire countryside. According to the reports, an incessant rain has been falling on a plot of land at Heflin no larger than 7 feet square for the past week, while everywhere around the atmosphere was perfectly dry. Residents of Heflin have been unable to give any explanation of the phenomenon."

Source:
"Incessant Rain Falling on Spot 7 Feet Square," 1910. *Atlanta Constitution* (October 28).

Alaska

1800 and After: Little Men

ACCORDING TO A LEGEND handed down to Lois Foster through her great-grandmother and grandmother, three little men came to Koyuk, an Inuit (Eskimo) village, in around 1800. The great-grandmother said that they arrived via a "silvery looking disc that sailed through the air." The grandmother, now quite elderly, claimed to have seen the little men during her childhood in the early 1910s.

She told her granddaughter that the little men were very strong, able to lift heavy logs and dead deer without assistance. They were stranded in the village because something had gone wrong with their disc and they could no longer go home. Eventually, they learned to speak the native language and so were able to communicate with villagers.

This story is probably a conflation of traditional Inuit beliefs in little people—beliefs in such entities are nearly universal—and modern UFO lore.

Source:

Chorvinsky, Mark, 1990. "Our Strange World." *Fate* 43, 1 (January): 17–19, 22–23, 26, 28.

1889 and After: City in the Mist

SOME ONE REVIVES EVERY FEW years the legend of the Silent City, seen as a mirage over the Muir glacier, in Glacier bay, Alaska, 150 miles above Juneau. What professes to be a photo-engraving of the city, with affidavit attached, was published and sold some years ago, and at least half a dozen white men profess to have seen the city, while the natives of the Alaskan coast cherished the tradition of its existence. The mirage is usually visible at about this time of the year [summer]. One man declares that he saw it on July 5, 1889, and two others, it is declared, have made affidavit that they saw it three days before that date. John M. White, a Virginian who after 10 or 15 years spent in various parts of the west has settled down in Philadelphia, solemnly declares that he saw it on June 21 some years ago.

"Mr. White's account of the phenomenon is circumstantial in the highest degree, and he joins to it a theory as to what the city is and where it lies. He declares that he studied the mirage for nine hours through a powerful glass as it was spread above the glacier on the side of Mount Fairweather. He affirms that the city is walled; that its houses are battlemented and the chimneys surmounted by chimney pots; that within the walls there is a tall monument surmounted by the sculptured figure of an Indian in full headdress and feathers. His glass revealed to him some of the inhabitants, men in knee breeches and jackets. The only beast visible was a donkeylike creature, with a body as large as that of a horse.

"The mirage appeared at first about 11:30 a.m. as a mist, and out of this rose the towers and battlements of the city as did those of ancient Troy. By noon the city was as clearly outlined as New York is from the Jersey heights.

"Mr. White rejects the various guesses that the phantom city is Antwerp, Montreal or Salt Lake City. Its architecture is unlike that of any other city he has seen. That it is a real city he is certain from the fact that he has seen three photographs taken of the mirage, one of which shows a tower rising amid the houses and a later one the same tower finished. He believes that it is a mirage of a city at the north pole on the edge

The mirage appeared at first about 11:30 a.m. as a mist, and out of this rose the towers and battlements of the city as did those of ancient Troy. By noon the city was as clearly outlined as New York is from the Jersey heights.

of the traditional open polar sea. He believes that when the sun is at its highest northern point, as it is on June 21, the mirage of the arctic metropolis is reflected to the point where it appears over the Muir glacier. The legend of the Chilcats of Alaska supports this theory. They say that many centuries ago, when Alaska was a warm and densely peopled country, there came from the north through the ice barrier a savage people fully armed who laid waste the region and put its inhabitants to the sword. These savage warriors he believes to have been the ancestors of the American Indians, and he is convinced they left behind a warm region about the pole, where the remnant of their people continued to develop and at length built the metropolis seen on St. John's day in mirage above the Muir glacier.

"The pictures purporting to have been made from photographic negatives of the mirage represent an ordinary modern city without walls or battlements, but with spacious, comfortable looking houses, surmounted with broad chimneys and interspersed with trees. In fact, they look like photographs of wash drawings made by an artist that was not too careful to follow the details of the legend."(1)

☞

"If any of the thousands of tourists and gold hunters in the Alaskan region shall happen to catch a glimpse of the beautiful and wonderful cloud city which seems to hover in the neighborhood of Mount St. Elias and of Glacier bay, they may consider that one of the rarest sights on this globe has been vouchsafed them. The

Italian Prince Luigi and his men saw it last summer [1897] when they climbed Mount St. Elias. In July, 1889, Professor Richard C. Willoughby saw and photographed the phenomenon. The same summer of 1889 Mr. L. B. French, a traveler in Alaska, was fortunate enough to behold the city mirrored in the clouds. He said it looked large enough to contain 100,000 inhabitants."(2)

☞

"Professor Otto J. Klotz, chief astronomer of the department of the interior at Ottawa, returned from Alaska by the Corona, and came to Victoria yesterday [October 3]. From here he goes to California, and returns to Victoria and goes back to Ottawa. During his visit to Alaska, Professor Klotz visited the famed Muir glacier, which has attained celebrity by accounts of a 'Silent City,' a mirage reported seen in the ice of the city of Montreal. The story of the mirage in Muir glacier Professor Klotz pronounces a humbug. He says a mirage could not be seen in the ice as described. It is [a] phenomenon of the air. Many fine reflections are obtained in Northern waters of wild scenery and fantastic formations of ice, but such a thing as the reflection of a city thousands of miles distant would be an utter impossibility, and for a mirage to be seen in water or ice would be equally absurd. He does not consider the atmosphere of the far North capable of producing such a phenomenon, which is generally the result of overheated air. The peculiarity of the mirage has been proved by science to be produced by air stratas of different heat overlying each other, and causing rays of light striking on some particular spot to be broken and sent back to the earth again, thereby enabling the place or object which the ray first touched to become visible at the place at which it was finally directed.

"Another ridiculous feature of the 'Silent City' story is that people uneducated in scientific points are led to believe that a mirage is a phenomenon generally seen in one particular place. This alone would be sufficient to stamp the story as a myth in the minds of all well read people."(3)

Sources:
(1) "City of the Mirage," 1895. *Newark* [Ohio] *Daily Advocate* (July 6). Reprinted from the *New York Sun.*
(2) "Mirage City of Alaska," 1897. *Newark* [Ohio] *Daily Advocate* (November 19).
(3) "A Scientist Explodes the 'Silent City' Bubble," 1889. *Morning Oregonian* (October 5).

1890: The Last Mammoth

IN THE OCTOBER 1899 ISSUE of *McClure's Magazine,* a contributor who signed himself H. Tukeman recounted a fantastic experience in the Alaskan interior. Nine years earlier, he wrote, he had seen and killed what may have been the last living mammoth.

The killing was a consequence of an accidental meeting, in Fort Yukon, with an elderly Indian identified only as Joe. In the course of conversation, Tukeman happened to show his acquaintance a scrapbook. Inside it was a photograph of an elephant. Joe claimed that just such an animal lived in a nearby mountain valley.

Intrigued, Tukeman hired a guide and made a trek to the site, where soon they sighted the creature standing in a river and eating the moss off trees along the bank. It was a woolly mammoth, an animal supposedly extinct for thousands of years. Immediately, the two men laid plans to kill it. They accomplished their ambition by creating a bonfire to attract the animal's attention. When it strolled over to examine the fire, the two pumped a volley of bullets into the mammoth until finally it died. As it was doing so, Tukeman suddenly realized what he had done, and a feeling of pity and guilt overwhelmed him as he reflected that he had killed a harmless prehistoric animal, perhaps the last of its kind.

They dug a deep hole and placed the mammoth's hide and bones inside, intending to retrieve them later. They then sat down to eat a meal of mammoth steak, which proved "not unpalatable, but terribly tough."

Afterward, Tukeman and his guide returned to San Francisco. He related his strange experience to a naturalist, a Mr. Conradi, who offered him millions of dollars for the remains. In a second trip to Alaska, Tukeman dug up the remains and had them shipped to Conradi. To protect Tukeman, Conradi told the Smithsonian Institution, to which he gave the hide and bones, that he—Conradi—had found the body frozen in the arctic ice.

Impressionable readers bombarded *McClure's* with requests for more information—even though the issue's table of contents had clearly labeled the story, titled "The Killing of the Mammoth," as fiction. A note in a subsequent issue stated that the author and the magazine had not intended to deceive readers. It added, "We doubt if any writer of realistic fiction ever had a more general and convincing proof of success."

Source:
Besse, Nancy L., 1980. "The Great Mammoth Hoax." *Alaska Journal* 10, 4: 10–16.

Arizona

1893: Headless Ghosts

I WAS ON MY WAY HOME last Friday night [August 18] some time, I think, before midnight. . . . I was walking slowly along the east side of Monroe [in Phoenix] not thinking of anything I now remember and was looking neither to the right nor the left. When I came directly in front of the Catholic church I became uneasily aware of a strange presence, though at that moment I saw nothing. I turned involuntarily toward the door of the church and what I saw I'll never forget.

"In the archway stood two female figures clad in white; they did'nt [sic] look shadowy as ghosts are supposed to appear but there was a strange plainness of outline luminous and uncanny. I could see the arms distinctly and the feet which seemed to be incased [sic] in white slippers. The hands were also unnaturally visible. It is probably twenty feet from where I stood to the archway and you know the moon had gone down some hours. It was neither very dark nor very light but I don't think that a substantial object could have been as clearly seen at that distance.

"Strangest and most horrible of all, both of the figures were headless and badly scared as I was. I remember I thought they resembled those models used in dressmaker's shops.

"It seemed to me I stood there half an hour but perhaps it was more nearly half a minute[,] for time you know passes unnaturally slow or fast under certain circumstances. When I collected myself I was standing in the middle of Fourth street. I'm no coward or superstitious fool and I began to reason. I didn't believe I had seen anything but tried to convince myself that either my eyes or my imagination had played a

trick on me. I determined to solve the mystery, for unless I did I feared that I would carry away with me a permanent belief in ghosts.

Strangest and most horrible of all, both of the figures were headless and badly scared as I was. I remember I thought they resembled those models used in dressmaker's shops.

"I went back with some hesitation to the opening which leads from the sidewalk to the church. There the frightful pair stood, plainer if possible in luminous outline and headlessness than before. I tried to make up my mind to go up to them, and did go to within may be [*sic*] ten or fifteen feet. I think I remained there two minutes. I remember that on the Tuesday before, there had been the ceremonies of the feast of the Assumption, and I thought it might be that there was some midnight ceremony, and that the actors were persons of flesh and blood, but that almost fiery outline of the figures, the missing heads and the still and dark church, were against my theory. I went away without turning my face, until I had got half way back to Fourth street. I decided to say nothing about the thing, and never did mention it until last Sunday, when I heard that Frank—had seen something strange in front of the church the next night. I saw him and we compared notes.

"He hadn't made as much of an investigation as I did, but he swears he saw for an instant, and might have seen longer if he hadn't been in a hurry to leave the neighborhood, two women in white, standing in the archway.

"I've gone past the church, on the other side of the street, every night since, but I haven't seen any more ghosts."

A press account noted that "the names of both gentlemen are on file" at the newspaper and characterized them as reliable.

Source:
"A Ghost Story," 1893. *Arizona Republican* (August 26).

1900: The Ghost Fires Back

THE MEXICAN RESIDENTS OF Fort Lowell are disturbed because of a new dweller in the midst of the ruins of the old fort. He came a few nights ago and vanishes in the day time. Almost every night some of the Mexicans see him, and describe the visitor as of unusual stature, wearing high boots, and the uniform of a soldier. Last Friday night [December 14] several shots were fired at him and all the dogs in the neighborhood were set upon the figure as it dodged about amid the mass of shattered walls of the abandoned fortress. It appears that only the Mexicans have thus far been able to see the figure. Reputable citizens have told the story of their adventures with the ghost and Saturday one of the residents of the Rillito came into Tucson to secure [a supply] of ammunition for the final attack upon the visitor.

"It appears that the ghost has appeared each night coming out of the ruins and fading away in the mass of adobe when pursued. Men have shot at it from all sides and it has thus far escaped. One night one of the residents of the fort was within twenty yards of the ghost, when it suddenly turned on him and fired a volley of stones at the pursuer. The Mexicans believe that the place is haunted and they are all willing to make affidavit to the effect that they have seen the ghost come out of the ruins at night and disappear mysteriously when pursued."

Source:
"Ghost at Fort Lowell," 1900. *Arizona Republican* (December 16).

Arkansas

1851 and Before: Hairy Giant

IT APPEARS THAT DURING MARCH last, Mr. Hamilton, of Greene county, Arkansas, while out hunting with an acquaintance, observed a drove of cattle in a state of apparent alarm, evidently pursued by some dreaded enemy. Halting for the purpose, they discovered, as the animals fled by them, that they were followed by an animal bearing the unmistakable likeness of humanity.

"He was of gigantic stature, the body being covered with hair, and the head with long locks that fairly enveloped his neck and shoulders. The wild man, after looking at them deliberately for a short time, turned, and ran away with great speed, leaping from twelve to fourteen feet at a time. His foot-prints measured thirteen inches each.

"This singular creature , , , has long been known traditionally in St. Francis, Greene and Poinsett counties, Ark., sportsmen and hunters having described him seventeen years earlier. A planter indeed saw him very recently, but withheld the information lest he should not be credited, until the account of Mr. Hamilton and his friend placed the existence of the animal beyond cavil.

"A great deal of interest is felt in the matter by the inhabitants of that region, and various conjectures have been ventured in regard to him. The most generally entertained idea appears to be that he was a survivor of the earthquake which desolated that region in 1811. Thrown helpless upon the wilderness by that disaster, it is probable that he grew up in his savage state, until he now bears only the outward resemblance of humanity.

"So well authenticated have now become the accounts of this creature, that an expedition is organizing in Memphis, by Col. David C. Cross and Dr. Sullivan, to scout for him."

Source:
"A Wild Man of the Woods," 1851. *Adams* [Pennsylvania] *Sentinel* (June 2). Reprinted from the *Memphis Inquirer*.

1890: Merry Spirit

A SPECTER OF A STRANGELY JOVIAL and friendly disposition is at present puzzling the people of Helena, Ark., who would be alarmed at the antics of the ghost were it not for the perfect good-nature and jolly ways of the fellow. He is a little, ugly man, dressed in a gray suit, with a soft black hat worn set back on his head, with a comical, pleasant face and quick, bright eyes. He will appear in broad daylight without warning and, as if evolved from the empty air, step up to some man, give him a hail-fellow-well-met slap on the shoulder with vigor, and immediately the man turns to gone completely, with his happy, genial laugh still ringing in the air. The man singled out never sees him on that occasion, though the ghost is plainly visible to the bystanders; he only feels the slap on the shoulder. When some one else is thus favored instead, however, he can see him then.

"Many attempts have been made to lay hands on him, but nothing but impalpable air rewards the eager grasp of the ghost's would-be captors. The merry little specter only shakes his head with roguish enjoyment at the discomfited

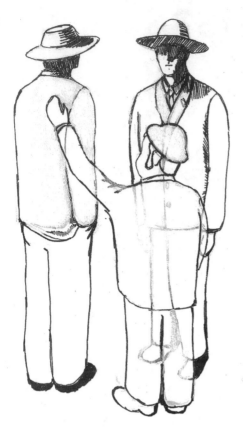

No clue to his identity has as yet been found, but, though a stranger in the flesh, all familiar with his appearance profess to becoming really attached to the friendly little specter. He never speaks, only smiles and nods his head when spoken to, with a frequent laugh at anything that amuses him.

and laughs loudly and heartily. No clue to his identity has as yet been found, but, though a stranger in the flesh, all familiar with his appearance profess to becoming really attached to the friendly little specter. He never speaks, only smiles and nods his head when spoken to, with a frequent laugh at anything that amuses him.

"The only variation that has taken place to his manner was on a day last week, when, just as the town clock struck 3 in the afternoon, he suddenly appeared to a group standing in front of Thorp's drug store, and looking at the clock, hung his head and drew aside his shirt and vest, pointing to a huge hole in his breast just over his

heart, from which big, heavy drops of blood were slowly welling. He turned on the group a sorrowful look, as if asking for comfort or help, and sighing heavily, disappeared. The next day, however, he turned up as jolly as ever. His movements are erratic, for he will be invisible for days together, but it has been noticed that he is never visible to a woman."

Source:

"A Merry Ghost," 1890. *Fort Wayne* [Indiana] *Sentinel* (June 28). Reprinted from the *St. Louis Globe-Democrat.*

1894: The Bells Toll

IN THE VILLAGE OF LINCOLNTOWN, which is settled principally by negroes, and which lies eight or nine miles south of this place, there is a little church, surrounding which is a mystery that is greatly perplexing and worrying the community. Every night there is to be seen in the belfry of the edifice a woman in white, who rings the bell three times in the most solemn fashion and who then disappears. How the woman gets there is what is puzzling the good people of Lincolntown, for the only approach to the belfry is a stairway to which entrance is gained by a single door, and not only is the door guarded every night since the commencement of the mysterious tolling, but the staircase is watched by 200 or 300 eyes, and at dark the belfry itself is watched and thoroughly searched and is found to be entirely empty. Besides, the belfry is only large enough to hold the bell itself, and when that is set in motion there is no footing for a person.

"The rope that is ordinarily employed in ringing the bell hangs all the time in plain view of the crowd and is perfectly motionless. The woman is also distinctly visible, but whether black or white it is impossible to tell. Even if the figure itself is a figment of the imagination, the ringing of the bell is not, as that is to be unmistakably heard for a quarter of a mile. . . . As to the identity of the ghost, it is generally believed that it is the restless spirit of a woman named Jonelle Lambkin, who, on account of some misdemeanor charged against her, was put out of the church here in spite of her continued reiteration of her innocence. Jonelle died about two months ago, alleging with her dying breath that she was a wrongfully accused woman and that the community would ultimately receive proof of this."

Source:

"A Ghost Rings a Bell," 1894. *Woodland* [California] *Daily Democrat* (February 14). Reprinted from the *Arkedelphia* [Arkansas] *Dispatch*.

California

1833: Giant Found and Lost

ONE OF MANY NINETEENTH-CENTURY legends and tall tales of buried giants comes from Lompoc. In 1833, as the story goes, Mexican soldiers digging a pit uncovered an enormous human skeleton. Twelve feet tall, the skeleton had double rows of teeth along both the top and the bottom of its jaw. Apparent offerings, including oversized stone axes, blocks with apparent writing on them, and carved shells, lay all around the figure.

Local Indians, it is said, took the find as a sinister omen and insisted that the bones be reburied immediately. Fearing trouble, the Mexican soldiers did just that, and the find, if indeed it was ever made, was forever lost.

Source:
Marinacci, Mike, 1988. *Mysterious California: Strange Places and Eerie Phenomena in the Golden State.* Los Angeles: Panpipes Press.

1851: Raining the Unimaginable

ACCORDING TO AN ACCOUNT published in the *San Francisco Herald* of July 24, 1851, meat—or at least a meatlike substance—fell out of the clear sky a few days earlier. The witnesses were soldiers from an army base at Benicia. They claimed that the rain lasted some two to three minutes and was confined to an area 30 yards wide by 300 yards long. A wind was blowing from the west, and no passing birds were to be seen.

On the ground they found pieces as small as a bird's egg and others as large as an orange. They brought samples to their commanding officer and to the base surgeon. The latter put them in alcohol, though not before weighing them and carefully examining one. That particular piece, in the surgeon's judgment, gave evidence of muscle fiber and sheathing, as well as part of a blood vessel.

Source:
Splitter, Henry Winfred, 1953. "Wonders from the Sky." *Fate* 6, 10 (October): 33–40.

1859: Apparition at the Forecastle Door

SOME EXCITEMENT WAS CREATED in the cabin of a vessel in this port [in San Francisco] last week [early March], by an apparition which stood awhile at the forecastle door, and waved its hand ominously towards a couple of men playing cards. It is described as the figure of a young looking man, with closed eyes. After a few minutes it disappeared, and when the men had resolution enough to seek an explanation of the strange appearance nothing more could be seen of it. If the men had been intoxicated, the circumstance might be attributed to some phantom evoked by John Barleycorn, but it is said that both were quite sober, and the affair has not yet been solved. No person looking like the apparition was known to be on board at the time."

Source:
"Still Another California Ghost," 1859. *Hornellsville* [New York] *Tribune* (April 14). Reprinted from the *San Francisco Bulletin,* March 7.

1860: "The Shape of a Huge Man"

LYONS' RANCH, NEAR SONORA, is haunted. The place was once the property of Jas. Lyons, who assassinated one of the Blakely brothers, some months ago, after having sold the ranch to them. The shade is presumed to be the murdered brother, Blakely. The barn seems to be the principal scene of its operations, and a number of miners who slept there have been driven from their lodgings by its remarkable pranks. According to their story, it appears to them in the shape of a huge man, about sixteen feet high, who arose from amidst the hay piled up in the barn, and tossed the bales around as if they were light as feathers in his grasp. The story goes, that this unwelcome visitor, on several occasions, chased the lodgers from the barn, making giant strides after the fugitives. On one occasion, one of the men fired at the ghost, but the ball had no effect. The upshot of the story is, his ghostship remains master of the field—or, rather, the barn. The story is solemnly told by the editor of the Sonora Age."

Source:

"A Ghost Sixteen Feet High," 1860. *Progressive Age* (March 7).

1868: Giant Snake on the Road

SOME TIME AGO WE PUBLISHED an account of an immense snake, reported to have been seen by several persons in the neighborhood of Young's Ranch, near Spring Valley. While its size may have been exaggerated by those who saw it, we never have had any doubts but that an enormous reptile was actually seen in that vicinity. As if to settle the matter, however, in regard to the reality of his existence, his snakeship has again deigned to make his appearance in public, in a different locality. As Mr. W. P. Peck, of our town [Calaveras], a gentleman whose reputation for truth and veracity is 'above suspicion,' and whose habits of life are such as to preclude the possibility of his carrying around imaginary serpents in his boots, was returning from Alexander & Co.'s mill, in old Rich Gulch, one day last

week, he saw the monster. The road leading from the gulch is exceedingly steep, and winds around the side of the hill, so that it is difficult for teams to pass each other when they meet.

"Mr. Peck was driving up the grade in a buggy, when he heard a peculiar hissing kind of a noise, which he supposed was caused by a loaded wagon coming down the hill with a break [*sic*] on. Being at the time in the most eligible place in the road to pass a team he endeavored to stop his horses for that purpose, but they roared and plunged so violently that he was obliged to proceed to avoid being precipitated over the embankment. The road is fringed with dense chaparral, and hearing a rustling among the bushes Mr. Peck cast his eyes upon the bank above him, when he saw a huge snake gliding into the brush. As it was just at dark he was unable to get a full view of the reptile, but the portion of him which he saw, some six or eight feet of his tail, 'was as large as a man's leg.' Mr. Peck met no team. The noise that he heard must have proceeded from the snake."

Source:

"The California Snake," 1868. *Hagerstown* [Maryland] *Herald and Torch Light* (October 28). Reprinted from the *Calaveras* [California] *Chronicle*.

1879: A Rain of Water Lizards

ONE MORNING IN AUGUST the editor of the *Sacramento Reporter* was summoned to the west end of the city, where he was told that lizardlike creatures had fallen with the rain. Once there, he observed thousands of the creatures, colored dark brown with bright spots, moving animatedly over several blocks. The fall seemed to have been particularly intense near the Opera House. As many as 200 of the animals were counted in a single puddle.

When taken out of the water, the animals lost their energy and became almost immobile. They had no bones in their limbs and had soft, pliable bodies. By the next day all those that were not still in water were dried up and dead.

Some theorized that the animals, compared to mud puppies (salamanders), had been there on the ground all along. Others said that seemed unlikely, since the animals could not survive outside water and this was the first rain in some time.

Source:
Splitter, Henry Winfred, 1953. "Wonders from the Sky." *Fate* 6, 10 (October): 33–40.

a hatchet and cut the monster to pieces. They proceeded to measure it, and, placing together the parts severed, it measured forty-three feet and seven inches, and was as large around as a man's leg."

Source:
Untitled, 1882. *Marion* [Ohio] *Daily Star* (January 16).

1882: Attacked by an Enormous Snake

A COUPLE OF CHINAMEN, while fishing in the Sacramento river near Chico landing, were attacked by a large snake, which coiled about one of them. The other Chinaman seized

1882: Flying Train, Flying Reptile

ONE OF THE MOST STARTLING snake stories that has been told in these parts for some time was related by the engineer and fireman

The train and snake came together, but the snake's tail was not where it should have been, and a portion of its lower extremities was clipped off. This seemed to put this flying snake on his mettle and he prepared for war. He wheeled around and gave chase to the flying train.

who came in last night on the Southern Pacific express, and was corroborated by the passengers. It seems that just after the train had passed Dos Palms, the engineer noticed, about half a mile ahead, what looked like a column of sand moving very slowly from east to west. At that time it was only a short distance from the track, and moving at such a pace that it was evident that train and column would come together. When the two monsters were but a short distance apart it was discovered that the column was not sand, but an animal of some kind.

"It was moving in almost a perpendicular position, the tail dragging on the ground and propelled by two large wings near the head. The bird, snake or whatever it was, seemed to be about thirty feet long, and twelve inches in diameter. By this time everybody, almost, on the train, had put their heads out of the windows and were banging on the platform to get a look at the monster snake. The train and snake came together, but the snake's tail was not where it should have been, and a portion of its lower extremities was clipped off. This seemed to put this flying snake on his mettle and he prepared for war. He wheeled around and gave chase to the flying train.

"The motion of the animal seemed to change in an instant, and he seemed to fly through the air two miles faster than chain lightning. In a few moments he, she or it, overtook the train, and began war after the latest snake style. The angry animal kept over the train and gave the train a lively thrashing, roaring like a cow in distress all the time. After breaking several windows and frightening the women and children almost to death, the monster called it off, followed by a shower of lead from the pistols of the passengers, which seemed to have no effect at all, if any of the bullets hit him. This is vouched for by every one who was on the train, and is given for what it is worth."

Source:
"A Flying Snake," 1882. *Brooklyn Eagle* (February 12). Reprinted from the *Los Angeles Times*.

1882: Ghost in a Circle

BODIE HAS A GHOST. A spectral woman[,] with a large basket in her hand, a white hood on her head, and clothed in white and black, appears at midnight at a certain locality, moves rapidly in a circle for a while, and then vanishes."

Source:
"Coast Clippings," 1882. *Reno* [Nevada] *Evening Gazette* (November 21).

1883: "A Large Mass of Glowing Matter"

A FEW NIGHTS AGO [early December], as Mrs. S. J. Merchant was in Brush street [in Oakland], she heard a swishing through the air, and was startled by seeing a large mass of glowing matter fall directly ahead of her in the roadway. Her superstitious fears aroused, she turned and ran down Thirteenth street. She rushed full tilt against an old gentleman, and the collision threw her to the ground, with him prostrate over her. They regained their feet, and she stated the reason for her headlong pace. The vicinity at the time was deserted by all save the two.

"The burning object threw a strong but wavering light. The old gentleman advanced cautiously toward the source of the light, followed at a respectable distance by Mrs. Merchant. He gained the corner, and his eyes fell on a strange phenomenon. In the road lay a ball of fire about two feet in diameter. It was blazing like a full[y] ignited chunk of coal in a grate. The heat from the ball was so intense it affected the observer's eyes, and he was forced to fall back. Mrs. Merchant had summoned courage and advanced to his side. Together they contemplated the mysterious spectacle. They noticed that the seething flames were gradually reducing the bulk of the ball, and that the heat was growing less and less strong. They continued to progress nearer and nearer to the object.

"Finally, sheltering his face with his cloak, the old gentleman approached the mass and thrust his cane into it. The part of the cane probed in was instantly consumed, and the investigator's hand was singed and slightly burned. He resumed his station on the corner, and before twenty minutes the flames had spent their force, and darkness once more reigned supreme. In company with Mrs. Merchant he went up to the spot again. To his surprise there was not a vestige left of the burning body. The ground was hot and blackened, but not a trace of the meteor, if meteor it was, remained."

Source:
"Was It a Lost Comet?," 1883. *Trenton* [New Jersey] *Times* (December 13).

1885: Drumming in the Night

MARIPOSA IS TROUBLED with a ghost which makes noise enough to keep a whole community awake o' nights. The sounds which it makes are like the drumming of a grouse, and cannot be located."

Source:
"Just Outside," 1885. *Fresno* [California] *Republican* (April 5).

1885: Crickets from Clouds

LAST WEDNESDAY EVENING [April 29] Antioch was visited by a singular phenomenon, it being no less than a regular shower of large black crickets. In the early evening, when the air was murky and threatening a thunder storm, and in fact when the thunder kept an ominous rumbling, swarms of these crickets seemed to fall from the clouds into Antioch, and the streets and sidewalks were soon black with them."

Source:
"State News," 1885. *Fresno* [California] *Republican* (May 2).

1886: Big Shower of Little Frogs

A CURIOUS PHENOMENON occurred on the 16th [of July], at Big Flat, Siskiyou county, during a heavy shower of rain. The ground became literally covered with frogs for many hundreds of acres, extending fully three miles in one direction. On the same day a similar shower fell on the bars below Cecilville, though on a smaller scale. In both cases the frogs were small, being about the size of crickets, and very lively."

Source:
"Pacific Coast Notes," 1886. *Fresno* [California] *Republican* (August 5).

1886: Ashes and Cinders

ON SUNDAY EVENING [September 19] about 4 o'clock, says the *Chico Chronicle*, the southwesterly portion of the town of Oroville was visited by a shower of ashes and cinders which lasted several minutes. The ground became nearly white from being covered with ashes. Many of the inhabitants became much alarmed, fearing that there was likely to be an earthquake. . . . The phenomenon is as yet unexplained."

Source:
"Here and There," 1886. *Placerville* [California] *Mountain Democrat* (September 25).

1886: Fireboy

TURLOCK COMES TO THE FRONT with a most peculiar mystery. It seems that a twelve-year-old boy, Willie Brough, living near Turlock, apparently sets fire to objects by his glance. On last Sunday [October 3] the phenomenon was first discovered, and the destruction of $9,000 worth of property by fire is laid to his charge—or, rather, eyes. He has recently been expelled from the Madison school near Turlock on account of his wonderful freaks. After Sunday's fire

Brough's family refused to have anything to do with him, believing him to be possessed of a devil. The boy was taken in by a farmer and sent to school. On the first day there were five fires in the school—one in the center of the ceiling, one in the teacher's desk, one in the teacher's wardrobe and two on the wall. The boy discovered all, and cried from fright. The trustees met and expelled him that night. One Turlock insurance agent has given notice that he will cancel all policies on property occupied by the boy. The neighborhood of Turlock is in a furor of excitement about the mystery."

Source:

Untitled, 1886. *Fresno* [California] *Republican* (October 8).

1888: Ball of Fire Too Closely Encountered

YESTERDAY AFTERNOON MR. Clark A. Stevens and a friend were seated in the front room of Stanislaus and I streets, watching the rain storm, when they were almost blinded by the sudden appearance in the air of a ball of fire which appeared to be about one foot in diameter. The blazing mass filled the room with light as it swiftly descended, and when about three feet from the ground it exploded with a report as loud as that of a cannon discharged. The report was heard by hundreds of people in various parts of the city as far distant as the courthouse. The effect of the explosion was to render not only Mr. Stevens and his friend partially insensible for several minutes, but all the other inmates of the house felt the shock to a greater or lesser extent. As soon as they had recovered from the shock sufficiently the men went into the yard to see if they could discover any traces of the explosion, but there was nothing to show that an aerial visitor had called except a strong smell of sulphur. The phenomenon was undoubtedly the result of the electricity in the atmosphere, but Mr. Stevens is not desirous of pursuing his investigations any farther."

Source:

"An Electrical Phenomenon," 1888. *Fresno* [California] *Weekly Republican* (November 23).

1888 and Beyond: Phantom Giant

FOR A WEEK AND A HALF in mid-July 2001, a search team of two dozen members led by land surveyor Larry Otter looked for what, if it existed, may have been the world's most massive tree, or at least the largest sequoia. They were following up on a legend of a giant tree that has been rumored in the area for more than 100 years. The biggest tree in the world on record, located in the same forest, is called General Sherman and measures 36.5 feet in diameter at its base.

Perhaps the first published account of the legendary giant tree, said to be in what would become Sequoia National Forest, appeared in an 1888 California newspaper. According to the article, two men who came upon it used a 4-foot-long rifle to determine the base's dimensions. It took forty-four rifle lengths before those dimensions had been exhausted. Local lore held that others had seen the tree. Studying the accounts, Otter became intrigued enough to organize an expedition consisting of experienced hikers, biologists, and photographers.

Over a ten-day period, the searchers endured rough terrain, climbing a mountainside that steeped at a 45-degree angle. One consequence was a number of injuries, all fortunately minor. A helicopter aided in the search but, like those who searched at ground level, found nothing. By the time the expedition had ended, Otter was declaring the phantom tree no more than a legend. Though he did not mention it, nineteenth-century Western newspapers are a notoriously unreliable source of information, especially when the subject is wondrous phenomena. Many such accounts are simple inventions conjured up for the amusement of readers.

Still, not all expedition members were convinced that the tree was not out there somewhere. A number vowed to go back into the forest when funding and resources were available again. Reporter Kerry Cavanaugh interviewed locals who insisted that the tree did exist; they had even seen photographs of it, and they thought they knew where it could be found. Doing them one better, a woman named Sue Ann Farley-Palmer claimed that the tree, or something just or nearly as big as the tree, existed on her weekend home property. Cavanaugh remarked, "Looks like Otter and his crew could have more evidence if they decide to continue the search."

Source:
Cavanaugh, Kerry, 2001. "Phantom Tree Stays Tall Tale of the Mountains." *Bakersfield Californian* (August 9).

1890: Raining Crabs

A MOST REMARKABLE phenomenon was witnessed on Morton street, just off Kearney, the other morning, it being nothing more nor less than a shower of small crabs, says a San Francisco exchange. A light shower was falling, and accompanying it were crabs by hundreds, ranging in size from that of a dime up to that of a good-sized California oyster.

"There is a barber shop on Morton street, and the bootblack employed there was washing spittoons on the sidewalk. He let out a wild whoop as he saw the crabs descend and called on all the saints in the calendar when he beheld his spittoons filling with the tiny crustaceans from the clouds. The barber shop was soon emptied of its occupants, who gazed wonderstruck at the strange sight, but were soon busy picking up the creeping creatures, which were as alive and kicking and active as infant crabs well could be.

"A similar visitation occurred at about the same hour on California street, between Sansome and Battery, but the crabs which fell there were few in number, while on Morton street they covered the sidewalk and gutter for a space at least twenty feet in length. This phenomenon is said to be not entirely novel, but this is certainly the first time it has been recorded in San Francisco."

Source:
"It Rained Crabs," 1890. *Stevens Point [Wisconsin] Journal* (February 8).

1890: Serpent of Land and Water

THE *RECORD-UNION* HAS rather a fishy tale from J. R. Williams, a well known rancher located near Pleasant Grove. According to his story the country about there is in great excitement over a reported sea-serpent. 'I have not seen the thing myself[,]' said he[,] [']and am not over-anxious to. But there are plenty of reputable citizens among my neighbors who have seen it. The serpent has been seen several times in the overflow east of Pritchard's lake and very close to the Spanish ranch; and again has been seen on the land, where it seems to be as much at home as in the water. Those who saw it in the water say that it is over forty feet in length and fully seven feet in diameter near the center of its body. It has a great flat head and a pair of jaws that, when distended, could with ease admit a full-grown man. Those who saw it in the water describe it as being a very rapid swimmer, carrying his head at times about eight feet above the surface of the water.

"'The serpent was last seen about a week ago on the land, a short distance from the Spanish ranch. Those who saw it then do not believe that it is over thirty feet in length, but aside from this description, it is the same as that given by the others. It doesn't appear to have any fins, or wing-like appendages like the sea-serpents we read about, but is more like a giant eel, except that it has a terrible head.'

"Mr. Williams, continuing, said a neighbor of his named Lyons, while out looking for some stock, came upon the tracks of the serpent. He followed it for fully half a mile, when he was seized with a premonition that he was getting too close to the reptile, and hastily 'dug out.'

"Several parties have been made up, said Mr. Williams, but when the time came to start out they all backed out. One man, however, a professional hunter named Pete St. Clair, who has the reputation in that neighborhood of being a dare-devil, started out a day or two ago accompanied by another man, with the intention of killing the serpent. They were armed with large bore Winchesters, and were confident of success. Mr. Williams is of the opinion, though, that they would have been better fixed had they wheeled a cannon along with them, or provided themselves with whalers' dynamite harpoons.

"Will St. Clair succeed? 'St. Clair is a pretty brave fellow,' said Mr. Williams, 'but I think he has tackled a tough job this time. He says he will camp on the enemy's trail until he lands him. He thinks his fortune will be made if he can succeed in capturing this monster, dead or alive, and I've no doubt such would be the case.'

"'Has the monster been doing any damage in the vicinity?' asked the reporter.

"'Yes—at least it is attributed to him. You see there are a great many sheep up there, and as a number of them have turned up missing lately, I've no doubt the serpent is responsible. The people, however, are not so much concerned about their stock as they are for their own safety. You don't see any more fishing around the lakes—particularly in Bartwell Lake, where they think the serpent has now located himself. But everybody is now waiting anxiously for news from St. Clair. While many of them are confident of his ability to kill the serpent, the majority of them—more particularly those who have seen it—believe that if he runs across the serpent he will never come back alive.'"

Source:
"The Sea-Serpent," 1890. *Woodland* [California] *Daily Democrat* (August 28).

1891: "What Is It?"

MR. SMITH, A WELL KNOWN CITIZEN of Northern Capay Valley, called on us to-day and tells us the following strange story. . . . [He] has always had a spotless reputation. Several days ago, Mr. Smith, together with a party of hunters, were [sic] above Rumsey hunting. One morning Mr. Smith started out early in question of game[;] he had not gone far when his attention was attracted by a peculiar noise that seemed to come from an oak tree that stood near by. Looking up Mr. Smith was startled to see gazing at him what was apparently a man clothed in a suit of shaggy fur. . . .

"Mr. Smith . . . called to the supposed man to come down. . . . The strange thing gave grunts of unmistakable anger. Believing that discretion was the better part of valor, our informant . . . went at once in a bee-line for the camp. After placing some distance between himself and the strange creature, the hunter turned around just in time to see it descend the tree. Upon reaching the ground, instead of standing upright as a man would, it commenced to trot along on the ground as a dog or any other animal would do.

"Smith then realized that it was no hermit he had seen, but some kind of monstrosity, such as he had never heard of, much less seen before. The hunter stood amazed and spell-bound for a moment, but soon gathered his scattered senses again and was soon making his best speed to camp, where in a few breathless words, [he] was telling his companions of what he had seen. . . .

"A hasty council was held, and the party decided to go in search of the monster, so taking their guns and dogs they were piloted by Mr. Smith. . . . They soon came in sight of the unnamed animal. In the meantime it had commenced to devour the contents of Mr. Smith's game bag that he had dropped in his hasty retreat. The creature would plunge its long arms or legs into the bag and pulling forth the small game that was in it, transferred it to its mouth in a most disgusting manner. An effort was made to set the dogs upon it, but they crouched at their masters' heels and gave vent to the most piteous whines. The whines attracted the attention of the nondescript [unknown creature], and it commenced to make the most unearthly yells and screams, at the same time fleeing to the undergrowth, some half a mile distant, upon which the whole party immediately gave chase.

"They soon gained upon the strange beast, and it, seeing that such was the case, suddenly turned, and sitting upon its haunchs [sic], commenced to beat its breast with its hairy fists. It would break off the great branches of trees that were around it, and snap them as easily as if they had been so many toothpicks. Once it pulled up a sapling five inches through at the base, and snapping it in twain, brandished the lower part over its head, much after the same manner a man would sling a club. The hunters[,] seeing that they had a creature with the strength of a gorilla to contend with, beat a hasty retreat to camp, which soon broke up, fearing a visit from their chance acquaintance.

"Mr. Smith describes the animal as being about six feet high when standing, which it did not do perfectly but bent over, after the manner of a bear. Its head was very much like that of a human being. The trapezie muscles were very thick and aided much to giving the animal its brutal look. The brow was low and contracted, while the eyes were deep set, giving it a wicked look. It was covered with long shaggy hair, except the head, where the hair was black and curly.

"Mr. Smith says that of late sheep and hogs to a considerable extent have disappeared in his vicinity and their disappearance can be traced to the hiding place of the 'What Is It.' Among those who have suffered are Henry Sharp, Jordan Sumner, Herman Laird and J. C. Treadle."

Source:
  "What Is It?," 1891. *Woodland* [California] *Daily Democrat* (April 9).

1891: "This Unknown Creature"

WE HAVE RECEIVED A communication from Mr. James E. Martin of Casey's Flat, giving us additional information regarding the strange animal seen in that vicinity some weeks ago. Mr. Martin is one of the best known citizens of this county and is well known in this city [Woodland][;] at present he is homesteading a quarter section above Rumsey. Much of his spare time Mr. Martin spends in trapping, and of course if any one would know anything of the 'wild man of the woods' it would be him.

"But to his story: About the sixth of this month, some of Mr. Martin's horses strayed from his premises. While out looking for them, he suddenly came across a trail showing that some kind of a dead animal had been dragged through the brush. The gentleman's curiosity was excited at such a large animal being made away with, so he followed it as fast as the jagged rocks and matted brush would allow him. Proceeding slow[ly] and cautiously, with keen eyes and steady nerves, he felt as though he was about to have a firsthand encounter with the beast that had done the killing. After going some distance, he came upon the partly devoured remains of a two-year-old heifer, which he recognized as the property of Mr. John W. Clapp. After the monster had satisfied himself on a large portion of the flesh, he had covered the dead carcass over with brush and dirt and had departed. Mr. Martin measured the tracks of the beast, and found them to be sixteen inches long and eight inches broad, with long claws, with which it had torn up the earth that covered the slain heifer.

"Our informant is not a coward by any means, but he suddenly remembered that he always felt better about that time of the day if he had his trusty Winchester with him, so, making his way homeward, he took his gun, and, mounting a horse, proceeded to Mr. Clapp's, where he told him what he had seen. Mr. Clapp returned with him, and together they proceeded to the spot. To say that Mr. Clapp was grieved at the loss of his fine bovine, now torn limb from limb, would be putting it mildly.

"Late that night, as Mr. Martin lay asleep, he was aroused by the piteous whines of his dogs, which are blood thirsty and ferocious animals. On opening the door, the dogs rushed in and skulked under the bed, where they shivered with fright and fear, and from which place they could not be driven either by threats or entreaties. Stepping outside to see what was the matter, Mr. Martin heard something moving away from his cabin, at the same time giving vent to some most unearthly screams that echoed from crag to mountain, and which finally died away in the lonely canyons. The gentleman asks for aid in the capturing of this unknown creature, and says that if he can but secure the necessary help, he will not stop until he has captured it dead or alive."

Source:
    "The 'What Is It?'," 1891. *Woodland* [California]
        Daily Democrat (May 13).

1891: Enigmas

PEOPLE AROUND HERE [Capay Valley] are afraid to travel the roads above here after midnight for fear of meeting the 'What is it' or getting struck by a meteor. Very likely this valley will be uninhabited in a few years as strange monsters and falling bodies are getting more numerous and bolder each day, and footprints of the monster are now seen as far down the creek as Madison."

Source:
    Untitled, 1891. *Woodland* [California] *Daily
        Democrat* (June 3). Reprinted from the *Capay
        Times.*

1891: "A Gigantic Lizard"

A YOUNG MAN OF Woodland, noted for his veracity and temperance, has a strange story to tell. Last Tuesday night [August 4], together with some companions, he went to the Sacramento River to fish. As they did not wish to begin until the early morning, they went to sleep beneath the blankets that they had brought with them. The young man . . . arose during the night to inspect some trout lines that he had set the evening previous. As he neared the bank, he could see that the water in the river was violently disturbed. He was much surprised at this, as he could see no boats either up or down the stream.

"As he stood looking at the water, he saw, so he says, a head of what appeared to be a gigantic lizard stick up out of the water. He ran back to camp and seizing a gun, retraced his steps to find the animal attempting to crawl upon the dry land. Raising his gun, he fired a shot at it, which evidently took effect, as the creature threw itself backward into the water with a loud splash, and commenced to swim down stream at a rapid rate, nearly a mile a minute, he thinks. The report aroused his friends, who rushed to the spot, only to catch a fleeting glance of the rapidly retreating monster."

Source:
    "Did He See the Sea Serpent?," 1891. *Woodland*
        [California] *Daily Democrat* (August 8).

1891: Boiling Water Falling

W. S. BARTON, a well known mining man and prospector, has returned to San Bernardino from an exploration of the famous Death valley. He states that of a sudden one of the members of the party became insane. In the trip south Barton stopped at Ibex, on the edge of Death valley. Here at midnight the thermometer stood at 145 degs. Over the valley . . . at 7 p.m., there occurred a remarkable phenomenon. Two clouds, one from the east and one from the west, met. An electric storm followed, the like of which has never been seen by living man, and for an hour the blaze was simply terrific. Then following the electric storm came a fall of boiling hot rain. This lasted for about ten minutes. The parties in camp were actually forced to cover themselves over in order to protect their bodies from scalding water. The thunder was something that no artillery force on the face of the earth, no matter how great, could equal."

Source:
    "It Rained Boiling Water," 1891. *Bismarck*
        [North Dakota] *Daily Tribune* (December 9).
        Reprinted from the *Toronto World.*

1892: Mounted Phantoms

LATELY CUPERTINO HAS BEEN visited by some spooks, or ghosts, or specters, or something of that sort. At least, several reputable citizens so declare. On their first appearance your correspondent was in San Francisco, and, of course, when told of the apparition, not having convincing proof, remained silent, but on Saturday evening [probably June 18] he saw with his own eyes the apparitions, and this is what he saw: Five figures dash by on horseback at a mad pace; two were females and three were males.

One of the males, who evidently was the leader, was a giant in size and was mounted on an immense charger. He was dressed in martial array and was of commanding mien. His companions I did not notice so closely, but they were all mounted on Indian ponies, and, as near as I could tell, his male companions wore no uniforms. The females had their long hair flying wildly about their faces, but from what glimpses I did get I should say they were Indians or Mexicans. Several others saw them, and some say they gave unearthly yells or warwhoops whenever they saw any one. This is their second visit, and as yet no one is able to explain the mystery.

"Joe Kelly, an old settler, who resides on Stevens creek, tells a legend about an American soldier who was lured to his death by a beautiful Indian maiden, the daughter of a chief of a tribe which once owned the country around Cupertino. Mr. Kelly claims that these apparitions were often seen around here in the forties and early fifties and says their present activity is caused by their resting place being disturbed by some of the buildings that have lately been erected. In the fight that took place when the soldier found he was betrayed before the Indians killed him, two braves, his treacherous sweetheart and another Indian maiden met their doom: hence the five specters that are now disturbing our peace."

Source:
"Ghosts Mounted on Ponies," 1892. *Woodland [California] Daily Democrat* (June 21). Reprinted from the *San Jose Mercury.*

1893: Astronomical Oddity

AT 11:30 P.M. A MOST PECULIAR phenomenon is visible in the heavens here [San Francisco] a little south of west. It appears to be like a very large star and is shooting out red and blue sparks. It is not very far above the horizon and is slowly sinking. It is also moving rapidly from south to north. The same phenomenon is reported from Santa Cruz, San Louis [sic] Obispo and other points in the state. Efforts are being made to communicate with Lick Observatory and other observatories

which are shut off from telegraph communication at this hour."

Source:
"Queer Sight in the Heavens," 1893. *Heber [Utah] Wasatch Wave* (April 4).

1896: Airships Sailing and Stolen

ON THE EVENING OF November 17, a light that the *Sacramento Evening Bee* characterized as an "electric arc lamp propelled by some mysterious force" sailed over the city in full view of hundreds of witnesses. It was flying at a low altitude and, like an intelligently directed vehicle, evaded buildings and hills that were in its way. Some of the observers, who claimed a particularly close look at the mysterious object, said they had heard voices "singing in a chorus, a rattling song, which gradually died away in the distance," as the *San Francisco Call* reported.

As sightings continued throughout the rest of the month and into December, an uproarious controversy ensued, with extravagant stories and rumors of the airship's secret inventors. Most, though not all, centered on the allegations of San Francisco attorney George D. Collins, who stated that he knew the inventor, whose interests he had been hired to represent. The inventor was said to be a wealthy man who had come to California from Maine seven years earlier. The controversy moved on to the issue of whether Collins had seen the invention up close at its testing site near Oroville. Press accounts had him saying he had done exactly that, and a fellow attorney said Collins had related him the same information privately. (The airship was 150 feet long, with "two canvas wings eighteen feet wide and a rudder shaped like a bird's tail," Collins allegedly revealed.) Then Collins vehemently denied ever having said anything like this. San Francisco's two rival papers, the *Call* and the *Examiner,* took opposite sides in the dispute. The latter snapped, "There is no wealthy man living here who came from Maine, or any other state, in the past seven years, and there is no one here that a diligent search can uncover who knows anything at all about a flying ship having been invented here or having left here for Sacramento." Suspicions turned to itinerant dentist E. H. Benjamin, a native of Carmel, Maine. (One source gives his first name as "Elmer," adding that when he appeared in Woodland a few years earlier, he "had neither a diploma nor a recommendation.") When tracked down, Benjamin insisted that his "inventions have to do with dentistry."

On November 24, another alleged informant told the *Oakland Tribune* that he knew someone, a fellow Mason, "who talked with the man who saw the machine" being tested near Oroville. No less than California's former attorney general, W. H. H. Hart, now took stage. The unnamed inventor had now hired him and fired Collins for talking too much. Hart then proceeded to talk more loudly and incessantly than Collins had. There were, he said, two airships, one in the East, the other in the West, and his job was to "consolidate both interests." "I propose to use [the California airship] wholly for war purposes," he grandly pronounced. "It will carry four men and one thousand pounds of dynamite," which would be dropped on Havana to support the Cuban resistance to the Spanish rulers. Later, citing his anonymous inventor source, he added that two machines had already been constructed and another was in the process. Of the finished ones, Hart told the *Call,* in its paraphrase of his remarks, "one . . . was of a large size, capable of carrying three weight, and another . . . was much smaller, capable of carrying one man, the machinery, fixtures and 500 or 600 pounds of other matter."

After a few days, as no evidence emerged to support these fantastic assertions, the story sputtered out, though sightings of ostensible airships continued (in the coming months expanding elsewhere in the nation) and were never entirely explained. As often as not, "airships" were no more than anomalous lights to which a vehicle presumably was attached.

For example, on December 4, as a freight train on its way from Davisville (now Davis) to Oakland passed between Dixon and Elmira just after 6:30 p.m., an observer in the caboose noticed two bright lights on the right side of the track and about one thousand miles in the air. One seemed "as large as the head light of an engine." After a few minutes he alerted the conductor, who was so puzzled by the sight that he called in two other witnesses. The four of them watched the lights as they followed the train and then surpassed it, vanishing to the southwest.

The sighting, reported in the *Woodland Daily Democrat,* was one of several in that area of largely rural central California. The local airship scare ended on a farcical note, something that could have come right out of a Mark Twain short story.

On December 8, the *Daily Democrat* ran as a feature story a letter written by a frustrated inventor who signed his name as H. Lytle. (Another source indicates that the "H." stood for Harry.) A headline writer crowned Lytle's testimony with the title, seriously or otherwise, "THE MYSTERY SOLVED. A Resident of Arbuckle Explains the Airship Phenomenon." Lytle opened his communication by noting that he had seen the airship on "Tuesday last" (December 1) as it flew toward the southwest over the Coast Ranges 15 miles west of Arbuckle and 3 miles north of Rumsey. The sighting sparked not awe but exasperation. He recognized it for what it was, and that was not a welcome sight. He went on:

"Three years ago I submitted the plans I had drawn of an aircycle, with explanations, to Mr. E. W. Brown, of Davisville. He at once saw that there was no doubt as to the successful operation of a machine constructed on those plans and at once advanced the necessary funds for the construction of the same.

"The machine consists of an aluminum tube twelve feet long and three feet in diameter pointed at the end like a cigar. Four feet under the tube and running parallel with it is a light steel frame similar to a tandem bicycle frame. Each end of this frame is connected with the tube. On the frame there are two seats, two handle-bars and two pedals, the same as on a tandem. The pedals operate the propeller and side wings. The propeller is used to force the aircycle ahead and the wings are used to ascend and descend as the operator chooses. There is also a rudder that sits behind the propeller the same as those on a steamship. This is operated by the handle-bars. This part is very simple. The difficult part is in obtaining the gas which is used in the tube and has great lifting power. This gas is obtained at an altitude of twenty-five miles above sea level. How this is obtained will not be stated here. Suffice to say that this is also simple when one has the instruments."

The ship was dubbed the *Nonesuch*. Brown and Lytle commenced trying short flights, and at Brown's urging, these flights, always under cover of darkness, became longer and longer. The two were elated at the aircycle's success, and they decided to risk an extended flight southward. To keep them from asking questions, Brown told his friends that he planned to take a short vacation in the countryside. On March 1, 1893, at 10:30 p.m., they set off. Lytle reported:

"We could make thirty miles per hour without any exertion, and, with the wind at our backs, seventy-five miles an hour was easy. We could always make better time at night, for, when we were surrounded by darkness, we did not experience that dizzy feeling that would come over us in the daytime.

"On the morning of the 2d day of March we landed near Los Angeles. Securing our cycle to a fence (which was necessary, for as soon as we dismounted there was an upward pull of 330 pounds) we started for a farmhouse that stood near by. We were just in time to enjoy a good breakfast, which we very much needed after our night's ride."

That, however, was the end of the good times. On their return, when they were about a hundred yards from the spot at which they had parked the *Nonesuch,* they saw two hoboes inside the ship. As Brown and Lytle rushed the craft and shouted at the tramps, the two men rose into the sky and were quickly gone.

The now-earthbound aeronauts had only $2.75 between them. Brown feared that if he telegraphed home for money, it would be hard to conceal the true reason he was not where he had said he would be. When Brown tried to pawn his watch, a suspicious pawnbroker refused to accept it. They finally decided that they had no choice but to walk back to Sacramento. At the conclusion of that no doubt exhausting

trek, they looked up a friend of Brown's. The friend loaned them $5 for train fare to Davisville. Lytle said:

"We never heard any more from the aircycle until a short time ago. I saw in the newspapers that an airship was reported at various places, but that no one had had the opportunity of examining it and that the actions were a mystery. I suspected that it was the Nonesuch which had returned. Those who are in charge know that if they land they will at once be arrested.

"I expect the operation will be asked why I didn't build another aircycle. The answer is that the instruments for securing the gas were on the Nonesuch when she was taken charge of by those two knights of the road."

Lytle surely has the distinction of being the only witness in history to assert that the UFO he observed was his stolen invention.

Sources:
"Aerial Navigation," 1896. *Woodland* [California] *Daily Democrat* (November 23).
"The Airship Again," 1896. *Woodland* [California] *Daily Democrat* (December 7).
"The Airship Craze Now Fading Fast," 1896. *San Francisco Chronicle* (November 26).
"Airships Now Fly in Flocks," 1896. *San Francisco Examiner* (November 25).
"The Apparition of the Air," 1896. *San Francisco Call* (November 24).
"As Large as a Big Whale," 1896. *San Francisco Call* (November 27).
"Claim They Saw a Flying Airship," 1896. *San Francisco Call* (November 18).
"A Clue at Last," 1896. *Oakland Tribune* (November 24).
"Collins Sticks to His Airship Story," 1896. *San Francisco Chronicle* (November 23).
"Coy Mr. Collins and His Airship," 1896. *San Francisco Chronicle* (November 24).
"Davisville Doings," 1896. *Woodland* [California] *Daily Democrat* (November 23).
"Floating in the Air," 1896. *Oakland Tribune* (November 23).
"Hart Stands by His Ship," 1896. *San Francisco Call* (November 26).
"Hart's Inventor Has Three Aerial Fliers," 1896. *San Francisco Call* (November 29).
"Have We Got 'Em Again?," 1896. *Sacramento Evening Bee* (November 23).
"Have You Seen It in the Sky?," 1896. *San Francisco Examiner* (November 24).
"It Flitted over San Jose," 1896. *San Francisco Call* (November 28).

"A Lawyer's Word for That Airship," 1896. *San Francisco Chronicle* (November 22).

"Mission of the Airship," 1896. *San Francisco Call* (November 25).

"The Mystery Solved," 1896. *Woodland* [California] *Daily Democrat* (December 8).

"Queer Things You See When—," 1896. *San Francisco Examiner* (November 23).

"The Scarecrow Fly-by-Night," 1896. *San Francisco Examiner* (November 26).

"Voices in the Sky," 1896. *Sacramento Evening Bee* (November 18).

1897: Airship Again

THE AIRSHIP CRAZE WAS revived here by the appearance of an aerial navigator in daylight. Many people in Acampo, three miles north of here, said they saw it the other day sailing over as plain as the sun. It seemed as big as a small house and looked like it was built of canvas. It went southeast. Some farmers also saw it the same day near here. The ship seemed to be under perfect control."

Source:

"That Airship Again," 1897. *Reno* [Nevada] *Weekly Gazette and Stockman* (January 28).

1897: Fast-Moving Light

THE AIRSHIP MYSTERY IS again attracting attention. A pale white light moving with great rapidity over this city tonight has revived the theme, and many who saw the phenomenon are confident that the problem of air navigation has been solved."

Source:

"Another Airship," 1897. *Fresno* [California] *Weekly Republican* (March 26).

1897: Vampire Attacks

IN AN INTERVIEW WITH a San Francisco newspaper, railroad brakeman John Santine expressed a most extraordinary complaint:

"For about three years and a half I have been fighting ghosts at my house. . . . A mysterious, uncanny intruder has kept me in a constant state of nervousness at night when I wanted to sleep. I thought for a long time it might be some 'varmint,' and set all sorts of pitfalls and snares to catch it. I had my bedroom, where I sleep alone, just filled with traps—mousetraps, rat-traps, 'figure 4s,' and deadfalls big enough to kill an alligator, but nothing ever came of it.

"As these singular nocturnal disturbances continued I came to the conclusion finally that I must be afflicted with a ghost. I bought a pistol and increased my watchfulness. At the least noise I would jump up and grab my pistol, but I never saw anything in the room.

"The curious thing about the affair was that after each disturbance, usually some time about the middle of the night, I would wake up to find myself covered with strange insects that died when exposed to the light. I bought all kinds of insect powder to beat the deuce—you just ask the druggist—but it didn't do any good. Then I tried putting two cats in the room, but not a rat or mouse could be discovered.

"I feel sure now that the creature is a vampire, which comes to my bed and lays itself on my neck, for what purpose I cannot tell.

"It was only last Monday morning, about 2 o'clock, that I awoke suddenly and felt an object resting on my left shoulder. In a twinkling it flitted away. I heard it pattering along the pillow as it went, and I am certain it was the creature, vampire or what you will, that has been haunting me so long. It always has the most horrid smell, just like something from the grave, and I think it the odor as much as anything which wakes me up.

"I have always left my window open at the top to have plenty of air, but it never occurred to me that the cause of my continual annoyance might find ingress in that way. Those vampires are awfully cunning creatures, and as soon as they see the least movement they are off like a flash. Sometimes the thing wouldn't come near for a week, so that though I tried to be on the watch all the time I could never catch it. I have fixed up a lot of snares, which hang across my room, and I hope to get it so tangled in them some night that I can shut the window before it escapes.

"I have awakened in the morning many a time with a sickness at the stomach, and I believe that it was that vampire's presence during the night that caused it. There are lots of cases of children that pine away under the care of a physician, and I believe that night visits from these hideous creatures are responsible for it. I don't know whether my case is an isolated one or not, but I think people should know about it and take necessary precautions."

Santine swore that he was not a drinking man.

Source:
"Haunted by a Vampire," 1897. *Delphos* [Ohio] *Herald* (March 13). Reprinted from the *San Francisco Chronicle.*

1898: The Night before Last I Had the Strangest Dream

A YOUNG WOMAN NAMED Bertha G. Allen wrote a San Francisco newspaper to recount a dream that eerily prefigures rumors of UFO crashes of the later twentieth century and beyond:

"It was very dim last night and as I hurried home from the factory the street lamps seemed brighter than usual. I said I hurried; yes, I almost ran. I saw nothing to frighten me; yet my fear of the darkness was irresistible.

"On turning down our dingy street I paused a moment for breath. As I did so I heard a whirring sound as of wind rushing suddenly and swiftly around a corner. Then I heard it again and again. I looked about, but saw not the cause of it. Surely it wasn't the wind, for the trees were standing still—as still as the stone statues across the street. Then again I heard the same queer sounds.

"In my amazement I looked toward the sky. Probably I thought that it could solve the mystery. Instead of seeing its usual blue color I saw it to be of a purple hue. How beautiful it did look, so royal, so stately!

"As I turned around I saw a white object rising slowly above the tree tops. There seemed to be something black on it, but I could not tell what it was. Then it sank from view again. Instead of being frightened, I was entranced with the purple sky and funny airship.

"When I again looked toward the place where I had last seen this queer affair, it was rising gradually in the air. It drew nearer, giving me a fine chance to see it. It was long, round and hollow. I distinguished the black figure to be a man, who sat in the center and guided it.

"It circled above me, shot through the air with lightning speed, and then again it would pause, as if to rest. I saw something else too. It had the shape of a coil, with one end flattened and hanging down. It looked as if it were made of fire. Not only did it circle in the air, but it followed the airship constantly.

"Then, almost too quickly to be seen, it swooped down upon this aerial contrivance. That was the last I saw of it, but it had done its work. Instantly the airship began to fall, fall, till it reached the earth. Not with a crash and a bang did it touch, but as if it had been a feather.

"I hurried to the spot, and what did I see? I saw the man only. He was dead. He was about three feet long and was wrinkled and shriveled.

"The next thing I remember was that I was lying on my own bed. This, of course, is not a true story. It is only what I dreamed the night before last."

Source:
"A California Girl Writes of the Wonderful Thing She Saw at Night," 1898. *Daily Nevada State Journal* (June 1). Reprinted from the *San Francisco Call.*

1905: Screaming Sky

A CCORDING TO AN ACCOUNT she would write nearly five decades later, Rose Bushnell heard the sky screaming early on the morning of April 20, 1905. "I am 58 now and still tremble when I recall that frightening experience," she said in 1953.

In the year in question, she lived in her grandfather's residence with her parents Mr. and Mrs. William Taylor, her brother, and her sister at Bull Creek Valley in Humboldt County. The house was large enough to accommodate guests, and the previous evening Mr. and Mrs. Charles Lounes had arrived with their three children. Sometime in the middle of the night, they awoke to screaming sounds.

"The screams came from the sky," Bushnell would remember, "and at first seemed to come from a great distance, fading away for about two minutes and then rising again, a little louder each time. They lasted for about two minutes at their loudest though we heard them for a period of fifteen minutes altogether. Part of the time the screams sounded like those of men, women and children together, as if in terrible pain. Then they sounded like men alone, then like women alone, then like children alone. Words cannot describe those horrible sounds."

Their neighbors, who lived a mile away, heard nothing.

Source:
Bushnell, Rose W., 1953. "True Mystic Experiences: Screams from the Sky." *Fate* 6, 12 (December): 57–58.

1909: Mystery Ship over the Imperial Valley

ALL IMPERIAL VALLEY IS EXCITED over reports of a mysterious airship which is taking nightly flights over Salton Sea. The monster ship has been seen by many persons in various localities, and the stories agree in general details. The airship was first seen by Imperial Valley people Tuesday evening [June 1] a short time after sunset. W. D. Conser, a merchant of Imperial, and his wife were among the first to observe the moving object in the sky. They were driving into Imperial from the country and stopped their team to observe the strange sight. At first it appeared to be stationary at a point directly over Salton Sea near the intake of the Alamo and New rivers. Then the airship began a rapid flight, swerving from a southern course to one directly northwest and apparently passing directly across Salton Sea at its widest point until it disappeared in the shadows of the San Jacinto Mountains.

"Others who saw the airship at Imperial were Mr. and Mrs. H. M. McCall and Mr. and Mrs. George Southwell, whose attention was called to the strange sight by a party of seven Mexicans standing in the road spellbound by the apparition in the sky. At Brawley a party of twenty men witnessed the flight across the sky. Securing a field glass they closely studied the machine. Its appearance was that of a basket fastened between two wide wings, and when the turn was made it is said the propeller could be plainly seen, while the railings of the basket stood forth clearly against the sky, but the observers were unable to distinguish any persons in the basket. The huge machine was handled apparently without difficulty and the turn was made smoothly and at good speed.

"Last night [June 2] observers were again able to witness the flight of the machine, it appearing at about the same hour as the day before. Otto Conser, son of W. D. Conser, with a party of several persons who had become interested in the observations of the night before, went out to a ranch north of Imperial to watch for the airship's appearance and were rewarded by seeing it go through the same movements as when first seen here. Many persons were on the lookout tonight, and towns of Coachella Valley have been notified to be on the lookout."

Source:
"Airship Flies over Salton Sea," 1909. *Los Angeles Times* (June 4).

1912: Mysterious Ocean Thing

ONE OF THE QUEEREST deep sea creatures ever seen in the vicinity was brought in a few days ago by a fisherman of Venice, Cal. It is five feet in length, black and green mottled, with a tail like that of a shark. It has a dorsal fin and four feet, shaped like those of a parrot. Its mouth resembles that of a Gila monster, while its head is a replica on a large scale of that of a California horned toad."

Source:
"Queer Catch from the Sea," 1912. *Ada* [Oklahoma] *Evening News* (November 29).

Colorado

1878: Creek Lights

LATE ON THE NIGHT OF July 15, near Colorado Springs, "a gentleman, passing along the Fountain... a short distance above the Half-way house, witnessed a very remarkable phenomenon just across the bridge on the Garden of the Gods road."

The unnamed observer saw what looked like a shower of falling stars show up a few feet above the creek. They plunged into the water, making a loud splash, followed by a continuous sound "resembling the breaking of glass bottles mingled with human cries and 'swear words.'" Presumably, the lights and the sound of a drinking party were two different, unrelated phenomena.

The light display had lasted only seconds but was a striking, beautiful sight, the witness said. The press account states, "The different colors which appeared were very bright and vivid—the parabolic curve made by reason of its bright stripes, presenting much the appearance of a rainbow."

Source:
   "A Singular Phenomenon," 1878. *Colorado Springs* (July 17).

1892: Massive Space Rock

A SPECIAL FROM NEWCASTLE, Colo., conveys the news that about 9 o'clock Tuesday morning [November 29] a stone weighing ten tons fell from the sky, striking the earth near the town. No one saw the stone fall, but the fact that it sunk [*sic*] deep into the earth and was in a heated condition when found leaves no doubt

as to where it came from. The stone has a color entirely foreign to any stone in that locality, being of a slate color."

Source:
   "Came in Free," 1892. *Decatur* [Illinois] *Daily Republican* (December 1).

1895: Luminous Snow

THE STORY OF A MOST remarkable snowstorm... is told by Lieutenant John P. Finley, one of the best-informed meteorologists in the country, who encountered the storm in making an ascent of Pike's Peak. He says the storm could be described as a 'shower of cold fire.' In reality it was so charged with electricity as to present a scene more easily imagined than described.

"At first the flakes only discharged their tiny lights on coming in contact with the hair of the mule on which the lieutenant was mounted. Presently they began coming thicker and faster, each flake emitting its spark as it sank into drifts of the snow, or settled on the clothing of the lieutenant or the hair of the mule. As the storm increased in fury and the flakes became smaller, each of the icy particles appeared as a trailing blaze of ghostly white light, and the noise produced by the constant electric explosions conveyed an impression of nature's power which Lieutenant Finley will never forget.

"When the storm was at its height and each flake of snow was like a drop of fire, electric sparks were shaken in streams from the lieutenant's fingertips, as well as from his ears, beard and nose, and a wave of his arms was like the sweep of flaming sword-blades through the air, every point of snow touched giving out its little snap and flash of light.

"This phenomenon, though rare, is by no means new to meteorologists, it having been recorded several times before. It has by some observers been treated as a sort of phosphorescence, but in the case above cited each flake appears to have been charged with static electricity."

Source:

"Electric Snow," 1895. *New Oxford* [Pennsylvania] *Item* (October 4). Reprinted from the *New York Tribune*.

1897: Light in the Spring Night

WHAT IS SUPPOSED TO BE the mysterious airship was watched by hundreds of people here [Salida] tonight [April 29]. The light it carries was seen coming from behind Tenderfoot hill. It seemed to go straight up into the air for a distance and then sailed over the city. When over the town[,] a bright light, resembling an arc light or searchlight[,] was displayed. After scrutinizing the town a few seconds, the white light was changed to one of a red color. It then disappeared toward Buena Vista.

"It was 12:10 a.m. when the light was seen over Salida, and at 12:30 it was displayed over Buena Vista. There was no wind, therefore the aerial mystery must have pretty good powers of propulsion to travel twenty miles in twenty minutes. The operator at Malta, a short distance south of Leadville, was on the lookout for the airship and soon reported that the light had suddenly changed its course and was going in a westerly direction and was soon lost to view behind the mountains."

Source:

"White and Red Lights," 1897. *Arizona Gazette* (April 30).

1897: "Genuine Flying Machine"

AN UNIDENTIFIED FLYING OBJECT, described in press accounts as a "genuine flying machine," was seen near Sterling in late August. G. A. Nenstein observed a large dark object in the southeastern sky as it flew on a level course and at a high rate of speed toward the northeast. He kept it in view until it was no longer visible.

Source:

"A Colorado Airship," 1897. *Portland Oregonian* (September 1).

Connecticut

1729 and After: Moodus Noises

IN A LETTER WRITTEN on August 13, 1729, the Rev. Mr. Stephen Hosmer of East Haddam, Connecticut, referred to "fearful and dreadful" sounds that unnerved locals. "I have myself heard eight or ten sounds successively, and imitating small arms, in the space of five minutes," he said. "I have, I suppose, heard several hundreds of them within twenty years; some more, some less terrible. Sometimes we have heard them almost every day, and great numbers of them in the space of a year. Oftentimes I have observed them coming down from the north, imitating slow thunder, which shakes the houses and all that is in them." He added that since the "great earthquake" of 1727, the sounds "have in a manner ceased. . . . There have been but two heard since that time and these but moderate."

Half a century earlier, the first European settlers learned of the noises from the native Wangunk Indians. The Wangunk called the area—at the foot of Mount Tom, in the south-central part of the state just east of the Connecticut River—Machemoodus, meaning "place of noises" (later, shortened to Moodus, it became the name of a village two and a half miles north of East Haddam). At first skeptical, the colonists soon enough heard the sounds themselves. The first printed reference dates to 1702 and suggests that even then the sounds were thought to emanate from underground. The sounds would come and go. Sometimes, after going unheard for years, they would resume at a furious pace.

A variety of legends and theories grew up around the noises. Traditional Wangunk lore held that they were the result of battles inside the mountain between good and bad witches. One white settler proposed that a subterranean passage linked the mountain south to the sea and that the tunnel carried wind and tidal sounds inland. Others suspected the spontaneous explosions of underground gases and minerals. According to one local legend, a man named Steele arrived in East Haddam around 1765, touting his discovery of a fossil he called a "carbuncle." It looked like a white stone but became so brilliantly luminous in the dark that at night the house in which Steele was staying "appeared to be on fire, and was seen at a great distance," the Rev. Henry Chapman reported to the editor of the *American Journal of Science* many years later. Steele, who had found the "fossil" on the mountain, claimed that the carbuncle was in some way responsible for the noises. On his departure soon after, he predicted that they would be heard no more.

As early as the late eighteenth century, however, educated opinion linked the sounds to earthquake activity, as the Rev. Hosmer's words indicate. (Indeed, there is some evidence that the sorts of seismic stresses that generate earthquakes may also cause luminosity in certain rocks.) During particularly dramatic manifestations of the sounds, buildings and ground would shake. In 1897, for example, according to an article in *Science* soon afterward, "there was a sound like a clap of thunder, followed for some two hours by a roar like the echoes of a distant cataract. A day later there was a crashing sound like heavy muffled thunder, and a roar not unlike the wind in a tempest. The ground was shaken, causing houses to tremble and crockery to rattle, 'as though in an earthquake.'"

Modern seismic-detection equipment has confirmed the connection between the sounds and earthquakes. By 1979, seismometers had established that earthquake activity, though usually so slight it could be discerned only by technology, and the sounds always accompanied one another. The seismic activity was contained within a relatively small, sphere-shaped area, some 760 feet in diameter, 4,560 feet beneath the surface. As yet, however, geologists cannot explain why such small earthquakes—many of them just 2 on the Richter scale—can generate such dramatic sonic effects.

Comparable seismic-linked sonic phenomena have been reported at a number of locations around the world.

Sources:
Brooke, Harrison V., 1975. "Thunder of the Mackimoodus." *Fate* 28, 10 (October): 70–79.

Corliss, William R., 1977. *Handbook of Unusual Natural Phenomena.* Glen Arm, MD: Sourcebook Project, 383–384, 385.

———, ed., 1983. *Earthquakes, Tides, Unidentified Sounds and Related Phenomena: A Catalogue of Geophysical Anomalies.* Glen Arm, MD: Sourcebook Project, 151–152

"Earthquake in Connecticut," 1840. *American Journal of Science* 1, 39: 335–342.

Ebel, John E., et al., 1982. "The 1981 Microearthquake Swarm Near Moodus, Connecticut." *Geophysical Research Letters* 9: 397.

Grant, Steve, 1999. "Could State Be on Shaky Ground?" *Hartford Courant* (July 29).

Helfferich, Carla, 1988. "Things That Go Boom in the Night," http://www.gi.alaska.edu/ScienceForum/ASF8/896.html.

"The Moodus Noises," 1897. *Science* 6: 834–835.

1855: "A Brilliant Red Ball of Fire"

A CORRESPONDENT TO THE *New-Haven Palladium* reported a sighting made on the evening of January 22: "a brilliant red ball of fire about two minutes in diameter." The object caught his attention as he was crossing the public square around 10 p.m. An astronomically so-phisticated observer, he placed the light "about eight degrees below the guards in Ursa Minor and near the small star Gamma in the belly of the Dragon."

For the first fifteen seconds the object was motionless, and then it began to travel slowly on an eastward, straight-line course with "at first a slight and undulatory motion. It passed below and about one degree from the star Benetuash in Ursa Major, and disappeared in the distance, not far from Denebola in the constellation Leo. Whole time visible ten minutes." Besides being in sight too long for a meteor, it did not throw off a starry trail in the fashion of a shooting star.

Source:
"Meteoritic Phenomenon," 1855. *New York Daily Times* (January 25). Reprinted from *New-Haven Palladium,* January 23.

1867: South Canaan Serpent

THE GREAT SNAKE [of South Canaan] has not been killed, nor has any one attempted to shoot it. An ordinary shot gun would have no more effect on it than a popgun on an iron-clad, and a man would be called very crazy who would attempt it. It has been seen every summer for the last fifteen years, in nearly the same spot. It was seen but a short time ago by a gentleman of the highest respectability, who is well known, and no one would think of doubting his statements. His name is A. Boardman, a skillful machinist. . . . Mr. Boardman was passing along the highway where the snake had been generally seen, not thinking anything about it, when suddenly his eye caught sight of something lying under a large elm tree which looked like the shadow of a large limb.

"Its shining brilliancy, and the real shadow of the tree being opposite from where the object lay, the thought instantly entered his mind, that it was nothing else than the big snake he had heard so much about, and so it proved to be, for it immediately started for the swamp. The animal having to cross the road, he had a fair view of his snakeship. A thrill of horror passed through his veins as the monster crushed its way through the brush, with almost the speed of lightning, into the swamp and was immediately lost to sight. Mr. B. thinks it would measure thirty inches around its body in the largest place; in fact, it appeared to be nearly of one size, except three feet from the end of the tail, which tapered considerably towards the end. Its length, he should judge, to be not far from twenty-eight or thirty feet. Its skin was so black and bright that it fairly dazzled.

"The farmers in the vicinity complain less of the depredations of the animal than formerly. On the borders of the swamp where it lives, graze large herds of cows, and it is among these that it gets its living. The owners of these cows have for some time past, wondered why some of their best milkers had failed to give their usual quantity of milk, as it is only in the morning the owners miss it, and they have come to the conclusion that the snake sucks them and thus gets its living. It is well known that snakes are fond of warm, new milk, and it is this kind of food, no doubt, that has caused this one to grow to such immense proportions.

"There is a reward of one hundred dollars offered by a private citizen for its capture alive, and fifty if killed. There is a strong pressure bearing on the town authorities to offer a reward large enough to secure its capture or destruction, and the subject will be brought before our annual town meeting, the first Monday in October. Hundreds who attended the camp-meeting lately held there, came more for the purpose of seeing where this monster snake lived, than to seek for the straight and narrow way. Few could be induced to stay on the ground at night, for fear it might make them a visit. . . . One man has determined to sell his farm and leave the place. His wife has already left, and says she will never live on the place again until the snake is killed. There is a perfect panic among the people, and what will be the result, time alone can determine."

Source:
"The Big Canaan Snake," 1867. *Hagerstown* [Maryland] *Herald and Torch Light* (December 18). Reprinted from a letter to the *Poughkeepsie* [New York] *Eagle.*

1882: Anomalous Ocean Disturbance

A STRANGE PHENOMENON was observed on Wednesday [March 8] at Barlett's Reef Light ship, off Saybrook, which resembled a tidal wave on a circumscribed scale. The vessel was lying perfectly still on a sea calm as a mill-pond, and not a breath of wind stirred the atmosphere, when all at once an area of about a quarter of an acre was lifted out of the depths to a height of 25 feet, churned into foam, and in a second threw the light-ship on her beam ends, almost swamping her. The phenomenon subsided as quickly as it appeared, but the billowy roll continued for a long time afterward. Capt. Edwards, who was on board at the time, states that he saw a similar occurrence about the same place nearly 16 years ago."

Source:
"A Marine Phenomenon," 1882. *New York Times* (March 13). Reprinted from the *New London* [Connecticut] *Day,* March 10.

1882: Comet Or . . . ?

THE GREAT SEPTEMBER COMET attracted attention in the town of Hartford when it exhibited odd behavior on the morning of December 13. A reporter wrote: "Just what was happening to disturb the comet's equilibrium does not appear, but yesterday morning a curious phenomenon was carefully observed, the explanation of which is not here attempted. At about 4 o'clock, while the sky was so thickly overcast with clouds that no stars were seen, at precisely the place where the comet has been visible at that hour of the morning, was seen a bright streak of light, of the same apparent length and breadth as the comet's tail, that was perpendicular to the south horizon. The phenomenon was just what might be expected if the comet stood on its head and pointed its tail at the pole star. A curious feature of the phenomenon is that the light was considerably brighter than that emitted by the comet for several weeks. Was it the comet?"

Source:
"Queer Doings of the Comet," 1882. *New York Times* (December 17). Reprinted from the *Hartford Courant*, December 14.

1886: Gnome

NORTH HAVEN, A QUIET VILLAGE six miles from this city [New Haven], is excited over the appearance of a ghost. It is described as about 3 feet in height, and wears a velvet suit and hat. Oliver McNulty, a laborer in Snears's brickyard, with four of his companions, ran across the ghost a night or two ago, when they followed up a moving light. The next day the men happened to be near the same spot and again discovered the gnome. McNulty struck it with his spade. He declares that he cut it in two, and that the severed parts reunited and then disappeared altogether. Old residents of the town say that they have seen it at different times for 30 years. McNulty's story is indorsed [*sic*] by the four men who were with him. The neighborhood is low, level, and damp, and the ignis fatuus is frequently seen there."

Source:
"A Connecticut Ghost Story," 1886. *New York Times* (January 12).

1886: Phantom Passenger

PATRICK EGAN, ONE OF THE best known residents of this town [Cheshire], tells a strange story of an adventure he had with a ghost a night or two ago. He says he was driving by the old Jenny Hill baryta mine when his horse suddenly shied to one side of the road. The horse needed all his care, and he had no chance to see what had frightened him for some time. When he had succeeded in quieting the animal he turned to look back upon the road and was astonished to see a tall, white, flimsy figure climbing into the buggy. Before he could move or think the ghost sat down on the seat beside him and sat there until the horse had trotted half a mile, when the strange visitor jumped out and vanished as suddenly as it had appeared. Egan says that he was not scared particularly because he had once before had a similar experience on the same road. At that time he resolved to say nothing about the ghost, but the second appearance was more than he could keep to himself. Whatever the thing was it had no weight, because it did not alter the trim of the buggy a bit. The road near the mine is dark and lonely, and what Egan really saw is greatly puzzling all Cheshire."

Source:
"Riding with a Ghost," 1886. *New York Times* (August 7).

1886: "Huge Sea Serpent"

THREE TRUSTWORTHY GENTLEMEN, who occupied a sail-boat, report that at 12:30 o'clock the other day, when half way between Westport and Southport, Conn., they saw a huge sea serpent, seventy-five to one hundred feet of the body of which was exposed, while the monster carried its head five feet out of the water."

Source:
Untitled, 1886. *Stevens Point* [Wisconsin] *Gazette* (October 23).

1888: "As Though by a Gigantic Plow"

A MOST SINGULAR SCENE was witnessed on Long [now Highland] lake during a storm recently. . . . The lake is three miles long and is divided into three bays. About 3:15 o'clock a vivid flash of lightning illuminated the scenery, followed by a terrific peal of thunder. The wind by this time was blowing with cyclonic force. Suddenly there came a roar, and far down the lake a huge flame of fire could be seen. The water for yards ahead was parted as though by a gigantic plow, and the billows seemed to rise at

Suddenly there came a roar, and far down the lake a huge flame of fire could be seen. The water for yards ahead was parted as though by a gigantic plow, and the billows seemed to rise at the side of this furrow for fully twenty feet.

the side of this furrow for fully twenty feet. The ball of fire seemed to force the water aside, and so deep did it go that the bottom of the lake could almost be seen as it passed through the narrows. The parted waters, with their singular propeller, advanced toward the head of the lake with great rapidity. When within one hundred yards of the short there came another flash of lightning, and the fire disappeared as suddenly as it had come. . . . It was many hours before the waters of the lake became calm.

"The residents along the lake who witnessed the strange phenomenon were greatly alarmed. They insist that the ball of fire was fully ten feet long, and half of the mass appeared to be buried in the waters of the lake. It was many hours before the waters of the lake became calm."

Source:
"Plowed by Fire," 1888. *Coshocton* [Ohio] *Semi Weekly Age* (October 9). Reprinted from the *Hartford Times.*

1892: Musical Thunder

A STARTLING AND MOST remarkable phenomenon occurred in Brookfield, Fairfield county, on Sunday night, Aug. 30, which will be remembered to life's end by those who heard it. About the time for the evening service, and when the congregations of the churches were waiting the beginning of worship, it began to thunder and lightning in the distance, and the shower appeared to be rapidly approaching until it was directly overhead.

"Suddenly there was a burst of musical thunder, sounding somewhat like a gong in different tones, and so marked were the musical notes as to be sweet and almost bugle like. As quick as a flash all the eyes of the congregation in the church were directed to the ceiling, and the suppressed cry of 'What's that?' could be heard all over the church. 'It's thunder!' was the exclamation from all. All were startled, although some were more frightened than others."

Source:
"Music in the Thunder," 1892. *Weekly Nevada State Journal* (February 6). Reprinted from the *Danbury* [Connecticut] *News*.

1906: Strange Flying Machine

FARMERS AND OTHER RESIDENTS of New Fairfield, Conn., a town five miles north of Danbury, were startled the other day by the appearance of a strange object in the air. . . . There did not appear to be any persons aboard, and no one observed anything like a basket or car suspended from it. The object first attracted attention, as reported by members of the family of Mrs. O. D. Taylor, by a noise like an automobile. It went through the air very swiftly, and in shape it resembled a big naval torpedo. It was near enough to be seen quite plainly."

Source:
"Airship That Startled Farmers," 1906. *Massillon* [Ohio] *Independent* (October 25).

1908: Groton Poltergeist

SCARCELY A MILE EAST of the Yale-Harvard crew's training quarters in Center Groton there have been strange doings in the past three days in the farm house owned by George R. Hempstead. The villagers are mystified at the happenings in this usually quiet household. The Hempstead house was built half a century ago, but had never before this had the reputation of being haunted. The Hempstead family consists of Mr. and Mrs. Hempstead, an adopted boy, Franklin, thirteen years of age, and a hired boy, Gilbert Edwards, a lad of sixteen.

"The first antics noticed by the Hempstead family took place on Wednesday evening [August 12], when spools of thread began to tumble from the work basket on the second floor down the stairs and sometimes apparently through the air directly at the feet of Mrs. Hempstead in the dining room below. Mr. and Mrs. Hempstead are prominent members of the Groton Ferry Baptist Church and are highly respected members of Groton borough. Several professional men, accompanied by newspaper correspondents[,] visited the Hempstead farm this morning to ascertain the truth of the stories about the farm house, and were cordially received by Farmer Hempstead and his good natured wife.

"'We cannot account for the mysterious goings on here the past three days,' said Mr. Hempstead apologetically, as he ushered his visitors into the sitting-room. The inquisitive delegation who had walked from Groton Ferry to the Hempstead farm did not have long to wait for manifestations, for spools of thread soon dropped to the floor, tiny marbles, which the boys had been playing with about the house[,] moved slowly from one room to another, and beans[,] which had also been used in a childish game, were seen to arise from the floor and seek new locations.

"Mr. Hempstead merely shook his head while the phenomenon, or whatever it might be termed, was in progress, while his wife looked at the deeply interested visitors with an expression that said, 'I hope you are satisfied.'

"'Are you not afraid, especially at night, on account of these mysterious doings?' ventured one of the party. 'Yes, of course I am,' she replied, as she shook her head. 'There has been no one harmed and what has happened has been of a trivial nature, so we might as well stay here as anywhere else.'

"'Those peculiar movements were first noticed Wednesday night for a short period early in the evening. During the night while everyone was abed nothing out of the ordinary was noticed, but early Thursday morning the beans, marbles, spools of thread and other light articles about the house resumed their antics, but again ceased at nightfall. I cannot account for the phenomena nor have I met any one who could explain the mystery.'"

Source:
"This Ghost Has Sense of Humor," 1908. *Trenton* [New Jersey] *Evening Times* (August 15).

1911: Worms in a Snowstorm

ABRAM C. SHELLY, AN AGED and stanch [*sic*] member of the temperance party, while walking along Torringford street the other morning during a snowstorm, perceived hundreds of live grub worms on top of the snow. He gathered a handful of them and brought them to Winsted to corroborate his statement. In a warm room the worms appeared as lively as in the summer. Shelly is certain the worms did not crawl up through five inches of snow, and the only way he can account for their presence on the snow is that the winds picked them up in the South and they came down in Winsted with the snowstorm."

Source:

"Snows Grub Worms in East," 1911. *Sheboygan* [Wisconsin] *Press* (March 9).

Delaware

1904: Anomalous Cloud

THE BRITISH SHIP MOHICAN while making for the Delaware breakwater encountered a strange phenomenon. A cloud of phosphoric appearance enveloped the vessel, magnetizing everything on board."

Source:

Untitled, 1904. *Iowa City Daily Press* (September 23).

Florida

1844: "Nearly Equal to the Size of the Moon"

ON WEDNESDAY SEPTEMBER 11th, at about 9 o'clock, was observed from the deck of the sloop Mount Vernon, Capt. J. P. Smith, and also by all his crew, what to all appearances was a star, but of such size and brilliancy, considering the sky was unclouded, and the sun was pouring down its rays with unusual luster, as almost to lead to the [belief] that it was some supernatural vision. This singular phenomenon is represented as being, in appearance, nearly equal to the size of the moon. It remained visible nearly the whole day, and disappeared only as the shades of night were fast approaching, when all expected to have a better and more distinct view of this brilliant and apparently erratic heavenly body. Capt. S. states it to have presented an appearance, in color, similar to that of the planets at night, only a shade whiter."

Source:
    "Singular Phenomenon," 1844. *Lorain* [Ohio] *Republican* (October 30).

1840s and After: Phantom Volcano

TO THE SOUTHEAST AND SOUTH of Tallahassee," C. L. Norton wrote in 1892, in *Handbook of Florida,* "there extends a vast belt of flat woods, merging into an almost impenetrable tangle of undergrowth and swamp. It is a famous hunting ground, and somewhere within its shades is the alleged Wakulla volcano. The curious inquirer is sure to hear the most contradictory statements regarding this mystery."

The "mystery" is barely remembered today, but it has never been explained to everyone's satisfaction. It was first remarked on, as far as is known, in the mid–nineteenth century, perhaps as early as the 1840s. It was also claimed that the original inhabitants of the area, the Indians, had known of it for a long time, though none of the accounts quotes an actual native informant. Most reports located it in the wild, all but impenetrable Pinbrook Swamp, west of the Wacissa River. During the day witnesses spoke of seeing a vast column of thick, black smoke; at night the smoke would turn into bright light, "as though a large house was burning and the flames were not quite visible."

Geologists have never documented the existence of an active volcano in Florida, a seismologically stable state that rests on a sedimentary formation. The area from Tallahassee to the Gulf Coast sits on limestone, which is not a volcanic rock.

One account, published in the *Tallahassee Floridian,* alleged that a party of "adventurous gentlemen" managed to hack their way through the dense vegetation until they got to a "strange country of volcanic appearance. Everywhere were seen great masses of rocks, often an acre in extent, all cracked and ragged as if upheaved from a great depth." Though no names are provided, the story seems credible, based on later discoveries. In any event, there is little doubt that many people saw *something,* in nearly every instance from a distance of an estimated 5 miles. At one point, according to some reports, Union gunboats on the Gulf mistook the smoke for evidence of a Confederate blockade runner hidden in the swamp, generating extended and futile shelling.

The phenomenon ended, it is said, in the early 1880s. It remained a subject of discussion and debate, however, for at least another five decades. During the dry winter of 1932–1933, William Wyatt and a local man who knew the area managed to trek deep into the swamp. In 1935, Wyatt recalled that "we found a number of sinks with piles of rock close by that seemed to have been blown out of them, the rocks being different to the kind usually found on the surface, for the edges were rounded off as though they had been subjected to great heat at some time or another. There were mounds or piles of rock as high as fifteen feet that looked as though they had been blown or pushed out of the earth by some gigantic hand, with small rocks close by, that again looked as though they had been melted."

Wyatt rejected the volcanic explanation, however, since the rocks were manifestly not of volcanic origin. He theorized that the smoke arose from underground peat fires. In 1934, State Librarian W. T. Cash came to a similar conclusion. Though superficially plausible, the theory fails to explain why witnesses invariably described the smoke as rising—and staying—in a column, rather than expanding into a diffuse mass. Though this is surely a geophysical, not a paranormal, enigma, its cause remains undetermined, no doubt because geologists have long since abandoned the search for it.

Sources:

Cash, W. T., 1934. Letter to Herman Gunter, State Geologist (November 1), http://www.vashti.net/volcano/archive/Historic/WTCash.htm.

Norton, C. L., 1892. "Handbook of Florida," http://www.vashti.net/volcano/archive/Historic/Norton.htm.

Shoemaker, Michael T., 1985. "Florida's Phantom Volcano." *Fate* 38, 4 (April): 47–50.

Wyatt, William, 1935. "The Wakulla Volcano," http://www.vashti.net/volcano/archive/Historic/Wwyatt.htm.

1878: "Dumfounded at Seeing Such a Monster"

WE ARE INFORMED BY Mr. Long, of Brevard county, Fla., that while driving his ox team near Ft. Drum, in that county, his ox shied and ran out of the road. Seeing something raise its head and a movement in the grass, Mr. Long, after stopping his team, went back to see what it was. Upon approaching the object he heard a great rustling and rattling, which convinced him that it was a rattlesnake, but he could not see it, because of the palmetto and high grass, until it threw itself into a coil and stood nearly as high as himself. He was almost dumfounded at seeing such a monster, and hastily retreated, but soon summoning up his courage, he advanced near enough to be within reach of the reptile with his long cow whip, which he knew how to handle.

"With this weapon he opened the conflict, which lasted nearly fifteen minutes, Mr. Long keeping out of reach of the snake, but still near enough to strike it with his cow-whip, which was about eighteen feet long. Finally Mr. Long began to feel sick and weak from the excitement, as well as from the musk emitted from the snake, and, putting in two[,] three or [more] rapid strokes with his whip, he retreated toward his cart, but fainted before he reached it. Upon coming to his senses again, he found that he had killed the snake. Mr. Long had no means of measuring its length but by his cow whip, which was 18 feet long, and the snake lacked about 2 1/2 feet of being as long as the whip. It had thirty-eight rattles and a button. He says it 'was as large around as a big blue bucket.' Mr. Long is one of the most reliable men of this section."

Source:

"Three Big Snakes," 1878. *Elyria* [Ohio] *Republican* (November 14). Reprinted from the *Fort Read Crescent.*

1881: "A Monster Unlike Anything Ever Seen or Heard of Before"

FISHING PARTIES AT GOOSE Lake, Florida, tell of a monster unlike anything ever seen or heard of before. J. Z. Scott first saw it, and says that it has a body between fifteen and twenty feet in length, and as large around as a common horse. It has a head like a dog and tail like a catfish. No fins or feet have ever been observed, though it seems to move with the motion of a fish, rather than that of a snake. All those portions of the body which have been exposed are covered with long hair of a dark color. It swims with astonishing rapidity, and will follow a lighted boat at night. Messrs. Aaron Terry and N. G. Osborne were out in a boat gig fishing, when the monster approached within striking distance, and they drove the gig deeply into it. The animal freed itself with a violent effort, twisting the prongs of the gig like so many straws."

Source:
Untitled, 1881. *Chester* [Pennsylvania] *Daily Times* (October 25).

1887: "As High as a Good Sized Man"

A FARMER NEAR ORLANDO, Fla., saw in the sand the trail of what he thought was a very big snake. He followed it, and after ten minutes' trailing came upon the largest serpent he had ever seen. It was engaged in swallowing a rabbit, and the farmer waited and watched the operation. After the rabbit had disappeared he walked forward to get a good shot at the monster, which, according to this story, at once reared up its head as high as a good sized man, and 'began racing back and forth before him, drawing nearer each time, hissing and darting out its tongue.' The farmer shot and broke the snake's back, and another shot killed it. It was a 'coachwhip' snake, of the boa constrictor family, and measured sixteen feet in length, and was four inches wide across the head."

Source:
"A Big Snake Story," 1887. *Gettysburg* [Pennsylvania] *Compiler* (September 18).

1890: Mermaid on the Line

A DISPATCH FROM Jacksonville, Fla., dated April 29 says: W. W. Stanton, mate of the schooner Addie Schaeffer, while fishing for bass 300 miles off St. Augustine, drew in his line and found entangled therein the strangest fish, if it is a fish, that has ever been caught. This strange creature is about six feet long, pure white and scaleless. The head and face are wonderfully human in shape and feature. The shoulders are well outlined, and very much resemble those of a woman, and the bosom is well defined and shows considerable development, while the hips and abdomen continue the human appearance. There are four flippers, two of which are placed at the lower termination of the body, and give one the impression that nature made an effort to supply the strange creature with lower limbs. Mr. Stanton confesses to quite a fright on first sight of his queer prize, which, on being drawn on board, gave utterance to a low, moaning sound, which might easily have been mistaken for the sobbing of a baby.

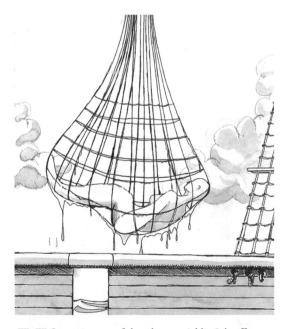

W. W. Stanton, mate of the schooner Addie Schaeffer, while fishing for bass 300 miles off St. Augustine, drew in his line and found entangled therein the strangest fish, if it is a fish, that has ever been caught.

"It is extremely unfortunate that Mr. Stanton did not succeed in keeping the creature alive, which he thinks might have been done, as the strange object lived two days after being taken. The schooner has been thronged all day by curious visitors, who express much wonder and astonishment at the strange object. Mr. Stanton, after visiting several ports and showing his queer creature, will donate it to the Smithsonian Institution. The fish or mermaid is in a large six foot glass in alcohol."

Source:
    "A Mermaid at Last," 1890. *Marion* [Ohio] *Daily Star* (May 13).

1894: Gulf Lights

CAPTAIN CORNING OF THE British schooner, from Porto Cabello, reports a peculiar phenomenon while his vessel was in the Gulf stream, on the west coast of Florida. The captain says he saw strange lights which apparently would rise from the water and ascend to a height of about 25 feet and then, after a report similar to an explosion, would go out. Captain Corning has no explanation to make of this strange phenomenon, but is positive that he did see the lights."

Source:
    "Cum Grano Sails," 1894. *Lowell* [Massachusetts] *Daily Sun* (February 8).

1895: "A Wicked Looking Head"

IN LATE FEBRUARY OR EARLY MARCH, individuals living along the Indian River near Titusville said they had seen an immense serpent in the water. Few believed them, but a sighting later in March had hundreds of witnesses.

At 9 a.m., people who had gathered to wait for a steamer to arrive noticed a big, apparently inanimate black object in the water some 75 feet from shore. Most thought it was a piece of wreckage, and two men, identified as Captain Simmonds and Fred White, boarded a boat and paddled out to investigate. They were about 25 feet from it when they realized that it was alive. In a moment they saw, according to a newspaper story, "a wicked looking head, with basilisk eyes." It hissed and darted in their direction. They stared incredulously at a 60-foot-long serpent, its body at its thickest the size of a barrel. "For about six feet along its back," it was said, "there appeared to be a row of fins. The body of the reptile tapered gradually to a pointed tail."

Hundreds of people observed the spectacle. Those who happened to be armed—apparently, this being nineteenth-century America, not a few—opened up on the animal, which responded by bringing its head 6 or more feet out of the water and hissing at its would-be killers. It then sank beneath the water, and it was gone.

On returning to shore, Simmonds and White were so shaken that onlookers provided them with "restoratives." The two men said they had seen its huge teeth and even smelled its bad breath. At noon a steamer came up the river to report it had passed the creature 30 miles to the south. According to the newspaper account, "The appearance of the monster has demoralized tourist travel on the Indian river and the house boats of the wealthy northerners have been deserted."

Source:
    "Mammoth Sea Serpent," 1895. *Atlanta Constitution* (March 29).

1899: Snake in the Air

POLICEMAN O'BRIEN OF New York has received a letter from a friend in Everglades, Fla., describing a monster seen there recently. O'Brien's correspondent calls the thing a 'flying snake.' He says it was first seen by the McCorkle brothers, whom O'Brien knows well, as they were walking through their orange grove. 'The snake rose from the top of an old orange tree,' says the writer, 'and started circling westward. It was about thirty-five feet long and had four wings, a skull like a puff adder, a bald pate, tapering tail, eyes that flashed fire, a tongue that was plainly venomous, and a look of dark blue annoyance.' O'Brien's correspondent adds that the sober citizens of the place had formed a hunt club and are making plans to bag the snake."

Source:

"Snake 35 Feet Long, with Wings," 1899. *Daily Iowa Press* (June 15).

Georgia

1869: Sheet of Flame

ABOUT THE HOUR of 1 P.M. yesterday, the 6th instant [i.e., October 6], the community [of Cuthbert] was startled by a terrific explosion in a direction apparently northwest from this, accompanied by a dense volume of smoke. One gentleman compared the report to a simultaneous discharge of . . . artillery, and distinctly saw the column of smoke which rose in the quarter from which the sound proceeded. The explosion was heard by two-thirds of our citizens, and some assert that the shock of an earthquake was plainly felt.

"Addison, an intelligent colored man in the employment of Mr. William H. Brooks, says he was at Beall's mill when the event occurred, and, in company with a white man, saw what resembled a sheet of flame descend from the heavens towards Lumpkin, northwest of Cuthbert, and heard at the same moment a terrific explosion.

"The true solution of the mystery may be found, perhaps, in the sudden projection from the moon or some other heavenly body, of a vast aerolite or metallic mass in a state of fusion which doubtless lies deeply imbedded in the bosom of mother earth."

Source:
"Astounding Phenomenon," 1869. *Petersburg* [Virginia] *Daily Index* (October 14).

1875: Lightning Figures

ON JULY 12, 1875, according to a contemporary issue of *American Journal of Science,* lightning struck a tree outside a house, splitting it apart. The strike was so close to the house that its occupants were rendered essentially unconscious. When they recovered, they noticed a most remarkable phenomenon: Every one of them had an image of the tree on his or her person.

The image was most dramatically imprinted on a child, age and gender unspecified. The child, who had been in the middle of the room when the lightning hit, "is impressed upon its back and exactly opposite upon its stomach," according to an adult male family member identified only as Mr. Simmons. "The entire tree is plain, and perfect *in toto;* every limb, branch, and leaf, and even the severed part, is plainly susceptible." The leaves were also correctly positioned, but they appeared not whole but in skeletal form. The marks faded in the following days.

Such phenomena have been reported before and since, though science does not recognize them as authentic.

Source:
"Images Produced by Lightning," 1875. *American Journal of Science* 3, 10: 317. Cited in William R. Corliss, ed., 2001. *Remarkable Luminous Phenomena in Nature: A Catalog of Anomalies.* Glen Arm, MD: Sourcebook Project, 244–245.

1880: "Wonders of a Meteor"

AN EXTRAORDINARY object, either a dramatic meteor or something more mysterious, sailed over Columbus at low altitude at around ten o'clock the evening of September 30. Rising from the southern horizon, it headed toward the northeast, "preserving a perfectly horizontal line in its journey. It was composed of three parts, which were perfectly developed balls of an equal size, and equi-distant from each other. The first ball drew out a tail which enveloped the two following balls and extended several yards behind them. This tail was exceedingly luminous, save at the extremity, which was somehow indistinct, having a nebulous appearance. Its motion was slow, and was visible to the observer for full fifty seconds. It did not fall to the ground like other meteors, but continued to course northeastward until lost sight of. It was indeed a brilliant and extraordinary phenomenon."

Source:

"Wonders of a Meteor," 1880. *Marion* [Ohio] *Daily Star*. Reprinted from the *Columbus* [Georgia] *Enquirer*.

1881: Ghost Machine, Phantom Engineer

WE LEARN THAT A mysterious apparition was seen one night recently, which was nothing more nor less than a phantom locomotive. The narrator, whose name we withhold, was walking the track of the Western and Atlantic railroad, about two miles from town [Dalton] one night last week, when he discovered the head light of an engine approaching around a curve. Strange to say, he heard no noise as the train came on, and presently he stepped from the track and waited for it to pass. He was still puzzled why no noise attended the engine's approach, and as it came opposite he noticed that the whole machinery had a ghostly, phantom-like appearance. At the throttle stood a pale, wild eyed engineer, while a spectre-like fireman was pulling the bell rope, but no sound came from the bell. All this he observed as the train rushed past like a shadow. He swears that the object he saw was a phantom train, of which there are several mentioned in railroad lore."

Source:

"What the Papers Say," 1881. *Atlanta Daily Constitution* (January 21). Reprinted from the *Dalton* [Georgia] *Citizen*.

1881: Fiery Whirlwind

AT THE Z. T. BAISDEN farm near Marietta, a strange whirlwind hit a 12-acre cornfield around noon on July 25. Four feet in diameter and sometimes 100 feet high, the dark black whirlwind sometimes spun off three minor whirlwinds. These would scamper across the field, ripping up the plants, then come back together with a loud crashing and burning, then shoot high into the sky.

The fires inside the big and little whirlwinds emitted a sulphurous odor that could be smelled as much as 300 yards away.

Three young women who were visiting Mrs. Baisden got within 150 feet of the extraordinary manifestation but were forced to retreat when burning sand flew into their faces. According to a local press account, "Mr. Baisden says that he cannot account for this strange phenomenon, and it certainly frightened all who saw it. The strange part was that it contained fire, yet did not appear to burn the corn that it did not tear up, and its sulphurous vapor sickened and burnt all who got close enough to get a full breath of it."

Source:

"Young Women in a Whirl-Wind," 1881. *Marietta* [Georgia] *Journal* (July 28).

1882: "Supernatural Monster"

A SUPERNATURAL MONSTER has been seen more than once in the swamp below Mr. Rufus T. Beacham's. Dogs and guns have so far proved useless in trying to bring him."

Source:

"Life in Georgia," 1882. *Atlanta Constitution* (August 18). Reprinted from the *Dublin* [Georgia] *Post*.

1882: "A Very Large Dog with a Lamp Perched upon Its Head"

AN APPARITION MADE its appearance at Mr. John A. Harvill's house last Tuesday night [December 5] that has thrown the neighborhood into a fever of excitement, indeed so intense has the excitement become and so eager are people to be convinced of the matter that many will doubtless go to Mr. Harvill's to see for themselves. On the night mentioned, Mr. H. and his wife, to whom he was married only a short time ago, were sitting around the fireside, when a noise resembling that of an old squeaking cart or wagon attracted attention. Not much heed was paid to the noise until it stopped in front of the gate, and although it had ceased to move, the noise continued without abatement.

"Mr. Harvill, thinking something strangely of the state of affairs, went to the door and there stood what appeared to be a very large dog with a lamp or torch of some kind perched upon its head. He hailed several times, thinking perhaps some of his neighbors were only playing off a trick upon him, but upon receiving no reply from his several demands shot at it several times; but to his utter astonishment and amazement, there stood the specter as steadfast as the rock of Gibraltar. Like the bard, he thought 'twas the weakness of his eyes that shaped the monstrous apparition. By this time, attracted by the reports of his gun, some of the neighbors of Harvill had assembled and a determination was agreed upon to make an attack. Having agreed upon the nature of the attack torches were procured and the advance began, but to the astonishment of all, when they approached with the light, the ghost began its onward march, accompanied by the same squeaking noise. While this story was being told our reporter, a gentleman who lives in the neighborhood of Harvill, says that he has seen the same object but that there was no noise with it."

Source:

"Ghosts in Laurens," 1882. *Atlanta Constitution* (December 9). Reprinted from the *Dublin* [Georgia] *Gazette*.

1883: A Snake Where No Snake Should Have Been

A GENTLEMAN LIVING in the lower part of this county [Taylor] while digging in a well last week, and having reached a depth of 54 feet, unearthed a large snake about three feet long of the moccasin species, which[,] after having been dug up, showed his revenge by making every effort to bite the person in the well, who[,] being closely confined in the well, escaped with his life, but managed to kill the snake. How the snake came to be so far under the earth is a mystery, as there was no stream nor pond of water in a mile of the place. The gentleman had been in the well all day and it was impossible for the snake to have fallen therein, but from all appearances the snake made his way from the top of the earth and had imbedded itself at this great distance. The soil through which it passed was mostly firm and hard, and its hole was first discovered about forty feet from the surface of the earth."

Source:

"Deep Down for a Snake," 1883. *Atlanta Constitution* (January 21). Reprinted from the *Butler* [Georgia] *Herald*.

1883: Invisible Rappers

IN THE LATTER PART of February, a house in Montgomery County fell victim to mysterious rappings whose cause seemed indeterminable. The house belonged to a man identified only as Mr. Adams.

He first heard them one day around the eighteenth of the month. He thought someone was hitting the side of his house with sticks, but when he went outside to look, no one was in sight. As he stood in the yard, the rappings started inside the house. Again, investigation turned up no source.

Over the next few days the rappings grew ever more intense. Frightened and exasperated, Adams virtually dismantled his house. But even after tearing out ceiling and walls, he was no closer to solving the mystery.

Source:
"A Ghost Story from Georgia," 1883. *New York Times* (March 1).

1883: "Aerial Cyclone on the Wing"

A SOMEWHAT NOVEL, not to say startling, cloud phenomenon was witnessed by early risers here [Albany] this morning. About twenty minutes to seven o'clock a large, long shaped, low, black cloud made its appearance in the eastern horizon, moving rapidly in a direction from northwest to southeast [and] requiring only a few minutes to pass entirely out of sight. As it passed across the early sun's disc, the sudden darkening occasioned by it, was very noticeable as well as a sudden movement of the air around in the direction of the cloud. Only a very few minutes, less than five in all, were allowed for observation before it was out of sight. It was surmised to be an aerial cyclone on the wing, and looking for a place to alight, and may be heard from further yet. It was the more conspicuously noticeable here from the fact that the horizon was perfectly clear and unobstructed by any other clouds at the time of its passage."

Source:
"A Striking Phenomenon," 1883. *Atlanta Constitution* (September 28).

1884: "Mysterious Luminary"

ABOUT ONE O'CLOCK [on July 17] a young lady in Mr. [W. P.] Ponder's house [in Monroe] was awakened by a light so bright that she was sure the house was on fire, and she ran down stairs and gave the alarm. Finding that the house was not burning, the family thought the woods were on fire; but a minute's investigation showed this to be a mistake. The cause of the brilliant light was discovered, and it caused the spectators to gaze with wonder and awe. Above the eastern horizon—a little north of east—about where the sun would be, if an hour high, was a brilliant light, apparently about the size of a full moon. It looked like a large star, its light white like that of the sun or of an electric lamp. At intervals it would expand and throw out luminous rays in all directions, making the surrounding brightness still brighter; and then it would immediately contract again to its former size. The light was as bright as sunlight, and objects could be seen as plainly as in the day-time. The sky to the south was especially bright. The body of light was apparently stationary. Mr. Ponder's family watched it about an hour, and then retired, leaving the mysterious luminary still blazing in the heavens. What is it?"

Source:
"Signs in the Heavens," 1884. *Dublin* [Georgia] *Post* (July 23). Reprinted from the *Monroe* [Georgia] *Advertiser.*

1886: A Spook in the Air

IN THE FAR EASTERN section of Atlanta, a party of local people was returning from an evening on the town when one of the group shouted, "Look there, oh, my God, look there." A press account two days later reports:

"She pointed into the air down the railroad track, and every member . . . looked in the direction indicated by the woman's finger. . . . In less than another second the entire party was running up the hill from the railroad at full speed, yelling as they went. Their noise attracted the attention of several persons and to the questions the crowd replied by saying that a spook was in the air. . . . Every body agreed that the something was a woman dressed in pure white. She appeared suspended in the air about ten feet above the ground and every second or two turned around[,] all the time uttering the most agonizing groans.

"Many . . . people would not believe the story and last night nearly a hundred visited the locality. The first batch went about seven o'clock, and in a short time they returned greatly excited and verifying the statements already current. This spread and was indorsed [*sic*] by others who went to the scene. By eleven o'clock the excitement was so intense that Captain Manly was sent for. He went through the ex[c]ited district and tried to ascertain the trouble. Every [person] he met,

She appeared suspended in the air about ten feet above the ground and every second or two turned around[,] all the time uttering the most agonizing groans.

confirmed the stories of the apparition, and had either seen it or heard of it. He could not induce one of them to go with him to the locality and went himself, but found nothing unusual. As he came away, however, he met a party of young men who told him that they had seen the same thing, but could not describe it."

Source:
"Spooks in the Air," 1886. *Atlanta Constitution* (September 16).

1888: Angels, Cross in the Sky, and a Day of Terror

THERE IS GREAT EXCITEMENT among the colored people and the ignorant whites along the Kenesaw Mountains, in Cobb and Gordon counties," a newspaper observed. It all started with the claims of a young African American woman, Lillie Marles, who lived in rural Calhoun and insisted that she had conversed with angels.

She had been directed to go to a hilltop to receive a revelation. When she reached the top, an angel descended and flew her high into the air so that she could view the whole world. "All

shall be destroyed," she said the angel said. "On the first Saturday in December next an earthquake will swallow up Calhoun. This will be accompanied by a whirlwind. Then a great fire will burn all that is left, and woe unto him who is not ready for the great day." There were other revelations about future events.

Hearers inclined to doubt were turned into fervent believers when many saw, or thought they saw, a rainbow-colored Greek cross suspended over Mount Kenesaw. After half an hour it dissolved, but it would return on several subsequent occasions.

Source:
"Future Events," 1888. *Coshocton* [Ohio] *Semi Weekly Age* (March 25).

1888: Subterranean Mystery Heat

IN EARLY SUMMER, a well on a Burke County plantation began to exude steam. The well was 100 feet deep and had 40 feet of water in it. Landowner Lawson E. Brown heard a roaring sound when the effect began. The water was boiling, and it continued to do so for weeks before ceasing for a short period of time. Then the boiling resumed at the same intensity.

Brown's tenant cleaned the well down to the bottom, thinking that perhaps something in the well was causing the problem. It wasn't, and the steam continued to rise. It was not reported how long this went on.

Source:
"Georgia Gossip," 1888. *Atlanta Constitution* (August 6)

1889: "Like a Terrible Cyclone"

A RUMOR IS CURRENT in Athens, Ga., to the effect that there is a rock in Clarke county, about six miles from town, upon which no man dares to tread. The superstitious say that about twenty or twenty-five years ago an old gentleman buried a coffee pot full of gold at the foot of this rock and has since died, and when a man passes that way his ghost appears and drives him off. Parties passing near the place have been run more than a mile by this invisible ghost. They say that when they come near the place, even though it be a perfectly still day, a noise can be heard like a terrible cyclone, and the tall pines which surround the rock begin to bow and many of them fall to the ground. One Sunday two gentlemen from Athens, who heard the rumors and doubted them, visited the rock with the intention of proving the reports to be false, but they did not stop long. They climbed to the top of the rock, when they became astonished by hearing a terrible crash. Hardly before they knew it a large pine tumbled to the ground right at their feet. While an examination of the tree was being made, which was twisted from the ground, another fully as large came down with a crash. The 'explorers' then 'skipped out.'"

Source:

"A Ghostly Guard," 1889. *Mitchell* [Dakota Territory] *Daily Republican* (February 19). Reprinted from the *Indianapolis Sun*.

1889: Hellhound on Their Trail

THAT STRIP OF COUNTRY lying between Colonel Mike Mattox's and Millstone [in Oglethorpe County] has always been considered as haunted ground, and a few years ago the good people living thereabout were greatly disturbed over the appearance in the public road of a strange looking wild animal that suddenly sprang up before them, or, if riding, beneath the feet of the horses, or under the wheels of the vehicles, but on being struck at with a whip would vanish from sight. This singular visitor only appeared at night, and was seen by a number of responsible white men, including Mr. A. G. Power, well known to our people, and we think also by Mr. Dock Mattox. It made periodical [*sic*] appearances for several months, when no more was heard of it.

"But recently this varmint, or whatever it is, has again began [*sic*] its pilgrimages, and has been seen several times recently, but this time only by American citizens of African descent. A few nights since . . . Perry Mattox, while near Mr. Mike Mattox's, suddenly noticed just in front of the wagon he was driving a strange little red colored quadruped, about the size of a fice dog, or perhaps somewhat larger. The mules noticed the thing as soon as the driver, and became terribly excited and frightened. They began to run away, but Perry managed to keep them in the road. When they slackened their gait[,] the thing in front would also hold up, so as to keep a certain distance ahead. When it reached a point on the road where there is an old grave yard, the apparition seemed to glide among the tombs and then vanish from sight. Perry was badly frightened and firmly believes that he has seen a spirit.

"Several other Negroes report having recently seen this thing, and they say that whoever strikes at it shall instantly drop dead. It seems only to pursue solitary travelers, and makes its appearance about midnight. Sometime since[,] it rushed between the legs of a Negro, and he says he could not feel it touch him. The thing then vanished from sight, to reappear in an instant about ten feet in advance of him, to again vanish in a grave yard. In fact, this little four-footed visitor seems to have its home among the numerous old burial grounds that dot the neighborhood."

Source:

"A Ghost in Goosepond," 1889. *Atlanta Constitution* (May 27). Reprinted from the *Elberton* [Georgia] *Star*.

1889: "A Woman Dressed in White and of Giant Size"

IT IS SAID THAT A GHOST was seen in the ridges of Walker county, Ga., by some parties returning from church. It appeared in the road, about as near as they could guess, 100 yards ahead of them. It was in the shape of a human when first seen—eyes looked like two great balls of fire, teeth as white as snow, hair almost trailing on the ground. As they neared the object it appeared as a woman dressed in white and of giant size. The party became scared, and, but for fear of being laughed at, would have turned and run. A few moments' consultation with renewed energy caused them to advance. They had moved but a few steps when the ghost commenced moving backward, all the time appearing larger to the frightened party. It moved on about 200 yards, when very suddenly it appeared to explode and throw its fragments in every direction, resembling the explosion of a coal oil lamp. Every fellow made for his respective home, scared within an inch of his life. No explanation of the apparition has been offered."

Source:
"Gigantic Ghost in White," 1889. *Denton* [Maryland] *Journal* (October 5). Reprinted from the *Cincinnati Enquirer*.

1890: Perpetual Rain

D. R. PARKHAM TELLS OF A curious phenomenon in Chattahoochee county—a place where rain falls perpetually. The spot is located on a little knoll in a thin wood on the Shipp place, two miles from Thad. Mr. Parkham says the discovery was first made last Thursday [March 14], and that rain has been falling steadily on the knoll since that time. The downfall covers a space of fifty square feet. This space is perfectly wet, and the leaves on the ground are full of water. Mr. Parkham says he visited the place with G. A. McBryde at noon Tuesday. There was not a cloud to be seen in the sky, and the leaves everywhere, except on the square, were as dry as tinder. 'I stood with the space between me and the sun,' said Mr. Parkham, 'and saw the raindrops coming down steadily from the sky. I held out my handkerchief and it was soon saturated with water.' Mr. Parkham says that everybody who hears about the phenomenon is skeptical, but that the many who have visited the place in the last few days have gone away convinced. No one has yet offered an explanation of the mysterious rainfall. Mr. Parkham suggests that some powerful unknown substance attracts the moisture from the atmosphere."

Source:
"A Curious Phenomenon," 1890. *Atlanta Constitution* (March 21).

1891: Snake-Headed Fish

A LARGE AND CURIOUS fish was caught in the creek near here [Calhoun] recently. It's [*sic*] head resembled that of a snake, and it had teeth like a human being. It is of a variety unknown to the oldest fishermen."

Source:
"A Fish with a Snake's Head," 1891. *Atlanta Constitution* (May 26).

1892: Fireballs All Around

MR. J. L. SHAWL, A very well known young man in our community, was out at a late-hour last Sunday night [presumably June 12]. He said there was a mysterious light visible in the heavens, and balls of fire fell around him. He was at a loss to account for the phenomenon and made tracks for home."

Source:
"Saw Balls of Fire," 1892. *Atlanta Constitution* (June 21).

1892: The Train Passed over Her

A LITTLE OVER A YEAR ago an engineer on the old Brunswick and Western road was frightened by an apparition of a woman in white that frequently appeared at a lonely station at night and made ghostly signals to the engineer. Recently another woman in white has appeared, and on several occasions has attempted to wave the train down. One engineer reports that the woman was seen standing on the track, just in front of the engine, which seemed to pass over her; and she was seen running towards the woods as the train dashed by. Evidently some crazy woman must be at large in that vicinity."

Source:
"A Woman in White," 1892. *Atlanta Constitution* (August 1).

1892: The Moon Vanishes

H. H. CROVATT, A SAVANNAH, Florida and Western engineer, reports having seen a wonderful phenomenon last night [November 24] at 8:30 o'clock while on his engine seven miles north of Savannah. He noticed that a portion of the moon was dark and thinking it an eclipse, called his fireman's attention to it. In another second half of the moon became dark and in about three minutes afterward [the moon] disappeared. He says he looked anxiously for it to reappear, but it was seen no more. He firmly believes that the moon was struck by a comet and knocked into smithereens. Crovatt has been [an] employee . . . for twenty-seven years. He vouches for the truthfulness of the foregoing, as also does his fireman, Green Chapple."

Source:

"Crovatt's Fiery Story," 1892. *Atlanta Constitution* (November 26).

1893: Mysterious Dog-Killer

A MYSTERIOUS ANIMAL attacked and killed dogs in the Atlanta suburb of West End in the first weeks of March. Over a three-day period, no fewer than six were slain, two of them mostly devoured. On three occasions, each of them at dusk, witnesses saw, or thought they saw, the killer, and twice it was fired upon, without apparent effect.

One witness said its outline was that of an enormous bear. Another said the creature looked like a large Newfoundland dog. A third swore it was a hyena.

The creature would come out of a woods located near a cemetery. Local people were terrified, and some refused to leave their houses after dark. Hunting parties trying to run down the strange beast were invariably unsuccessful.

Source:

"The People Scared," 1893. *Atlanta Constitution* (March 18).

1893: Invisible Rock-Thrower

IN OAK GROVE district, near Hammond postoffice and just fourteen miles from the city, the little settlement in that section has been thoroughly aroused. For over a week, there has been some rock throwing about the home of George T. Reeves, a respected and honest farmer of that country, and despite the fact that from fifty to 100 armed men have scoured the woods about his home in all directions, the mysterious rock throwing continued until Monday [October 16]. The neighborhood about the home of Reeves has been worked up to such a pitch that if the miscreant is captured he will be dealt with severely by the party capturing him.

"Tuesday one week ago a daughter of Reeves was out in the cotton field picking cotton. With her was an adopted child of Reeves that had been left an orphan. While the two girls were at work a rock fell near the daughter of Reeves. She turned to the other girl and told her not to throw again. The other replied that she had thrown no rock. Both girls went back to work, but it was not long before another rock was thrown. This time it fell near the adopted daughter of Reeves. She became angered and demanded of her sister why she threw the stone. Her supposed assailant denied having thrown the rock.

"Both girls quarreled for several minutes and returned to their work. Again a rock fell by the side of the first girl. She then became thoroughly aroused, thinking that the girl—her adopted sister—had only pretended that a rock had been thrown at her in order to find an excuse for throwing a second stone. The mother appeared on the scene while the quarrel was at the hottest, and suggested that some negroes in a field a couple of hundred yards away might have thrown the rocks. Mrs. Reeves then went and began to watch. While she was on the alert to discover the person throwing the stones, a fourth rock fell by the two girls, who were standing together. They saw that they had been mistaken in accusing each other, and the daughter of Reeves called for the 'coward to step in sight.'

"A search failed to reveal the whereabouts of any one. That night a number of rocks, some weighing as much as a pound and a half, were thrown against the house and upon the roof.

"The next day, Wednesday, Farmer Reeves was at work stacking fodder. He was leaning over a stack when a heavy stone struck him on the shoulder near the neck. He looked about him but no one was in sight. A few minutes elapsed and a second stone fell near him. This continued for some time, and even when Reeves walked to his home rocks fell about him, sometimes very nearly striking him, and at times falling either behind or in front of him. Again that night the house of Reeves received a bombardment of stones.

"About Thursday the story of the falling rocks had been heard in nearly all the settlement homes, and the curiosity of the people was thoroughly aroused. They began to come to the home of Reeves and the stones still fell. Some one suggested that it was some phenomenon of nature, and that the Reeves farm was receiving the remnants of the end of a comet's tail. But this suggestion didn't go at all.

"Sunday the men of the neighborhood gathered at Reeves's house, armed to the teeth. They were all not only determined to find out if any one was throwing the stones, but were exceedingly anxious to discover what or who it could possibly be. Sentinels were posted about the house for a distance of a quarter of a mile to watch for any suspicious characters. As soon as the outposts were established the men divided into posses and began to search for the embryo David who was casting pebbles about so recklessly. During the time that the men were in and about the house no rocks dropped or were thrown.

"The men had hardly been gone thirty minutes, however, when the stones began to fall. First one dropped fairly upon the residence of Reeves. Another fell near the front porch. A third struck the side of the house. Two old men and three or four women were at the residence of Reeves, but they had no means of communicating with the armed men.

"Sunday night the stones fell again, but no one could be discovered.

"If there has been only one man casting the stones about he must be a well-trained man with the sling-shot with plenty of time on his hands.

"On the day before yesterday morning (Monday) just about sun-up the throwing commenced with renewed energy and it only stopped at 11 o'clock. What the rock-throwing means, or who can be perpetrating it, the entire community about the Hammond postoffice is in a befuddled condition. An effort will be made to get a detective on the scene.

"Mr. W. J. Waits was in the city yesterday and confirmed the reports about the rock throwing and told the foregoing story. He said: 'I was in the home of Mr. Reeves when the party left to hunt down the rock thrower. Guards were posted about. They had been gone about half an hour when four rocks fell. I can't account for it at all, but we are pretty badly roused up in the neighborhood and we are determined to unearth the miscreants, whoever they may be.'"

Source:
"An Embryo David," 1893. *Atlanta Constitution* (October 18).

1894: Barking Monster

ONE DAY IN AUGUST hunters along the Alapaha River in the southern part of the state heard loud barking sounds. They naturally assumed that they came from a nearby, though unseen, dog. When the barking went on and on, they suspected that the dog had treed a game animal, and that attracted their interest as sportsmen.

"Approaching the spot carefully," the Atlanta press reported, "they, presently, came in sight of something that made their hair stand on end, so great was their surprise. There, lying on the sandbed, was an immense fish, the body and tail being perfect, but, instead of the regulation head, the monster's head was shaped like an English bull dog with great rows of teeth glistening in the sun, and, all the time[,] the creature was baying as if about to attack something. Before the hunters had recovered their senses and thought of using their guns, the fish glided into deep water and disappeared, the same dismal barking being heard once or twice after it went under."

Source:
Untitled, 1894. *Atlanta Constitution* (August 27).

1895: "The Most Loathsome-Looking Animal That Ever Inhabited This Earth"

A GEORGIA NEWSPAPER quoted the alleged testimony of an unnamed hunter said to have seen "the queerest freak on record" in a swamp near the Savannah River, apparently in early February:

"It was about 8 o'clock in the morning, and I was standing with my gun behind some trees, about a hundred yards from the outside, waiting to get a shot if possible at a passing deer. Suddenly I heard a noise in towards the river that made my hair stand on end. It was not such a loud noise, nor did it sound very dangerous—but it was peculiar; I had never heard anything like it before, and I hope I may never hear it again. It was a sound somewhat like the quacking of a duck and the hissing of a snake combined. That is as near as I can describe it, and yet that does not give any idea of how it sounded. It was stronger and louder than either, and yet that is the impression that it gave me—either that it must be a monster duck or a huge snake; and then I thought it must be both.

"The sound seemed to issue from a thick place surrounding a kind of lagoon. I kept my eyes fastened on the spot with cocked gun in hand. I had not long to wait. In a few seconds I heard a kind of splashing in the water and peering through the bushes I saw, about a hundred yards away, what seemed to be the head of an enormous duck. But I thought surely it was the king of all ducks—for the bill was at least a foot long and as black as could be. It was still making that blood-curdling, half-hissing, half-quacking noise, and seemed to be wading or swimming slowly along in the mud and water.

"Before I had time to think—even if my brain had been in any condition for such work—the creature raised itself up a little and I saw the blackest, ugliest, most loathsome-looking animal that ever inhabited this earth, I do firmly believe. Its body was between three and four feet long and was also black. When I first caught a glimpse of it I thought I must have been mistaken in the head and that it was an alligator coming out to sun himself; and yet I had never seen an alligator like that before.

"Coming on up nearer to where I crouched in terror behind a tree I soon had an opportunity to see the thing in all its strangeness and ugliness. There was a little knoll where the puddle of water ended, and—horror of horrows [sic]—what I had thought was an alligator stepped up on this little elevation, and I then saw that it had only two feet. As near as I could judge, its legs were about a foot and a half long, and it stood there like some huge blackbird in the night, with its bill stuck downward and still emitting that unearthly kind of noise.

"To say that I was paralyzed with fear would hardly give you an idea of my condition. If the creature had seen me and started in my direction, I am sure I would never have been alive when it reached me. It never occurred to me that my gun would have been any protection; I was so completely terrified by the appearance of the unnatural looking thing that I couldn't think of anything else. I believe I would rather have braved a thousand alligators than that bird, or beast, or whatever you might call it.

"It stood there, I suppose, about a minute, and I had a good opportunity to examine it with my eye. Its body was tough and scaly, like an alligator's, and the tail went off to a point. It had legs like a turkey or duck, only they were larger and stronger; its feet I could not see on account of some bushes. It kept turning its bill up and down and around, but try as I could I never did locate its eyes. They were the features I was right then most interested in not seeing, and I suppose they were so black that at that distance I could not make them out.

"I can never describe the awful sound the thing made with its mouth. It made my blood run colder every time I heard it. After a short while, that seemed an age, the creature gave a kind of spring from the ground, and before I could realize what it was doing it went up into a large tree and sat on the lowest limb. As it did so, I could hardly believe my eyes when I saw two dark wings spread out from its side and strike the air with a heavy sound that made my heart sick. I had not noticed before that it had wings, but wings they certainly were, although I could see no feathers. As soon as it had poised itself on the limb its wings were drawn so closely to its body that it was impossible to detect where they were. I looked at it for a second or two, and then as its back was towards me, I thought it was a favorable opportunity to get out of the swamp, as I was hungry anyhow. I stole quietly out and never took a long breath until I had left the swamp a full half mile behind."

Source:
"A Strange Animal," 1895. *Atlanta Constitution* (February 11).

1895: Balls of Lightning Fire

FRIDAY NIGHT [OCTOBER 25] a strange phenomenon presented itself to a party of young ladies and gentlemen of this city [Albany] who went to a dance in the eastern section of the county. Several times to and from the dance[,] balls of fire, springing from tiny sparks, would gather together and dash under and around their vehicles, in many instances brilliantly illuminating the roadway and badly frightening the horses and the superstitious of the party.

"A gentleman from Harralson county who is down in this section selling his crop of apples had the same experience between the city and Dawson a few nights ago. As he was riding leisurely along in his covered wagon two immense balls of fire seemed to rise as if by magic and roll across the road underneath the feet of the gentleman's horses. So sudden and unexpected was the phenomenon that the team came near running away and tearing up the wagon."

Source:
"Balls of Fire," 1895. *Atlanta Constitution* (October 28).

1897: A Gentle Shower from Nowhere

AT MACON'S ROSE HILL Cemetery, on one bright and clear October morning, the sexton stared in disbelief as rain fell on a section that had once housed the bodies of Union soldiers from the Georgia theater of the Civil War.

A member of a crew of workmen digging at the site had alerted him to the strange occurrence. The worker had seen it while passing the area and, to be sure his eyes weren't fooling him, even stepped under the shower. He got wet, though not soaked, since this was a gentle shower. Once the sexton saw the shower for himself, he called the mayor's office and asked that a reputable citizen or city official be dispatched to investigate.

The phenomenon continued for at least four days. At one point, Captain Monroe Jones, chief of the fire department, and an *Atlanta Constitution* reporter went to the site. The reporter wrote, "The most remarkable part of the report of the sexton is that it is perfectly true. Away over the hills of Rose Hill cemetery on a shaded slope overlooking the winding river, there is a corner where once slept the union soldiers of [General James H.] Wilson's army who fell in battle around Macon. With one or two exceptions the graves of the union soldiers were relieved of their bones a number of years ago, when the ashes of the dead were removed to the Union cemetery at Nashville. . . .

"Numerous callers have visited the spot in the cemetery today. In fact, Sexton Hall has shown so many there he is thoroughly fatigued with the exercise. The phenomenon cannot be accounted for by those who have witnessed it. That it is rain cannot be doubted, for the drops can be seen falling from above the tops of a tall black-gum and hickory tree, between which the shower seems to make its way to the ground. During the entire day the rain has been falling between these two trees near the foot of the lone hill in the cemetery, all the while the sun fairly beaming down as bright as could be on the clearest, fairest of sunny autumn days. . . .

"This incident revives another story about Macon's beautiful cemetery. A few years ago Undertaker Dennis Keating discovered that at a certain spot in the cemetery behind a hill, there could be heard the noise of picks digging away within the bosom of the earth. Mr. Keating, who is one of the best known citizens of Macon, is ready to make affidavit to this at any time."

Source:
"Rain for Four Days on One Spot," 1897. *Atlanta Constitution* (October 26).

1902: Goat Ghost

People living in the vicinity of Oconee cemetery [in Athens] are talking of a strange phenomenon. Several people say that of late at a given hour in the night there arises out of the grave of a woman an airy figure that gradually takes the shape of a goat with long horns. The transition occupies several minutes, and as soon as the figure is well-defined it speeds to the Oconee[,] plunges into the current and is seen no more. Several attempts have been made to solve the phenomenon, but without success."

Source:

"Spook with Horns Seen in an Athens Cemetery," 1902. *Atlanta Constitution* (July 12).

1910: Elusive Illumination

In the late summer and early fall a mysterious ball of light began appearing nightly in Dyas, in western Monroe County, on property belonging to James Holloway. It would show up around 11 p.m. and remain in view until daybreak. A press report states, "The ball of fire generally remains about 30 or 35 feet above the earth, and constantly travels over about a 20-acre field near Mr. Holloway's house. Sometimes it dances very rapidly along the ground, and all of a sudden rises to a distance of thirty or forty feet above the earth."

Observers who tried to approach the object would see it vanish, then moments later reappear several hundred yards away. Curiosity-seekers—thousands in all—came from surrounding counties to witness the curious phenomenon.

Holloway claimed that this was not the light's first appearance. Thirty years earlier, it had performed identical stunts in the same field.

Source:

"Ball of Fire by Night Mystifies Dyas Citizens," 1910. *Atlanta Constitution* (October 16).

Idaho

1874: Beautiful Ghost, Bad Manners

IT IS CURRENTLY reported that Belle Chamberlain's ghost visits the brick house on Idaho street, above the Central hotel [in Boise City]. There is nobody living in the house, but several parties have undertaken to sleep there, and all report strange noises, the opening of bolted and locked doors, and a beautiful angel tripping through the room, turning tables, chairs, and otherwise disarranging things. Several of the boys have been running silver bullets to shoot the evil spirit, but others are laying plans to secure the charmer without doing her any violence."

Source:

"Curious," 1874. *Idaho Statesman* (November 14).

1894: Shower of Salt

FRIDAY [PERHAPS APRIL 6] was a balmy spring day. Late in the afternoon a drizzling rain blew up, carried on a wind directly from the south. It was of a peculiar whiteness, and after it had passed every one who had happened to have been out in it and who wore a dark suit of clothes or a dark hat were [*sic*] covered with thousands of tiny white specks. Later it was noticed that every window in town looking to the south was also covered with white spots.

"'It has been raining mud,' said every one who noticed the phenomenon at first, but later some of the curious tried tasting the spots. They had a distinctly salty flavor, and analyses made by druggists proved that they were salt.

"The question now is, Where did the salt come from? There can only be one answer— from the Great Salt Lake, nearly 300 miles south in Utah. It must have been a warm day over the Great Dead sea when a strong south wind swept over it, catching up the salty vapor and sweeping it north to Idaho and finally bespattering the clothes and windows of the good people of Pocatello with the salty spray."

Source:

"A Saline Shower," 1894. *Middletown* [New York] *Daily Argus* (April 11).

Illinois

1854: Ball-Lightning Swarm

DURING THE HEAVY rain of Friday night [May 12], a very remarkable phenomenon was observed by a large number of persons in this city. In all directions, the atmosphere was filled with illuminated balls, resembling fire, which floated through the air, only a short distance above the earth. They varied in size from that of a man's double[d] fists to that of a marble. They were not extinguished by the rain, but were brightest and most numerous, when the storm was the severest. Their appearance in the black midnight is represented to have been exceedingly beautiful. We presume they were electrical balls, but how generated or occasioned we do not pretend to explain."

Source:

"Fire Balls—Singular Phenomena," 1854. *Daily Alton* [Illinois] *Telegraph* (May 15).

1859: Apparition's Cold Glare

THERE IS A married man in Galena, Ill., who insists that every night, about 10 o'clock, a ghost in woman's garb appears in his chamber, looks at him with a stare that appals [*sic*] him, till he turns in terror to his wife, who, it seems, cannot see the apparition. He bolts the door and fastens his windows, but all is of no use, the intruder comes."

Source:

Untitled, 1859. *Gettysburg* [Pennsylvania] *Compiler* (June 6).

1867: "A Vast Monster, Part Fish and Part Serpent"

IN THE SUMMER of 1867, the *Chicago Tribune* reported some alarming news: "Lake Michigan is inhabited by a vast monster, part fish and part serpent," a fact that "no longer admits of doubt." It noted a previous sighting by crews of two small boats off Evanston, a suburb bordering Chicago to the north along the lake. "The animal is between forty and fifty feet in length, its shape serpentine, the size of its neck about that of a human being and the size of his body about that of an ordinary barrel."

The monster had now reappeared, according to the *Tribune,* seen this time below Hyde Park, on Chicago's south side, about a mile and a half from shore. The witness, Joseph Muhlke, "an intelligent German who gains his living by fishing," was out on an early morning. As the sun rose, he still had caught nothing, even though he was in an area where he had enjoyed much success in the past. Finally, he decided to pull closer to shore and try his luck there. So he drew in his lines and started to pull in the anchor. But at that point his boat began rocking in an east-to-west motion. The cause was not the wind, which was blowing from the south, but even as the motion grew, he was still unable to figure out why this was happening.

Feeling uneasy, he pulled harder on the anchor. Then he heard a loud, unsettling sound to the east, something between a heavy puffing and a muffled vocal expression. Muhlke turned around to look. About a quarter-mile away he could see a dark oval shape in the water, resembling an overturned boat. At first it was not moving, but then it began to expand, then burst through the water surface to a height of 3 or 4 feet. At that moment something else rose out of the water, about 20 feet closer to him. It was the head of a large animal, and at nearly the same moment the tail jutted out of the water.

As the *Tribune* told it, "About two-thirds of the monster was out of the water. Thus far the animal had made no forward motion, and manifested no disposition to do so, the only signs of activity displayed being a gentle motion of the head, north and south, as an occasional uplifting of a long neck out of the lake, and a few splashes of the tail upon the water."

Muhlke said the animal was a bluish black color on its upper surface and grayish white on its lower. Its head was slightly larger than a human being's and resembled a seal's in some respects. He could see no teeth and only a portion of the neck, which appeared rough with bony plates running along it as on a sturgeon's back. The *Tribune* stated that Muhlke thought "there were either fins or legs, toward the head and under the water, as there was a constant wash of the water on either side of him [i.e., the animal], near that point, as if he was sustaining his huge bulk by the motion of such appendages. A few feet forward of the tail there was a well developed fin of a greenish hue, corresponding with the dorsal fin of the sturgeon. The entire fin had a lateral motion, and the various spines of which it was composed had an individual longitudinal motion, so that sometimes the fin almost closed up like a fan. Immediately beneath this was an anal fin, possessing the same characteristics, but different in shape, being very long and the spines of equal length."

Two legs could be seen in front of this fin. They were apparently jointless and so flexible that they could be drawn up against the belly when they were not being used. Its tail was of "great size" and "very unsymmetrical in shape, with something resembling long hair covering its entire upper surface, the under surface being diversified with sharp ridges, radiating to the outer edge."

Though this fantastic story is related straightforwardly, the creature does not much sound like anything that could ever have existed, unless

Muhlke saw a very large sturgeon and imagined some un-sturgeonlike details. It is at least as possible that the *Chicago Tribune,* in common with other American newspapers of the nineteenth century, liked to feature outlandish yarns for the edification of its readers. Other lake-monster hoaxes saw print in the *Tribune* from time to time.

Sources:
Hall, Mark A., 1993. "Lake Michigan Monsters." *Wonders* 2, 2 (June): 22–32.
"The Lake Monster," 1867. *Chicago Tribune* (August 7).
"A Lake Serpent," 1867. *Chicago Tribune* (August 6).
Stonehouse, Frederick, 2000. *Haunted Lakes II: More Great Lakes Ghost Stories.* Duluth, MN: Lake Superior Port Cities, 171–174.

1869: Millions of Somethings

NORTH OF TAYLORVILLE, late on the evening of June 4, a shower fell. The next day the ditches, streams, and pools were found to be crawling with strange snakelike creatures—millions of them, according to a press story. They were described as "from one and a half to two feet long, and of three-fourths of an inch to an inch in diameter. This diameter is very slightly lessened at the head and tail. The tail is flat, like that of an eel, but has no caudal fin; indeed, there is no fin at all. The head is in the shape of that of an eel, but the mouth is that of the sucker. The eyes are small, and the ears are simply orifices. Immediately, behind the head, on each side, is a flipper, like that of a turtle, say three-fourths of an inch to an inch in length, including the limb, which has a perfectly developed joint. In color, these snakes, or whatever they are, of a dark hue. . . .

"The number of these creatures is beyond all estimate. They swim in every branch and puddle of water. Their mode of progression, in addition to the undulatory motion of a snake in the water, is by the use of the flipper described above, and they swim entirely under the water or with the head and a few inches of the body above the surface, thus indicating that the flippers are not absolutely essential to motion. They are perfectly harmless."

The animals, believed to be amphibians, were unfamiliar to anyone in the area.

Source:
"A Shower of Snakes,"1869. *Atlanta Constitution* (June 12).

1870: Artifact from a Lost Civilization?

AS THREE MEN DRILLED AN artesian well one day in August 1870 in Lawn Ridge, the pump brought up an object that apparently had lain more than 100 feet below the surface. Found in the drill residue, it was a small metal coin or medallion with a number of curious features.

The discovery came to the attention of a prominent American scientist of the time, Alexander Winchell, who investigated the find and later wrote an account of the mystery in his book *Sparks from a Geologist's Hammer* (1881). He secured a statement from one of the men, W. H. Wilmot, who detailed the circumstances of the find and related just what the object had lain under: 3 feet of soil, 17 feet of yellow clay, 44 feet of blue clay, 4 feet of dark vegetable matter, 18 feet of hard purplish clay, 8 feet of bright green clay, 18 feet of mottled clay, and 2 feet of ancient soils. Just under this last, in a foot of yellowish clay, the coin/medallion had been encased for—well, just how long was the crux of the question.

Composed of some sort of copper alloy and having the general size and shape of a roughly cut U.S. quarter, the object looked as if it had passed through a drilling mill—the edges bore evidence of cutting. Each side had artwork and hieroglyphics-like writing on it, not carved—according to Smithsonian researcher William E. Dubois—but etched in acid. The sketches were fairly crude. One depicted the profile of a woman with something on her head, either a crown or a headdress made of quills. She had an arm raised, while the other clasped a small child. The image on the other side was more ambigu-

ous. A figure faces outward, though not all of its body is shown. It has the ears of a mule or a jackrabbit, suggesting it is either a quadruped or—the modern eye, anyway, might infer—someone in a bunny suit.

The Smithsonian asked coin expert Dubois to analyze the find. In a lecture to the American Philosophical Society, he rejected the notion that it was the result of a hoax or practical joke. Far from being a recent concoction, he argued, it was clearly very old; as he put it, "the tooth of time is plainly visible."

In 1876, speaking to the Geological Section of the American Association for the Advancement of Science at its convention in Buffalo, New York, Professor Winchell expressed his conviction that the coin/medallion was an artifact of great age, disputing a countertheory by Professor J. R. Lesley that it was only a practical joke by some passing French or Spanish explorer. Lesley thought he had cracked a code which revealed a date on the object: 1572. Winchell replied that Lesley's reading owed more to imagination than to anything independently verifiable. Moreover, in his view, the hoax explanation was nonsensical on its face. Why would somebody drop a coin into a very deep hole, in the dim expectation that somehow it would be recovered hundreds of years into the future?

No one seems to have considered a more likely explanation, that the coin was a much more recent prank, planted at the recovery site by someone who knew of the drilling operation. It is true that no one confessed, so far as is recorded, but hoax artifacts purporting to demonstrate the presence of advanced ancient civilizations were virtually a cottage industry in nineteenth-century America.

Undeterred by these considerations, however, amateur historian and archaeologist J. R. Jochmans has expressed the most extraordinary interpretation possible: "What conclusions can we draw about the mystery coin? A lost civilization once existed on the North American continent which worked in copper and other metals; possessed art and writing; attired themselves with crowns and other clothing; knew of and perhaps domesticated several animals including the horse;

utilized acids for etching in a manner that is still not understood today; and perhaps the most disturbing, possessed forms of machinery for the cutting, rolling and processing of metal pieces." Jochmans believes the artifact to be somewhere between 100,000 and 150,000 years old.

Source:

Jochmans, J. R., 2002. *Ancient American* 7, 43: 9.

1870: "Immensely Awful Sky Battle"

FROM A LETTER WRITTEN to J. N. Loughborough on June 13, 1901, from Oakland, California. The correspondent is identified only as "Pastor Schultz":

"In the month of September, in the year 1870—I do not remember the exact day of the month, but it was in the dark of the moon—I was visiting with my uncle in northwestern Illinois, in the locality then known as Green Vale, now Stockton Post-office. One evening, my uncle and I were up until nearly midnight, and before retiring went out-of-doors, and noticed a great red light, lighting up all things lighter than a bright moonlight. The sky was perfectly clear, and as we looked into the heavens to see what caused this great light, we saw a large bright circle of golden red, with streaks of red from the edge of it all around. These outer circles seemed to be moving upward, and finally the object we were viewing assumed the form of a beautiful crown, apparently as large as a good sized washtub. From the body of the crown there went up sharp prongs about eighteen inches in length. All was vivid red. There were no jewels on this crown, but all was of one color. This view lasted about ten minutes after we first saw it. I do not know how long it may have been in the heavens, before we noticed it. It opened up as it appeared and disappeared from view, leaving all in dense darkness for a moment. Then it was immediately replaced by a streak of red light about four feet wide, extending over the entire heavens from east to west, with prongs out on the lower or south side of the band. These prongs were about two feet long, all pointing to the southwest. The prongs were also of deep red. The band then looked much like a timber saw, as all of these sharp points were pointed one way.

While we were looking wonderfully upon this, there arose from the north, as far up as the sun would be when two hours high, a perfect army composed of thousands of men, fully equipped, arms shouldered, and then, from the south, another army just like the first arose.

"While we were looking wonderfully upon this, there arose from the north, as far up as the sun would be when two hours high, a perfect army composed of thousands of men, fully equipped, arms shouldered, and then, from the south, another army just like the first arose. Both of these armies were clad in deep red, and marched toward one another. When they reached the red band—and that extended, as I said, from east to west—they leveled their guns, and both sides fired at once. We heard no report, but saw the barrels, and with the breach of the guns began beating one another, using their guns as clubs. I was in the Civil War, 1861 to 1865, and saw skirmishes where this very mode of warfare was resorted to, but this scene in the heavens was the most awful battle one could ever want to witness. It was immensely greater than anything I ever witnessed in the war. In this sky battle the clubbing went on until there was not a person left standing. All were killed and prostrate on the ground. We heard no noise, as before stated, but saw the thick smoke and confusion of the battle.

"The battle being thus over, the whole scene again disappeared, but only for a short time, when there arose from the east a bright flaming red light, extending from east to west across the whole heavens. This band of light was about one rod wide, and went from the east to the western horizon over the zenith of the heavens. This wave of light lasted about five minutes, revealing a terrible scene of dead bodies and broken guns, covering the whole earth as far as the eye could extend. This whole scene, with its varied changes, lasted from thirty to forty-five minutes.

"When I was in Missouri some years later, hunting land, I met a man to whom, in the course of our conversation, I quoted from Scripture the text about there being 'signs in the heavens,' when he at once described to me the very scene which I have here related, he having also seen it in Missouri at the same time I saw it in Illinois."

Source:
Loughborough, J. N., 1904. *Last-Day Tokens*. Lodi, CA: Pacific Press, 23–24.

1882: Fire-Breathing Phantom

A GHOST HAS FOR the past two weeks been doing business in the woods near Pontiac, Illinois. He is white above the waist and black below, and flames appear from his mouth and nose. A party of ten boys and girls paid him a visit one night, and he came out of a log and stampeded them. An expedition, organized by a correspondent of the Chicago Times, penetrated to his headquarters but was unable to obtain an interview."

Source:
Untitled, 1882. *Atchison* [Kansas] *Globe* (February 7).

1883: Lake Monster in Distress

THE FIRST SEA SERPENT of the year has been seen. He wasn't exactly a sea serpent either, for he was seen in Lake Michigan, off Fort Sheridan, and Captain Brinkerhoff and Lieutenant Blauvelt of the Fifteenth United States Infantry are his vouchers. Captain Brinkerhoff said of it: 'The creature poked its head up, and we saw it plainly with our naked eyes and through our glasses. The head was very dark above and light colored underneath. We could not see the features distinctly, but it looked like an alligator's head. It appeared to be disabled in some way and began to struggle.'"

Source:
"The First Sea Serpent," 1883. *Marion* [Ohio] *Daily Star* (April 22).

1883: Lights, Sounds, Grisly Scene

QUITE AN EXCITEMENT has been created in this county [Bond] in regard to a haunted house upon what is known as the James Nolan farm, about two miles southwest of Pocahontas. Large crowds visited this place every night, coming as far as twenty miles or more from every direction, and are rewarded by seeing lights moving about in the vicinity of and through an old, unoccupied house, disappearing occasionally, soon to reappear. Some go as far as to assert that strange sounds are heard emanating from the house, and others are firm in the conviction that they actually saw a man carrying from the house a woman who had been beheaded. An unusually large crowd went from here last night, and many are expected from points between Pocahontas and St. Louis, some even coming from the city. The probability is that an ignis fatuus has been playing pranks on the boys, and the superstitious have drawn on their imaginations for the other characters necessary to make up a first-glass ghost sensation."

Source:
"A Haunted Farmhouse," 1883. *Decatur* [Illinois] *Daily Republican* (June 20).

1883: Green Frogs in the Falling Rain

DURING THE PREVALENCE Wednesday night [August 1] of a drenching rain-storm, immense quantities of small green frogs about an inch long fell on the Mississippi levee [at Cairo]. In an area of about half a mile, the decks of two steamboats moored there were almost covered, while the shore and fences were literally alive with the crawling, hopping creatures. Nothing of the kind fell in the city. The visitation was only in and about the levee embankment. The phenomenon, although before observed in this section, is attracting considerable attention."

Source:
"A Shower of Frogs on the Mississippi Levee," 1883. *Waukesha* [Wisconsin] *Daily Freeman* (August 3).

1883–1884: Lake Michigan Monster

IN LATE 1883, WHILE towing a mud scow in Lake Michigan, just off Chicago's shore, Captain Dick Brewer happened to notice something that at first he thought was a log. At second

glance it seemed not to be a log at all but an unusual object. Brewer picked up his field glasses and continued his observation. Within moments the object was moving at a fast clip toward the boat. One hundred feet away it stopped, raised its head, and dived under the surface. Brewer had seen it well enough to estimate its length at eight feet. It had a huge head—"a cross between the heads of a horse and a cow," in the words of a press account—with two big eyes set close to the forehead. Long black hair covered its body. It had a large, gaping mouth with sharp teeth. All in all, it looked so ferocious that Brewer feared it would attack his vessel. He directed crew members to watch all sides, but the creature was seen no more.

A few days later, two fishermen in a small boat saw the creature swimming alongside them. One was so terrified that he took ill. There were other, unnamed witnesses to subsequent sightings.

In December 1884, two tugs in the lake, again not far off from Chicago, had their own sightings. The first happened in the afternoon, when the cook of the *Miller* was filling a water barrel. Suddenly something broke from under the surface by the port bow—the same or an identical animal that had figured in the previous winter's excitement. The cook shouted, and the captain and the engineer raced to the spot, where they, too, saw the creature. It disappeared under the water, then resurfaced a few hundred feet away, leisurely rolling about. Because the tug was towing another vessel, Captain Weiman could not give chase, but he watched the animal until it sank into the waves.

An hour later, another tug, *Success,* hauling two boats two miles into the lake from Chicago's Lincoln Park, came upon the animal. The witness was deckhand Ed Burke, who was staring out on the water when the animal shot above the surface with a roaring sound "as loud as a circus lion." It then darted, in what Burke took to be menacing fashion, toward the tug. Hearing the sound, Captain Everatt turned to glimpse the animal before it disappeared. A newspaper story noted, "Everybody who has seen the thing says it bellows like a bull and can be heard half a mile away."

Source:
"Can This Be True?," 1884. *Marion* [Ohio] *Daily Star* (December 20). Reprinted from the *Chicago Inter-Ocean*.

1886: Mirage Train

A MOON-LIGHT MIRAGE was lately witnessed in Illinois. The moon was shining brightly, but a dense fog hung over the flat lands near St. Joseph, and the passengers in a railway train saw a phantom train suspended in air under the fog bank. The apparition was visible for several minutes."

Source:
Untitled, 1886. *Stevens Point* [Wisconsin] *Gazette* (June 12).

1888: Phantom Train Wreck, Ghostly Trainmen

ON THE 16TH OF OCTOBER, 1888, I was employed as a night telegraph operator at ——, Ills., with hours from 7 p.m. to 7 a.m. On this evening I had been reading . . . and had just laid aside the book to reflect upon the work, when I found it was nearly 12 o'clock. I answered a call that one of the 'boys' 'sprung' on me. After finishing up this work I looked out to see if everything was clear for the passage, saw that all switches were in position, all signals displayed, and was about to turn to some other business, when I saw approaching what appeared to be a train from the west. I looked in the other direction and saw another train approaching. This surprised me, as I knew that there were no regular trains due until 12:35, and I had not heard of any extras on the road that night. As they came nearer I saw that the one from the west was a stock train, and the one from the east was a light train of only the engine, caboose and one freight car.

"Then I noticed that I could see through the cars, that they were not solid as an ordinary train. In a moment they had come together right in front of the window of the office, within ten feet of the chair in which I was sitting. There was not the slightest sound, but I saw the engines strike and stop; saw the cars piling up; saw the engineer try to jump as he was caught by the cars and pinned against the boiler head; saw a car double against the one that had caught the engineer in the same manner that the blade of a knife doubles as regards the handle; saw a brakeman caught between the cars that doubled up; saw one of them slide over or across the other, forced by the cars behind; saw the nameless appearance of the man after this action; saw a car fall against the water tank and tip it over; saw a side rod break and go through the cab; saw a portion of the boilerhead or front detach and come with terrible velocity toward the window; felt the shock as it passed through the window and by the chair in which I was sitting; saw the surviving trainmen as soon as the car stopped begin to carry the dead toward the door of the waiting room, through which they passed without opening the door or making a sound.

"I started to go into the waiting room, when I stumbled upon the body of a man lying upon the floor. I looked carefully, saw that he bore on his coat the pin of O.R.T. [Order of Railroad Telegraphers], thereby proclaiming himself as an operator. As I looked I recognized the face of an old acquaintance, and an operator from whom I had not heard for a long time by the name of Frank Willard. While I looked there came into the office two men, who picked up the body, carried it to the waiting room, where I saw them lay it beside those of several others; but as I stepped forward to see who they were the entire apparition vanished. I looked out of the door, but there was no sign of a wreck, the tank was as usual, but I noticed a very strong smell of smoke.

"I then went into the office; looked at the clock, which marked 12:03 a.m. I sat down at the key, called up the train dispatcher, asked if there had been a wreck near that place in the last year, and he replied that the night man had been killed there about a year ago in a collision. I asked my relief in the morning the name of the night man who was killed there, and he replied Frank Willard, and gave a good description of the man as I had known him. He also showed me a copy of the country paper, containing an article referring to the wreck and giving the names of those killed, and noting the fact that the wreck took fire and was burned, with the exception of a couple of cars that had been near the water tank when it was tipped over, and were so wet that they would not burn.

"I do not, or rather did not, believe in the existence of ghosts; but I think that in this case there is proof that on occasions the spirits of the deceased visit the places of their leaving this life, and appear as they did at the moment of departure."

This story is inconsistent with any findable record of a train wreck in Illinois in 1887—though one did take place that year. Two miles east of Chatsworth, one of the great railroad disasters in American history occurred, when a passenger train on the Toledo, Peoria, and Western Railroad hit a burning bridge just before 1 a.m. on August 12. The bridge collapsed, demolishing eleven of the fifteen cars. Eighty-one passengers perished, and 372 were injured. The tragedy attracted national attention and inspired a ballad, now forgotten but sung for some years afterward, "The Bridge Was Burned at Chatsworth."

Source:
Cohen, Norm, 1981. *Long Steel Rail: The Railroad in American Folksong.* Urbana: University of Illinois Press, 170, 272.
"The Phantom Train," 1890. *Decatur* [Illinois] *Daily Dispatch* (July 12). Reprinted from the *Detroit Tribune.*

1890: Neither Venus nor Meteor

AFTER DARK ON THE evening of July 15, Chicago residents were puzzled to see a brilliant object in the western heavens that, as the press reported, "shone like the illumination of a monster electric light." Though it occupied the spot ordinarily taken by Venus, it had a deep red color uncharacteristic of that planet. If that had been all there was to it, this would have been a curiosity, not a mystery. But after a stationary half hour, during which amateur as-

tronomers all over the city had trained their telescopes on the light, the "planet" suddenly shot toward the north, flew on a straight-line course for a few seconds, then descended gracefully before fading from sight. Unlike a meteor, it left no trail behind it.

Source:
"Bright and Curious," 1890. *Oshkosh* [Wisconsin] *Daily Northwestern* (July 31).

1890: "The Bridge Is Burned!"

NO. 15 IS THE BIG Four fast express which runs into Chicago over the Illinois Central tracks from Kankakee. The train is pulled by an Illinois Central locomotive, of which Mr. [Horace L.] Silver is the engineer. For forty-three years the veteran has been handling the throttle of Illinois Central engines. For forty-three years Mr. Seaver has been a spiritualist—not one of the table raising, bell ringing kind, but an intelligent believer that spirit bodies exist. He says he has had innumerable evidences that a spirit hand guided his engine through fearful dangers and happy escapes. Whenever he climbs up in his cab he knows that the spectral engineer is sitting beside him, ready to extend the hand of warning in time of need.

"Mr. Seaver was in the cab, gazing far out along the track, one dark night, wondering how many more trips he would make before his good spirit deserted him. In the train were a thousand old soldiers going to a reunion at Champaign, Ill. The throttle was out to the last notch and the speed more than sixty miles an hour. Suddenly the engineer heard a soft voice whispering in his ear: 'The bridge is burned! The bridge is burned!'

"As quickly as possible Mr. Seaver set the air brakes and stopped the train. In the coaches the thousand old soldiers were sleeping. The conductor hurried forward to the engine. 'What do you mean by stopping this train out here?' he demanded angrily. 'You would better go along the track and find out,' said the engineer quietly.

"Only a few feet ahead of the engine was the river, and over the river hung the charred remnants of the big bridge, which had burned only a short time before. The veterans were saved. This happened in 1890, and Mr. Seaver was hailed as a hero all over the country. 'Something unseen did it, not I,' said the engineer modestly."

Source:
   "Ghost Runs His Engine," 1908. *Indiana* [Pennsylvania] *Evening Gazette* (December 10).

1891: Balls of Fire in the Street

THIS WAS THE HOTTEST day of the season [August 10]. A storm came up, however, about 4 p.m., which cooled the atmosphere, and during the progress of which a strange phenomenon occurred. About 6 o'clock a bolt of lightning tore through the black clouds and struck with a deafening crash at the corner of Hermitage avenue and Polk street. Two big balls of fire, each as large as a half bushel measure, fell in the middle of the street, and for a moment blinded the fifteen or twenty persons who were standing in the vicinity. Several men were hurled to the ground and a little girl was slightly burned."

Source:
   "Curious Phenomenon," 1891. *Bismarck* [North Dakota] *Daily Tribune* (August 11).

1892: Collision?

IN EARLY FALL, A dramatic story came out of Springfield, Illinois, where it was said that "a bright body resembling a large star was seen moving with astonishing rapidity toward the moon, which it struck and then was seen to burst like a bomb, darkening the light of the moon for an instant. It is thought by some that a large meteor came within the power of the moon's attraction and fell into the moon."

If such an event had occurred, it would have been widely witnessed by observers all over the world. Whatever this story is or is not about, it is not about a collision with the moon.

Source:
   "A Meteor Hits the Moon," 1892. *Ogden* [Utah] *Standard* (September 28).

1893: Suicide Reenacted

A GHOST HAS MADE its appearance at New Berlin, in Sangamon county, and the people of that locality are greatly excited over it. About two weeks ago W. B. Smith, a wealthy and well-known stock-raiser, living near that place, made an assignment, and at the same time Charles Kinney, who was associated with Smith, committed suicide by shooting himself through the head. For several nights peculiar noises have been heard in and around the house in which the unfortunate Kinney killed himself. Smith's furniture was stored in the house, and his creditors[,] becoming suspicious that he was taking it away, decided to watch the house.

"Accordingly Saturday night [May 6] two men stationed themselves near the house to await developments. Presently there was a dim light in the house and a noise was heard within. The sentinels, putting a ladder against the portion, climbed to the second story and cautiously peered in at the window, through which a light faintly glimmered. They were shocked to see sitting on the side of the bed a man who resembled Kinney. While they were watching him he placed one hand to his head; there was a sharp report of a revolver; the man fell over and the light went out. The watchmen, in terror, ran down the ladder and called up a number of neighbors.

"Soon a good-sized posse, headed by John Munger, started for the house to capture the ghost. They explored the dwelling from cellar to garret, but found no light, no man, no goblin, and no trace of the tragedy which two veracious men had witnessed a few minutes before. The incident is the talk of the country folks for miles around. The two men who claim to have seen the apparition are said to be reputable and truthful citizens and not addicted to strong drink."

Source:
   "A Ghost Story," 1893. *Decatur* [Illinois] *Daily Republican* (May 10).

1893: "Mounted on a Magnificent Horse"

LEXINGTON, MCLEAN county, is becoming notorious by the appearance, nightly, of a ghost which appears mounted on a magnificent horse and is not confined to any particular spot. Sober, truthful citizens are said to have not only seen but have shot at the phantom. No one up to date is able to explain the mystery."

Source:

Untitled, 1893. *Decatur* [Illinois] *Daily Republican* (August 4).

1894: Ball-Lightning Destruction

DR. E. C. TOWNE reports that during last night's [September 9] thunder storm a large ball of fire fell in a vacant lot near the business portion of the [Chicago] suburb of Austin. A hole several feet in diameter and of considerable depth was torn in the ground[,] and the earth for twenty feet around was seared and cracked. The fall of the fire ball was accompanied by a terrific peal of thunder and vivid lightning."

Source:

"An Electrical Phenomenon," 1894. *Ogden* [Utah] *Standard* (September 11).

1896: Colored Snow

CHICAGO WAS VISITED BY the most singular meteorological phenomenon last night [February 18] that has ever come under the observation of the local weather observer. Black snow, yellow snow, and brown snow fell in blinding clouds over the entire city, and reports from suburban towns brought the news that the various colored snow storm was not an exclusive Chicago production."

Source:

"Colored Snow," 1896. *Fort Wayne* [Indiana] *News* (February 19).

1896: Giant Reptile

ON JUNE 6, CARL Smithson, who farmed 7 miles southeast of Tolono, heard a strange sound emanating from his barn. He recognized its source—a small Jersey calf—but he had never heard it making a cry like that before. Racing inside to investigate, he went to the calf's stall and was shocked to find that a monstrous snake had swallowed the calf's leg up to its knee.

The snake was enormous—as much as 18 feet long and the thickness of a beer keg—and far too large for Smithson to deal with it on his own. He ran for help and returned soon afterward with some neighbors, but by this time the snake was gone.

"The serpent has been seen by several persons in the neighborhood," a newspaper said. "A company of armed men will make a thorough search for the giant reptile."

Source:

"Illinois Boasts a Snake Story," 1896. *Decatur* [Illinois] *Daily Republican* (June 12).

1897: Airship over Chicago

HUNDREDS OF PERSONS in Evanston, Niles Center, Schermerville, South Chicago, and in the city of Chicago proper last evening saw gliding through the heavens an object the like of which they had never beheld before, and which all agreed in declaring must have been the storied airship of Kansas and none other. Various descriptions were given of the strange object that set nearly 800 people agape in Davis street, Evanston, but the descriptions agreed on points of major interest. The object described sweeping through the heavens bore tan-colored lights, according to all accounts. All declare that the brightest light was white, and seemed to be backed by a reflector that could be turned, swinging its searching rays from side to side through the night air. Behind this some discerned a small red light, which others failed to observe; but again all agreed that still farther behind the big white headlight could be seen a smaller white light and a green light, side by side. The latest glimpse of the traversers of darkness was caught by South Chicagoans at about 9:30 o'clock, when numerous persons state they saw an object like that already described approach the land from out over the lake [Lake Michigan], and, after reaching a point some distance inland, turn slowly to the northwest and fade away into the night and darkness."

Source:

"Airship over Chicago," 1897. *Oshkosh* [Wisconsin] *Daily Northwestern* (April 10).

1897: Airship Wrench

JOHN WEATHERS, A WELL-to-do farmer of Oswego township, came to the city yesterday greatly excited over the air-ship of which so much has been published. He says that the aerial mystery, which is about forty feet in length, and oblong in shape, lighted on his farm Monday night [April 19], and that the occupants borrowed a monkey-wrench of him with which to adjust their electrical devise [*sic*] of the large searchlight. It arose in the course of an hour, and went in a northerly direction. Mr. Weathers is ready at any time to show to any doubting Thomas the monkey-wrench which the air-ship engineer used as proof of the truth of his story."

Source:

"Show Monkey-Wrench," 1897. *Fort Wayne* [Indiana] *News* (April 21).

1897: Verses concerning "That Airship"

THERE'S A WILD, weird something that sails in the air,
With wings like the "Piasa bird."
Look upward at night from almost anywhere,
You'll see it since April the third.
The people at Elgin declare it has eyes,
As red as the sun in the fall.
Two Evanston lovers (who always are wise)
Declare it has no eyes at all.
The wise ones at Rockford have seen it arise
Like a cloud from the west, and they tell

How the mayor and council, with wide open eyes,
Stood breathless, as bound with a spell.
It raced with the engine that hauls the Fast Mail
Down the Burlington track yesterday.
The engineer sat, with his face deathly pale,
And watched the thing vanish away.
At Alton, perched high on the "Piasa" hill,
Last night there was seen a huge form.
It screamed like a panther, with voice strange and shrill
That rose high above the wild storm.
It's headed for Washington now, I am told,
And I doubt not, like dasher of churn,
It will hasten reports on affairs growing cold,
Awaiting the Dingley bill turn.
And then let us hope it will take a deep breath
And sail on to Cuba and Greece,
To frighten the Spaniards and Turks half to death,
Establishing honorable peace.
We need such a fowl, with a nest full of eggs,
To hatch out a brood that will fly
And cause lawless people to take to their legs,
When they catch the strange bird's warning cry.

Source:

"That Airship," 1897. *Olean* [New York] *Democrat* (April 27). Reprinted from the *Chicago Inter-Ocean*.

1900–1935: The Universe according to Zion

WILBUR GLENN VOLIVA'S kingdom was a small town in the state's far northeast, on Lake Michigan's western shore, just 3 miles south of the Wisconsin border. Zion came into existence in 1900, founded by a fanatical Scottish preacher, John Alexander Dowie, whose most recent ministry had been in Australia. Dowie, head of something called the Christian Catholic Apostolic Church, despised nearly all things of the world, and his Zion would be free of such pestilences as physicians, pharmacies, pork, tobacco, liquor, unions, competing churches, dances, and more. Dowie's right-hand man was the equally stern and humorless Voliva, who soon grew restless under his boss's rule. Dowie's mismanagement of the community brought financial ruin and widespread discontent. In 1906, Voliva engineered a coup, declared himself "general overseer" of the church and leader of the Theocratic Party, and became emperor of Zion.

Born in rural Indiana on March 10, 1870, into a religious household, he was an ordained minister (in the New Light Christian Church) by the age of nineteen. He attended various Bible colleges, punctuated with pastoral and evangelical activities. By 1898, he had become a member of the Disciples of Christ, but that changed quickly after he read Dowie's *Leaves of Healing,* a weekly magazine. In 1899, he joined Dowie's sect.

After overthrowing the Zionist founder, Voliva spent the next years fighting enemies, satisfying creditors, and consolidating his power. Through persuasion and intimidation he got church members to turn over their property to him and soon owned a good portion of the town. Those who had once been sympathizers joined forces with those who had never admired him to form the Independent Party, which battled unsuccessfully against Voliva's forces. With the Theocrats in control of the city council, the city parks were turned over to Voliva as his personal property. He instructed his police—and they were now *his* police—to make sure no children of Independents trespassed on his property.

Writer Robert Schadewald, an authority on Voliva's strange career, reports: "In 1914 Voliva's minions began passing the strictest blue laws in 20th-Century America. Tobacco was banned in Zion and when trains stopped in town, police boarded them to arrest smokers. Likewise forbidden were movie theaters, pork, alcohol, doctors, drug stores, unions and secret societies. Women could not cut their hair, expose their necks or straddle a horse. Zion detectives lurked in nearby Waukegan lest some Zionite should take the train south for a movie, a smoke or a taste of demon rum. The laws, of course, applied to both church members and Independents."

On August 16, 1914, Voliva preached a fiery sermon blasting the usual targets—doctors, scientists, and biblical scholars who had arrived at conclusions different from his own—but now adding a new one that would forever ensure him a place in the history of pseudoscience: the notion that the earth revolves around the sun. By the next year, he was openly championing a flat, pancake-shaped earth. That, however, was not all. He said, according to Schadewald: "I believe the earth is a stationary plane; that it rests upon water; and that there is no such thing as the earth moving, no such thing as the earth's axis or the earth's orbit. It is a lot of silly rot, born in the egotistical brains of infidels. Neither do I believe there is any such thing as the law of gravitation. I believe that is a lot of rot, too. There is no such thing! I get my astronomy from the Bible." Instead, he proposed that a dome encloses the earth and that within it are the stars, which are much smaller and closer than astronomers contend. Voliva offered $5,000 to anyone who could prove him wrong, at least to his own satisfaction. There were no takers.

It was no longer possible to teach anything but Volivan astronomy and geography in Zion's parochial schools, directed by close Voliva associate Anton Darms, an "apostle" in the church. In articles in the church's magazine Darms laid out the case for Voliva's version of Bible science. Voliva supplied the polemics, lashing out at astronomers and evolutionists as active agents of Satan.

All of this brought Zion worldwide attention, most of it less than admiring and laden with ridicule. Voliva was unfazed. "I can whip to smithereens any man in the world in a mental battle," he boasted. "I have met any professor or student who knew a millionth as much on any subject as I do." On another occasion, in the midst of a courtroom exchange, he declared, "Every man who fights me goes under. . . . The graveyard is full of fellows who tried to down Voliva. This other bunch will go to the graveyard too. God almighty will smite them."

In the late 1920s, the Illinois legislature investigated Voliva's finances and uncovered enough questionable dealings to urge prosecution. Though none was forthcoming, the revelations did Voliva's reputation even in Zion no good, and power began to slip from under his iron thumb. The Great Depression that came soon after further eroded his influence as his city slipped into poverty and despair. In 1935, Voliva was booted out of his position as the church's general overseer, and the church abandoned its flat-earth doctrine.

In his last years, Voliva wintered in Florida. His diet consisted of Brazil nuts and buttermilk, which he was convinced would give him a century's life span. On October 11, 1942, at seventy-two, he died in Zion. A few years later, a nonadmirer, science writer Martin Gardner, summed up his life thus: "Voliva's drives were two in number—a desire to defend a religious dogma, and a paranoid belief in his own greatness so far removed from reality as to border on the psychotic."

Sources:

Gardner, Martin, 1957. *Fads and Fallacies in the Name of Science.* New York: Dover, 16–19.

Schadewald, Robert, 1989. "The Earth Was Flat in Zion." *Fate* 42, 5 (May): 70–79.

1902: Ghostly Woman in Black

THE GOOD PEOPLE of Bushnell, Ill., have been much disturbed of late by a ghost which appears in the form of a woman. She usually presents herself to human view robed in deepest mourning. Only once has she been seen in white and then in what appeared to be material of flowing textures. On all other occasions she has shown herself in long black robes and a mourning veil over her face. She does not seem to confine herself to any particular place, but she has been seen on different streets and at all hours of the night.

"She suddenly appears before the belated citizen in a noiseless manner and as suddenly disappears. On two or three occasions she has given chase, and the frightened individuals declare it is with only the fleetest running that they have been able to keep out of the way. And on one occasion an attempt was made to capture the spirit, but after a chase of several squares she suddenly disappeared from view, and further chase was abandoned. Many stories are rife as to the true nature of the apparition. Many superstitious persons believe it to be the spirit of a woman who died recently after months of great suffering and that her spirit has returned to harass the city. It has resulted in keeping a great many children off the streets after dark and a few older ones too."

Source:

"Black Robed Ghost Haunts Illinois Town," 1902. *Fort Wayne* [Indiana] *News* (June 5).

1902: Well-Dressed Phantom Female

MISS DELLA GEARHEART, prominent in society at Alto Pass, had an encounter with a ghost while driving a mile west of town at 9 o'clock last night. Hearing her name spoken, she stopped her horse, and leaning out from the buggy asked what was wanted. She then beheld the form of a tall, handsomely gowned girl, who after gliding to within a few rods of the buggy turned and without another word vanished apparently into the air. Miss Gearheart, who is a strong, healthy young woman, was prostrated by the incident."

Source:

"Encountered a Ghost," 1902. *Decatur* [Illinois] *Herald* (August 8).

1907: Watseka Wonder

A STORY IS BEING circulated from an Illinois town which worries some of the railroad men in that state. The story is that a phantom freight train on the Chicago & Eastern Illinois has been seen to whir through Watseka at the speed of sixty miles an hour. A heavy freight train in charge of Conductor Fox and Engineer Hazzard was occupying the side track there when the phantom raced through.

"Operator Eckerty, who was on duty, saw the train and made an entry in his train book. When he reported the train to the dispatcher the latter reprimanded him, stating that no such train was on the division and for him to wake up. When the crew corroborated the story of the operator and described the train rushing by regardless of the stop signal, the engineer at the throttle and a fireman shoveling coal, the dispatchers began to take notice and asked for further particulars.

"When Operator Burnett, of the Topeka, Peoria & Western, which crosses the Chicago & Eastern at that point, also reported seeing the strange train[,] the phantom began to take on the appearance of a reality. All crews are on the alert to see if there is a repetition of the strange phenomenon."

Source:

"Phantom Train Flashes By," 1907. *Fort Wayne* [Indiana] *Journal-Gazette* (March 11).

had thundered by ten minutes before, and had been 'O.S.'d' by the next station north. Silence wrapped the station and Cammack and his bride sat gazing at the fog from the lowlands that drifted past the front windows. Out of this fog gradually materialized the ghost. All in white, with trailing drapery that waved in eccentric manner, the ghost approached the window. Its fiery eyes glared through the pane as it tapped with bony fingers on the glass and beckoned to Cammack. The latter's hair stood straight up. . . .

"Unfortunately, Cammack had no revolver, and he did not fancy sallying forth armed with the stove poker. So he and his wife gazed with eyes almost starting out of their heads at the uncanny visitor. The latter rapped once more on the glass, and as the close wrapped drapery fell away from the face the watchers saw the grisly outline of a fleshless skull. With a wild laugh that was half screech and that echoed back from the bluffs, the ghost dissolved into the atmosphere."

Source:

"Operator Sees Ghost," 1909. *Edwardsville* [Illinois] *Intelligencer* (January 27).

1909: Ghost with Fiery Eyes

Mr. and Mrs. [H. P.] Cammack live in the 1700 block on North Main street [in Edwardsville]. They are Newly weds, having been married on Christmas day at Decatur. . . . Last evening she decided to keep her husband company in his lonely vigil as third track night operator at Edwardsville Junction. For the benefit of those who have never become intimately acquainted with the new Junction it may be said that it is the most desolate, lonely and God-forsaken spot this side of No Man's Land. There is no wagon road or footpath to it, nothing but a network of railroad tracks, with steep bluffs on two sides and the Cahokia creek flowing past the door.

"At 12:30 this morning the ghost appeared. Number Eighteen, the last express for Chicago,

1910: Airship or Balloon?

An airship was seen passing over Decatur last night shortly before 7 o'clock. Don't ask for particulars, for those who saw it are unable to say whether it was a sloop, brig, barque, schooner, side wheeler, twin screw propeller, gasoline launch, armored cruiser or celestial greyhound. No signals could be seen. If she carried any lights they were not visible at that time in the evening.

"The airship passed to the west of Decatur, going north. It really passed over no part of the city. From the angle at which it was seen it must have been pretty far away. From North Main street it seemed to be but little above the tree tops. If the aerial craft was sailing at a comfortable height above the earth this would mean that it was several miles west of Decatur.

"Those who observed it in Decatur thought it was large. Of course, it might have been a small balloon and much nearer than it seemed. The spectators were unable to say whether it was an airship or a balloon. They are firm in their belief that it was either a very large balloon or an airship. It may be added that the men who saw the sky-craft are not persons given to dreaming dreams or seeing visions. They are not even in the habit of seeing double. They really saw something sailing through the sky. Whether it was a toy balloon launched from the university campus or a real airship passing over Harristown or Wyckles is a point on which they are not willing to commit themselves positively."

Source:
"Saw Something Sailing in Sky," 1910. *Decatur* [Illinois] *Review* (April 14).

Indiana

1838: The Monster of Devil's Lake

A PIECE APPEARED IN the *Alton* [Illinois] *Telegraph* in September written by an author identified only as "A Visitor to the Lake." The "Visitor" reported: "In the Northern portion of Indiana there are many beautiful little lakes, which give great interest to a country somewhat open. About 25 miles from Logansport, and in the vicinity of Rochester, there is one of these lakes about two miles in length, and half a mile in width, and of unknown depth. Soundings were once tried with a line of 13 fathoms, but with no effect.

"There is an ancient tradition of the Pottawattamie Indians relative to this lake, which has been handed down from generation to generation, and is now received by the white man with confirmed credence. The precise time at which the tradition was first received among the Indians cannot be determined—probably not a long time after the emigration of the Pottawattamies across the 'hard waters' of the north, some centuries since, to this district of country, which was then occupied by the Miamies, by whose grant the Pottawattamies became possessed of the lands. It appears that the tradition does not owe its origin to the superstitious fears of the red men, but that some gigantic creature inhabited the lake, and does at present time, is beyond the possibility of a doubt.

"This lake is called by the Indians 'Lake Man-i-too,' or the Devil's lake, and such is the terror in which it is held, that but few Indians would even dare to venture in a canoe upon its surface. The Indians will neither fish nor bathe in the lake; such is the powerful conviction that 'Man-i-too' or the evil spirit, dwells in the crystal waters. It may elicit a smile from the incredulous to assert gravely the fact that some very extraordinary creature claims monarchy of this beautiful lake. But the existence of a monster in this lake is not an object of more surprise to us than the remains of the Mastodon, whose teeth measure 18 inches—and which were found about two miles from town, in the prairie through which the canal runs. Were there no assurances from men entitled to credulity [*sic*] that a monster has been seen within a few days in the Lake Man-i-too, it might be supposed the strange story originated in the superstitious fears of the Aborigines.

"When the Pottawattamies' Mills were erecting some ten years since, at what is called the outlet of the lake, the monster was seen by those men known to Gen. Milroy, under whose direction the Mills, I believe, were erected. There are persons in Logansport who questioned closely those who lately saw the mysterious occupant of the lake, and are now convinced of this tradition of the Man-i-too being founded upon something more substantial than the basis of fish and snake stories generally.

"But two weeks since[,] some men by the name of Robinson were fishing in the lake, when they beheld with surprise the even surface of the water ruffled by something swimming rapidly, and which they suppose must have measured 60 feet. The Robinsons are respectable men, whose fears are not easily excited; yet such was the terror that this nondescript caused, that they made a hasty retreat to the shore, much alarmed. Since this circumstance took place, and but a few days since, Mr. Lindsey, who is well known here, was riding near the margin of the lake, when he saw, at a distance of 200 feet from him, some animal raise its head three or four feet above the surface of the water. He felt the security of the shore, and viewed the mysterious creature many minutes, when it disappeared and reappeared three times in succession. The head he described about three feet across the frontal bone, and having something of the contour of a 'beef's head,' but the neck tapering, and having the character of the serpent; color dingy, with large yellow spots. It turned its head from side to side with an easy motion, in apparent survey of the surrounding objects. Mr. L. is entitled to credulity [sic]. So convinced are many of the existence of the Monster, that some gentlemen in town have proposed an expedition to the lake, and by the aid of rafts to make an effort to capture the mysterious being which is a terror to the superstitious, but which becomes an object of interest to science, the naturalist and philosopher."

Historian Donald Smalley has concluded that the story and follow-up items were a hoax engineered by John Brown Dillon, editor and publisher of the *Logansport Telegraph*.

Sources:
Smalley, Donald, 1946. "The Logansport *Telegraph* and the Monster of the Indiana Lakes." *Indiana Magazine of History* 42, 3 (September): 249–267.
A Visitor to the Lake [probably John Brown Dillon], 1838. "The Devil's Lake." *Alton* [Illinois] *Telegraph* (September 19).

1839: Wild Child

STRANGE AS IT MAY appear, it is currently reported and very generally believed that a wild child, or lad, is now running at large among the sand hills round and in the vicinity of Fish Lake. It is reported to be about four feet high, and covered with a light coat of chestnut-colored hair. It runs with great velocity, and when pursued, as has often been the case, it sets up the most frightful and hideous yells, and seems to make efforts at speaking. It has been seen during the summer months running along the lake shore, apparently in search of fish and frogs, and appears to be very fond of the water, for it will plunge into Fish Lake and swim with great velocity, all the time whining most piteously."

Source:
"A Wild Child," 1839. *Adams* [Pennsylvania] *Sentinel* (December 30). Reprinted from the *Michigan City* [Indiana] *Gazette,* December 4.

1870: Rock Attacks

AT THE BENHAM SALT well, fifteen miles from Leavenworth, on the Jasper road, there is a frame dwelling house, erected a couple of years since, and now occupied by Mr. Hi. Benham. This house is now the great 'mystery.' For about six weeks it has been struck with rocks, during all hours of the day and night, and notwithstanding the fact that a close watch has been kept, no explanation of the curious occurrence or phenomenon has been discovered. The house is not in a very secluded spot, being directly on a public road. Some of the rocks weigh as much as four pounds, and come with considerable force, striking on the roof and all sides of the house. One of my informants, a gentleman of undoubted integrity, was there on business the other day, and while there four rocks came against the house. Some four or five men were watching at the time. There is also knocking on the door as if for admittance, but on opening it, no person is found.

"Mr. Benham has even shot through the door at the supposed intruder, but with no result. He was plowing in a field near his house one day, and left his team and went to the house. While there he heard some one plowing, and on going back to the field, found that a couple of furrows had been plowed, but no person was to be seen. Every one thinks it is some malicious person, but all say that no person could conceal himself in the vicinity of the house and throw rocks for five hours, without being seen, while it has been kept up for weeks with a watch day and night; regular 'pickets' being on duty . . . without making any discovery whatever. Hence the mystery. I give no theory, but merely the facts, as I believe them to exist."

Source:
J., 1870 "The Crawford County Mystery."
Fort Wayne [Indiana] *Daily Democrat*
(September 2).

1870: "The Spirit of a Beautiful Little Girl"

IT APPEARS THAT A LADY of repute, residing at Fredonia . . . has been haunted for years by the spirit of a beautiful little girl. The most remarkable circumstance connected with the matter is that the phantom was seen by others: that several attempts have been made to seize it, and that when pursued, it always melts into space. A gentleman residing in the same house with the haunted lady has not only seen the spirit, but declares, further, that he heard the noise of its feet upon the stairs, and could not be persuaded that it was not his own child until he looked into the cot of the latter. A committee of investigation will take the mystery in hand in a few days."

Source:
"Ghost Sensation," 1870. *Petersburg* [Virginia] *Index* (September 10).

1871: Through the Looking Glass

ONE OF THOSE REMARKABLE phenomena, which appear to set the laws of nature at defiance, and which the most profound scientists find themselves unable to explain, has just occurred at the farm-house of Mr. L—, in the suburbs of this city [Fort Wayne]. The family is one of the oldest and wealthiest in this community, are members of an orthodox church, and have never inclined toward Spiritualism.

"On last Thursday night [November 16], Mrs. L— went into her parlor at half past ten o'clock. Setting the lamp on the mantel, under a large oval mirror, she re-arranged the furniture which had been somewhat disordered by a party of young folks who had been visiting at the house in the afternoon. Having 'set the room to

rights,' she turned to take up the light, when her gaze fell upon the looking-glass, and she felt herself glued to the spot by the sight therein presented. Standing beside her was a woman, surpassingly beautiful, with black, wavy hair, blue eyes, and features of classic regularity. She was arrayed in a filmy, floating garment of white, and in her left hand she had a roll of parchment. Mrs. L— was too terrified to move, but at last, recovering her self-possession, she called to her son, in the adjoining room. At the sound of her voice the apparition disappeared.

"On last Sunday, however, another member of the family was found lying senseless on the floor before the mirror, and on being restored to consciousness, declared that while looking in the glass, she had seen a dark-eyed, dusky-faced man gazing over her shoulder, and in the fright which the apparition had evoked she swooned away.

"The house has long had the reputation of being haunted, and the dismal cedars with which it is surrounded, together with the fact that it stands in a lonely wood—isolated by stretches of hall and valley from any other habitation—would lead the timorous to look upon it as anything but a desirable residence."

Source:
"Truth Stranger Than Fiction," 1871. *Fort Wayne* [Indiana] *Daily Sentinel* (November 22).

1879: Chased by a Dragon

A NUMBER OF SNAKE stories have been going the rounds of the press, a large majority of which are rather 'fishy' than snaky, but for a genuine story, the facts of which can be easily ascertained, as all the parties concerned are well known and will corroborate it in every particular, the following 'lays over the deck:'

"Jacob Rishel, a farmer living twelve miles northeast of this city [Fort Wayne], in Jackson township, last Saturday evening [August 16], while returning from work had occasion to pass through a field of high grass. He had just reached the farther side of the field when he heard a loud noise behind him. Looking back he

Here, for the first time, he had a good view of the beast, a huge reptile almost forty feet in length, with head erect, and, most singular of all, it had a pair of horns or feelers projecting one from each side of the head, looking somewhat similar to the tentacles of a devil fish [octopus], but were about three feet in length.

noticed the tall grass waving about and being agitated in a very violent manner, having the exact appearance of a small whirlwind, only the grass was not twisted but was rather crushed and broken down, leaving a swath about eight feet wide behind it. Mr. Rishel was naturally alarmed, as it was then almost dusk and he was nearly half a mile from any house. His first supposition was that it was a whirlwind, and, as it was coming directly towards him at almost lightning speed at a distance of not over 100 yards, he started to run as rapidly as possible at right angles from the course the storm was taking. Imagine his consternation to see the— whatever it was—also change its course, and again head directly for him and only a short distance behind him—so close, in fact, that he could distinctly see that it was something more than 'wind' and nothing less than a huge reptile or monster the like of which he had never seen before. He realized now that it was a race for life, and started, again at right angles.

"The reptile changed its course with him; but Mr. Rishel saw that he had gained a yard or two by the maneuver. He continued running a short distance, not daring to look behind him, and then again changed his course and by a succession of doublings soon found himself in a corner of the field where dodging was no longer possible and his pursuer so close upon him that, as Mr. Rishel says, 'I could smell his breath.' A reaper was standing in the corner of the field preparatory to cutting the grass on the morrow, and Mr. Rishel ran around it, with the reptile close behind him. The reaper . . . stood about four feet from one fence and six feet from the other. In running around it he made a decided gain and got in the rear of the 'snaik.' Here, for the first time, he had a good view of the beast, a huge reptile almost forty feet in length, with head erect, and, most singular of all, it had a pair of horns or feelers projecting one from each side of the head, looking somewhat similar to the tentacles of a devil fish [octopus], but were about three feet in length.

"The reptile followed Mr. R. around the reaper, but owing to its length and the short space between the machine and the fence, it was comparatively easy for him to keep out of its way. He, however, was fast becoming exhausted, and knew that he could not keep the race up all night, which appeared probable, as the serpent showed no signs of relinquishing the chase, but with head erect and mouth open, thrashing its feelers around in a terrible manner, in the meantime emitting a most horrible hissing sound, more like a roar than a hiss, it was making its utmost endeavors to reach and strike the terrified man. Even in his fright Mr. R. realized that although he might get over the fence in safety his pursuer would do the same, and having him again in an open field the race, in his present exhausted condition, would soon be terminated. Seeing a scythe hanging on the fence, by which he passed each round, a wild hope flashed into his brain that by the means of that he might yet save himself, and he at least resolved to make one desperate effort in that direction. He says that he knows he suffered a terror and fear never experienced by mortal man before; yet he never lost his presence of mind in the slightest degree. Passing close to the fence he grasped the scythe, and while running unloosed it from the [sheath] and threw the latter out of his way. Slackening his pace he suffered the reptile to approach quite close to him, and then suddenly whirling, he struck at his pursuer with the long scythe blade, and fortunately succeeded in severing one of its 'horns' close to the head. With an unearthly noise the reptile leaped forward and almost succeeded in reaching its intended victim. Encouraged by his first attempt, he again struck at the animal, but missed it, as he did again and again while passing around the machine the second time. After his first stroke, and being on the side toward the field with the serpent on the side next to the fence, the latter, now thoroughly enraged, attempted to spring over the cutter bar of the machine directly at Mr. Rishel . . . but its body being in a curve, and not having momentum enough[,] it sprang with its body directly against one of the guards of the bar, which was elevated about ten inches from the ground and, being sharp, penetrated the skin about four feet from the head, and being a sort of a harpoon shape, held the snake firmly."

"Mr. R., seeing his advantage[,] summoned all his remaining strength and courage, and with his providential weapon by a well directed blow completely severed the head from the body. Mr. Rishel realized what he had done, and the suspense being gone the reaction came on. He staggered a few feet and fell to the ground in a dead faint, while the monster was thrashing and writhing in its death throes a few yards away. Mr. R. laid [sic] there for about two hours and recovered just as his friends, becoming alarmed at his prolonged absence, came up to him with lanterns on the search. He recited his terrible encounter, and the snake was laid out and found by actual measurement to be 34 feet and 3 inches in length and about as thick as a barrel, and says that had the snake not been dead and measured before his own eyes he would have been willing to take his solemn oath that it was fully a hundred feet in length. This may account for some ridiculous and absurd snake stories which are afloat. The tentacles spoken of were about forty-two inches long and about three inches in diameter where they joined the head. The head was remarkably small for the size of the snake and was flat, something of the nature of a flat-head snake. The color was precisely like that of the garter snake on a large scale with a dark green stripe running down the back. The snake was skinned and the skin sent to Chicago where it will be stuffed and placed on exhibition.

"Mr. Rishel was ill for several days afterwards and an attack of brain fever was anticipated, but this new calamity was happily averted. Today, however, was the first time he has been out of the house since the encounter. His hair, which was already beginning to turn slightly gray, is now as white as the driven snow.

"He stated to the *Sentinel* reporter who interviewed him that he had not intended to say anything about the matter, and that his friends had promised to maintain the strictest secrecy in regard to it, but that during the past two days several persons had asked him about it, and as he supposed it would all come out any how, he would rather make the statement himself than to have any exaggerated report of the affair published."

Source:
"A Tale of Terror," 1879. *Fort Wayne* [Indiana] *Sentinel* (August 20).

1881: Phantom Boat

OFFICERS ROHLE AND O'Connell claim that in the gray hours of the morning they saw a blood-curdling apparition in the shape of a white boat rowed by a shadowy spectre, which glides along the canal near the Bloomingdale mills. Several times the officers said they pursued the ghost, but it disappeared through a fence and was not to be seen more. This story comes from veracious men who tell it in all soberness. The effect will be to effectually frighten the boys and girls of Bloomingdale, and perhaps cause them to remain at home of nights."

Source:

Untitled, 1881. *Fort Wayne* [Indiana] *Daily Gazette* (May 30).

1883: Ghost in the Railroad Yard

SOME EXCITEMENT HAS been created here [Elkhart] by the report that a ghost nightly makes its appearance in the railroad yards. It is stated that the apparition has been seen by a number of persons, but a fear that the story would be received with incredulity has had a tendency to keep the matter suppressed. However, last night [March 9] while conductor Shaw was making a coupling, it made its appearance right in front of him. Startled, he held up his lantern, which the ghostly visitant extinguished as it advanced, when Shaw unceremoniously fled. It will be watched for to-night. As a man was killed there by the cars some years ago, the superstitious think it is his spirit roaming around."

Source:

"A Ghost Story," 1883. *Newark* [Ohio] *Daily Advocate* (March 10).

1883: Hairy Female Biped

WALKING THROUGH A wooded area near Lafayette one July day, Mrs. Frank Coffman, "the wife of a well known farmer," noticed something to her right. It was an extraordinary creature, female in contour, with long black hair blowing in the wind. Short gray hair covered its body. It was breaking twigs from a sassafras bush and eating the bark, at first oblivious to Mrs. Coffman's presence.

As the witness stood paralyzed with terror, the figure turned around and saw her. It glared in what Mrs. Coffman took to be anger, and then it raised its arms and let out a hideous shriek. It ran into the forest—presumably on two legs, though the account isn't clear on the point. Finally, the farmer's wife recovered sufficiently to run in the opposite direction. As she approached her house shouting, her husband heard her and caught her just as she collapsed.

After carrying her into the house, he grabbed a rifle and alerted the neighborhood. A party of a hundred men and boys, with hunting dogs, searched the site and soon came upon the creature. They set off in hot pursuit, sometimes getting close, but in each case the animal was able to outwit them. A contemporary newspaper reported, "For fully half a mile of the chase she was never ought of sight. Her feet touched the ground but seldom. She would grab the underbrush with her long, bony hands, and swing from bush to bush and limb to limb, with wonderful ease. She seemed only endeavoring to keep just beyond the reach of her pursuers, until, coming to a swamp, she disappeared as suddenly and effectively as an extinguishing light, and no searching served to ascertain her whereabouts."

The press could only speculate that the creature was either a "wild woman" or an escaped gorilla.

Source:

"Beast or Human Being?," 1883. *Atchison* [Kansas] *Globe* (July 19).

1884: "A Sudden Burst of Ruddy Light"

UNACCOUNTABLE LUMINOUS appearances are reported by several persons in this city who saw them a number of times during the past two or three nights. Our informants, all unscientific observers, who have occasion to be up and about at night, agree substantially in their descriptions of these phenomenon [sic]—a sudden burst of ruddy light, springs up as if from behind intervening buildings, irradiating the sky and locality for a moment and quickly subsiding. The observers say they first thought that the light was due to some burning building, and expected to hear the alarm sounded. The nights were quite dark, the sky overcast and the weather rainy, and as the mysterious appearances do not seem to be identical with any of the recognized electrical displays pertaining to the meteorology of this region, it is in order for some scientist to relieve public curiosity on this subject."

Source:
"Atmospheric Phenomenon," 1884. *Fort Wayne* [Indiana] *Daily Gazette* (December 16).

1886: Figures in the Night

FOUR MILES SOUTHEAST of Wabash, about a quarter mile from a small settlement consisting mostly of well-to-do retired farmers, stood an old, decaying frame house. Late one mid-January night a passerby identified as Dr. Watson was startled when his horse abruptly stopped and started to buck. As the doctor struggled to reassert control over the animal, he happened to glance toward the half-open door of the abandoned house. There, he spotted a black-clad man, his coat and vest thrown back enough to reveal a white shirt. The figure was swaying back and forth in an unsettling fashion. The horse bolted forward, and as it and its rider fled the scene, Watson saw the figure disappear instantly.

On another night a few days later, elderly farmer Jefferson Brown, driving by the house, noticed a woman standing in the doorway. A black dress covered most of her body, except for part of what looked like a white undergarment in front. Even more weirdly, she seemed to be standing in mid-air, "with hands uplifted as though in supplication."

Brown and two friends later observed a ghostly boy and heard creaking sounds inside

the house, as if someone were walking on its rotting floor. Rather than investigate further, they ran away and did not return.

Source:
"A Strange Apparition," 1886. *Trenton* [New Jersey] *Times* (July 15).

1887: Ghostly Giant

IN TIPPECANOE COUNTY, Ind., a farmer was riding home through a piece of woods at night when he noticed a ghost, fully 11 feet tall. He whipped up his horse, but the ghost kept right along until he reached the edge of the woods, when he disappeared."

Source:
Untitled, 1887. *Gettysburg* [Pennsylvania] *Compiler* (December 27).

1888: Flying Fish

DURING A HEAVY STORM Wednesday night [August 1] a large number of fish, of a variety unknown here [Seymour], some of them four inches in length, fell in this neighborhood. The occurrence excited a good deal of curiosity, but no one has been able to explain the phenomenon."

Source:
"Fish from the Clouds," 1888. *Newark* [Ohio] *Daily Advance* (August 4).

1890: Fainting the Dead Away

AROUND MIDNIGHT IN early June a young hired man—his name is spelled variously as "Kumfer" and "Kumfler"—was returning to his employer's farm southwest of Fort Wayne when, to his immense surprise, he encountered a white-robed woman with long hair and a beautiful face. He recognized her from stories current in local lore about a ghost. Terrified, he promptly fainted.

On returning to consciousness at some undetermined time later, he was relieved to find that the ghostly woman was gone. Too weak to walk, he passed out until neighbors came upon him and carried him home. For the next few days he remained bedridden under a doctor's care.

Source:

"Scared by a Ghost," 1890. *Fort Wayne* [Indiana] *Gazette* (June 10).

Circa 1890: Cry from the Sky

THE YEAR OF OLIVER Lerch's terrifying vanishing is given variously as 1889, 1890, or 1900. All that seems certain is that it never happened, though it has been told as true for decades, possibly as long ago as the early years of the twentieth century.

The heyday of the story was in the 1950s and 1960s, when "true mystery" potboilers were popular. Their proximate source was Joseph Rosenberger's "What Happened to Oliver Lerch?" in the September 1950 issue of *Fate,* a digest devoted to allegedly authentic unexplained occurrences and phenomena. In Rosenberger's telling, the incident took place on Christmas Eve 1890 after Oliver Lerch, twenty, left the family house, scene of a merry Christmas party, to retrieve water from a well. "Some minutes later, perhaps five," Rosenberger wrote, "a horrible cry for help, so terrifying that it could be heard above the singing, split the serenity of the occasion."

Everyone rushed outside. There was no sign of young Lerch, but his voice was sounding from overhead. For five minutes it cried out, sometimes as if close by, other times as if distant. According to Rosenberger, searchers disagreed on whether he was shouting "It's got me!" or "They've got me!" In any event, Lerch was never seen again. All that remained were his tracks in the snow, suddenly ceasing 225 feet from the house, halfway to the well.

Subsequently, the tale got retold, usually by those, such as M. K. Jessup (*The Case for the UFO* [1955]), who thought a flying saucer had snatched Lerch and presumably taken him to another world. Others speculated that he had fallen into another dimension. By the 1960s, like all legends, the Lerch story had evolved into variations with different names and geographical locations attached to them. An account in a 1966 paperback by Brad Steiger placed the incident in Wales in 1909 and identified the victim as eleven-year-old Oliver Thomas.

In 1979, a skeptical inquirer, Joe Nickell, secured a confession from Rosenberger, who acknowledged, "There is not a single bit of truth to the 'Oliver Lerch' tale. . . . It was all fiction for a buck." He went on to assert, more dubiously, that he had made it up himself. In fact, the story was in circulation well before Rosenberger's account appeared. As early as 1932, responding to a letter from British writer Harold T. Wilkins, the *South Bend Tribune*'s managing editor dismissed the Lerch yarn as "purely imaginary. We frequently hear of this supposed incident regarding the Lerch family, but have never been able to locate such a family." Other investigators have learned that the supposed witnesses were as imaginary as Lerch was and that weather records refute the claim that snow was on the ground in South Bend at Christmas 1890.

The story's origins almost surely lie in Ambrose Bierce's short piece "Charles Ashmore's Trail," published in 1893. Though it is fiction, Bierce narrates it as if it were true, with precise details about date (November 9, 1878) and place (Quincy, Illinois). In this account, young Ashmore leaves the family home in the evening to gather water from a stream. When he fails to return, his father and sister search for him and see his tracks in the snow, which stop halfway to his destination. Four days later, his distraught mother, on her own way to the springs, hears his disembodied voice crying out as if from a great distance. Though his words are clearly articulated, his mother later cannot remember what they were. Subsequently, other family members hear them until they grow fainter and after a few months are heard no more.

Sources:

Jessup, M. K., 1955. *The Case for the UFO.* New York: Citadel.

Nickell, Joe, 1980. "The Oliver Lerch Disappearance: A Postmortem." *Fate* 33, 3 (March): 61–65.

Nickell, Joe, with John F. Fischer, 1988. *Secrets of the Supernatural: Investigating the World's Occult Mysteries.* Buffalo, NY: Prometheus, 61–65.

Park, T. Peter, 1998/1999. "Vanishing Vanishings." *The Anomalist* 7 (Winter): 158–178.

Rosenberger, Joseph, 1950. "What Happened to Oliver Lerch?" *Fate* 4, 5 (September): 28–31.

Steiger, Brad, 1966. *Strangers from the Skies.* New York: Award Books, 33–35

Wilkins, Harold T., 1948. *Mysterious Disappearances of Men and Women in the U.S.A., Britain and Europe.* Girard, KS: Haldeman-Julius Publications, 4–5.

1891: Immense Rattler

JAMES GRAHAM, AN old-time Virginian . . . who resides in Scott county, was in this city [Columbus], and tells a big snake story, about killing a monster rattlesnake unlike any ever before seen in Indiana. He says he was engaged in pulling bark, last week, and came in contact with a den of rattlers, and after killing ten large ones, as large as are generally seen in this locality, he started up the hillside, where a monster snake, which looked as large as all the balance, lay basking in the sun.

"The rattler, on seeing him approach, set up his notes of warning. Graham retreated, and, procuring some hickory bark, made a lasso or harpoon by tying his bark spud to it and hurled it at his snakeship, striking the huge reptile back of the head. On measurement after death it was found to be nineteen feet in length, and had thirty-nine rattles and a button. The skin was stuffed and will be sent to a museum."

Source:

"Latest Snake Story," 1891. *Marion* [Ohio] *Daily Star* (May 28).

1891: A Shroud with Fins

AT 2 A.M. ON SEPTEMBER 5, according to an account in the *Crawfordsville Daily Journal*, as two men hitched a team of horses to an ice wagon, one of them, Marshall McIntyre, suddenly felt uneasy. Something caused him to look upward. There, at an altitude of 300 or 400 feet, was an eerie phenomenon without definite shape or form. It was pure white and perhaps 18 feet long and 8 feet wide, resembling nothing so much as a shroud with fins. A brilliantly glowing "eye" was located at what appeared to be the front. The object, or whatever it was, emitted a wheezing, agonized sound. The thing made flapping motions like a sheet in the wind, and sometimes it squirmed as if in agony.

It circled around a nearby house, then stopped and hovered. The two men retreated to a barn and watched it from there. It sailed eastward until it reached the city limits, at which point it returned to the house. McIntyre and his coworker, Bill Gray, debated whether to awaken the occupants, a family named Martin, but decided to let them sleep in peace.

Two days later the newspaper reported that others had seen the weird phenomenon the same night as McIntyre and Gray. One was a Methodist pastor, the Rev. George W. Switzer. Switzer had stepped out after midnight to get a drink of water from the backyard well. Like McIntyre, he had suddenly experienced an unsettling sensation that caused him to direct his eyes skyward, where he saw what looked like a "mass of floating drapery" coming in from the southwest. It moved rapidly, twisting like a serpent. Switzer alerted his wife. The couple watched it glide east of the church then descend until it was out of sight as if to land in the yard of a neighbor's house. Walking out into the street, they saw it ascend into view. It then circled the town and was still doing so when the Switzers, too exhausted to maintain their vigil, retired.

On September 8, the *Daily Journal* reported that two other witnesses, John Hornbeck and Abe Hernley, had also seen the strange visitor. Unlike the others, they had been able to identify it. They had followed it around town through the early morning hours until they "finally discovered it to be a flock of many hundred killdeers." The newspaper speculated, "These birds were evidently passing over the city and becoming bewildered by the electric lights had lost their way. Their white breasts and wings gave the flock their ghostly appearance and the sound of agony was their plaintive dismal cry. Messrs. Hornbeck and Hernley were quite close to [the birds] as they swept near the ground and are certain that they were not mistaken."

In later years, after UFO reports became widely known, investigators found that night-flying birds with breasts reflecting ground light did cause some ostensible UFO sightings. Thus, though nineteenth-century newspapers are notoriously unreliable, the conclusion attributed to Hornbeck and Hernley has at least surface plausibility. Nonetheless, the incident has been chronicled—even though the "object" does not sound much like a UFO as ordinarily understood—as an early encounter. The first to do so was the famous anomaly chronicler Charles Fort (1874–1932), who was initially skeptical, suspecting that no such thing had ever happened. Nonetheless, he found that the Rev. G. W. Switzer did indeed exist. "I wrote to him," Fort recorded, "and received a reply that he was traveling in California, and would send me an account of what he had seen in the sky, immediately after returning home. But I have been unable to get him to send that account. . . . The problem is: Did a 'headless monster' appear in Crawfordsville, in September, 1891? And I publish the results of my researches: 'Yes, a Rev. G. W. Switzer did live in Crawfordsville, at the time.'"

Subsequently, Vincent H. Gaddis, a writer interested in the sorts of anomalous phenomena that had so fascinated Fort, wrote a piece on the incident for the Fortean Society magazine *Doubt*. Gaddis, who had worked as a journalist in Crawfordsville in the 1930s, claimed that the "monster" had returned to Crawfordsville on the evening of September 5 and had been witnessed by hundreds. At one point, it supposedly swooped low over a group of onlookers, who felt its "hot breath." In the 1960s, Gaddis would coin the phrase and concept of the since-discredited "Bermuda Triangle."

Sources:

Fort, Charles, 1941. *The Books of Charles Fort.* New York: Henry Holt, 638.
Gaddis, Vincent H., 1946. "Indiana's Sky Monster." *Doubt* 14 (Spring): 209–210.
———, 1968. *Mysterious Fires and Lights.* New York: Dell, 34–35.
"Mr. Switzer Saw the Spook," 1891. *Crawfordsville* [Indiana] *Daily Journal* (September 7).
"The Spook Explained," 1891. *Crawfordsville Daily Journal* (September 8).
"A Strange Phenomenon," 1891. *Crawfordsville Daily Journal* (September 5).

1892: Phantom Prankster

THE LITTLE TOWN OF Nappaner, Ind., is all torn up over a ghost which it is said has been appearing for the last two months. Those who have seen it report that it has been playing all sorts of pranks unseemly in a ghost. Among other things reported is that it approached a boy, took a saw from his hands and disappeared. Also that it put three sets of harness on one horse, and tied three horses together by their tails, also, that it stacked all the farm implements in a pile in a barn and scattered a lot of meat about a yard. The ghost is said to have the form of a man and vanishes and reappears with startling and uncomfortable frequency."

Source:

"A Remarkable Indiana Ghost," 1892. *Woodland* [California] *Daily Democrat* (January 22). Reprinted from the *Philadelphia Ledger*.

1892: Strange Snow

AN EXTRAORDINARY meteorological phenomenon occurred in the eastern part of this county [LaPorte] by the recent fall of about one inch of strange looking snow. It was of darker color than ashes and looked like mill middlings or shorts. This snow when melted makes a muddy water, and when allowed to settle deposits a fine sediment, which to the naked eyes presents four different appearances, viz., two powders (one of a gray and the other of a black color), among which are mixed ragged flakes that look like sawdust and others that resemble scales of mica or copper filings. When seen through a microscope the gray powder appears to be the debris of myriads of broken down, semitransparent cells and fibers. The black powder, which is about five times as coarse as the former and about one-tenth in quantity, appears to be made up of little pear shaped, buglike animalculae.

"The copper colored scales are of hard substance, and when magnified become translucent and appear to be of a fibrous, cellular structure, of a purple or bloodlike color, in which are set the little black, buglike creatures before described. This is considered the original life substance from which all the other is derived. The little ragged, sawdust-like flakes are but the former in course of disintegration and look like white, fleshy cellular tissue in which the black objects are set like seeds in a fig. The substance can easily be found, as it forms a uniform dark crust, like stratum, with a considerable depth of ordinary snow both below and above it. The fall extended over quite an area of country and has attracted a good deal of attention. The matter is no doubt worthy of a thorough scientific investigation."

Source:
"An Extraordinary Shower," 1892. *Woodland* [California] *Daily Democrat* (March 31). Reprinted from the *Pittsburgh Dispatch*.

1892: "White Robed Visions Walking"

A FEW SHORT WEEKS AGO the dreadful word reached the city [Crawfordsville] . . . that a Monon passenger [train] had jumped the rails just north of town, and left in its track death and destruction. . . . The story of the awful, ghastly procession, with its blanched faces, as it came slowly into town that memorable afternoon is well remembered. . . . Since that fatal day a quiet melancholy seems to have hovered over the unfortunate spot. Even horses and cows shun it. Birds turn their flight as they approach it. Only the morbidly curious have ventured near it.

"Last night a leading physician of this city, who was present a few moments after the awful catastrophe and was an eye witness to the terrible, heart-rending scene, was called from his warm bed to see a sick man who resides not far from where the wreck occurred. He told this morning of his midnight trip which . . . gave him a scare that he will never recover from. He says that just as he approached the foot of the short hill where the wreck occurred he heard the distant rumbling of an approaching train. [He checked] his horse as he waited, and in a moment a north bound Monon passenger [train] rushed by on its way to Chicago. Before the bright lights of the many windows had disappeared and the distant rumbling of the wheels had died away, while the long line of heavy, black smoke still hovered over the hill, a sight met his gaze that almost paralyzed him with fear. With the greatest difficulty he held his horse, wild with fright and plunging and snorting to break away. Before him, on the hillside, where the fatal coaches had rolled down, he saw two figures clothed in white. They would rise from the ground, walk about and hold up their white arms in supplication.

"'I never,' continued the doctor, 'believed in ghosts. But there were two right before my eyes. I didn't feel like I was exactly scared, but I was possessed with a sensation that is indescribable. It was an awful moment. I can yet see those white robed visions walking about on that hillside. I only remained a moment, but it seemed to me I was there an age. My horse fairly flew up the hill, over the track and homeward bound. I don't believe I could get him near that spot again, even in daylight. If you doubt what I say just go out there tonight at 1:30 and I'll venture to say that you will witness the same sight that I did.'

"Since the horrible wreck of January no [fewer] than three accidents have occurred on this spot. Only last week, while John Rogers, a young farmer living north of the city, was walking along the track he noticed a distortion of the rails before him and was fortunate enough to stop the train before it dashed to pieces."

Source:
"Where a Train Was Wrecked," 1892. *Omaha Sunday World-Herald* (April 17).

1892: Serpent of Horse-Shoe Pond

REPORT, FROM WHAT seems to be a perfectly reliable authority, comes to the city of a strange aquatic monster resembling a sea serpent, which has been seen in Vincennes township, about six miles below the city. This monstrous reptile or animal, which is described as bigger and longer than a telegraph pole, resembles a huge snake in appearance and in its movement. It now inhabits Horse-shoe Pond, near the home of Mr. Daines, a highly respected farmer, whose veracity can not be questioned. He has seen the sea serpent, or whatever it is, on several occasions within the last few days, as has [sic] also his wife and several hired men and neighbors.

"Mr. Daines describes the monster as having a head shaped like that of a sea lion or that of a large dog, and fully as large as the head of a mastiff. Its . . . body is long and serpentine, and is fully sixty feet in length. Its color is black on the back and sides. It inhabits the water and does not seem to venture any distance on shore. It glides through the waters of the pond with that easy and graceful movement peculiar to a snake swimming. When in the pond it holds its head up out of the water. Often its hideous head is held six or eight feet from the surface though it usually holds its head about four feet out of the water. It often is seen lying at rest upon the top of the water, with its head just projecting above the surface. When any one approaches[,] it at once lifts its head as if to listen.

"When approached it becomes alarmed and swim[s] away; if pursued it flees with wonderful rapidity. Mr. Daines, who has seen the serpent, has several times attempted to kill the horrid creature, but failed. He shot it five times, but the balls, which evidently struck it, did not seem [to] harm the monster. It is probable that Mr. Daines will come to the city to collect a crowd of men to go down there armed with Winchester rifles, to kill or capture the sea serpent."

Source:
"A Sea Serpent," 1892. *Vincennes* [Indiana] *Commercial Weekly* (April 22).

1892: Serpent of Big Swan Pond

SOME WEEKS AGO Isaac Daines, who lives south of the city, reported that he had seen some strange monster resembling a sea serpent in a swamp near his home.

"Recently the same strange monster was again seen in Big Swan Pond, ten miles south of this city, in Vincennes township, by several parties. Among those who claim to have seen the strange and hideous creature are Robt. L. Hedges, William Wood, James Durham, Wildman Nolton, Lee Turnmeyer and other men of good repute for veracity.

"The description given corresponds with that made by Mr. Daines. The monster resembles in form a huge snake. It is long and serpentine in shape and movement. Its head, however, is white and similar to the head of a dog in size and shape. It carries its head high above the water when swimming. Its throat is white, its back is black and its sides are spotted or mottled, red and yellow, like the side of a larger water snake. The estimated length, as seen, is variously estimated, the average length being twenty or twenty-five feet long.

"What this strange aquatic creature can be no one seems to know. Those who have not seen it believe it to be a large water-moccasin—a snake peculiar to swamps[—]and express the opinion that the reported size has been greatly exaggerated. Yet all who have seen the monster express an opinion entirely different. They do not know what it is, yet they all agree in the belief that it is no water-moccasin but a much larger and different looking thing altogether, especially about its head[,] which resembles the head of a dog more than a snake."

Source:
"Sea Serpent Seen Again," 1892. *Vincennes* [Indiana] *Commercial Weekly* (June 17).

1893 and Before: On the Banks of the Wabash, and Elsewhere

IT IS ALLEGED THAT AS LONG ago as 1881 August White, a farmer, while seining [at Cedar Bass Lake, a resort,] caught some sort of a monster in his drag that immediately tore a hole through the same and escaped. Two years ago a rowboat was capsized off Cedar point by being struck by something swimming very fast near the surface. Last summer the little pleasure steamer City of Kokomo was pulled several feet by something becoming entangled in her anchor line.

"Therefore when Attorney Beeman, Auditor Nosmin, Sheriff Vanderweele and George Scoville, the attorney who defended the murderer of President [James A.] Garfield, went fishing in May of this year, they were prepared for a big catch. It was with the expectation that their fondest hopes were to be realized that Attorney Beeman felt a powerful tug on his line. He let out about 1,000 feet and then sought to check his catch. But the fish wouldn't 'check.' He pulled out all the line and started for deep water with the boat. The line was made fast, and the fish gave them a free ride for half an hour. Finally all became quiet. Then Beeman, assisted by Nosmin, began to pull in their prize. The fish came in sluggishly, showing no resistance until within 10 feet of the boat, when he suddenly rose to the surface, whirled around and darted off. As he turned he struck the rear end of the boat with his tail, smashing the stern into a thousand pieces and precipitating Beeman and Nosmin, who were standing in the rear end of the boat, into the water, whence they were rescued with considerable difficulty.

"All hands agreed that the animal was 40 feet long and 3 feet thick. Its head was huge and pointed, its color greenish black, and it was devoid of any visible fins. The story winds up with the significant suggestion that 'bathing in the lake will be less popular this summer than heretofore.'

"Not to be outdone by Cedar Bass lake, the raging Wabash next put in a bid for notoriety and cited three young women—Misses Eva Douglass, Cora Nave and Cora Kilander of Huntington—to support its claim. As the young ladies were driving along the banks of the Wabash one Sunday they heard a splashing out in the river and saw, to their great amazement, what appeared to be a huge sea serpent. Its head was above the water; its body submerged, but two or three feet of tail were above the water's surface, and it kept up a splashing with its tail, moving its head from side to side in serpentine manner.

"The ladies had all seen sea lions and unite in describing its head as being like [unlike?] that of a sea lion. Its head was as large as the head of a child 12 years old. They watched it for several minutes, and then it disappeared. They drove on and it again appeared, opposite the buggy, swimming as before, for several minutes."

Source:

"Monsters of the Deep," 1893. *Marion* [Ohio] *Daily Star* (August 26).

1893: Snake of Fire, Ropes of Ice

THE HOME OF MARK Weston, situated near Alexandria, a small town southeast of this city [Delphi], is now the center of attraction for hundreds of curious people, drawn by news of the most remarkable phenomenon ever recorded in this section. The people who have visited the scene gaze in awe upon the startling work wrought by some indescribable power. The story was graphically told by Mr. Weston to your correspondent today, and is verified by witnesses.

"He says: 'Just after dark[,] night before last[,] I had occasion to go out to the barn to look after the horses. A public highway passes within two hundred yards of my house, and the barn is built about twenty rods from the house due south and somewhat nearer the road. I started from the house in the direction of the barn and had gone perhaps half the distance when I noticed something playing along the ground that looked like a tremendous fiery snake. The object crossed my path, and as it did so I felt the air grow much colder, and a peculiar moaning sound arose, like the sighing of the wind through the trees, only it was loud enough to drown a man's voice when he would shout. Then I felt something come over me like electricity, and I became motionless, as though I had grown fast to the ground.

"'I was terribly scared but I never lost use of my hands or legs through fear, though there was something peculiar in the air that simply paralyzed me. When the thing had got perhaps fifty feet from me going west it turned and came back, and as it did so the moaning sound changed to a shrill whistle, something like a locomotive would make, and when it got just in front of me it took a course directly away from me and toward the barn. It traveled very rapidly and looked like a large, ragged streak of fire, perhaps thirty feet long and eighteen inches in diameter. The thing reached the barn, and in almost an instant ran directly up the front of the building and onto the roof. I expected every moment to see the barn burst into flames. But it did not.

"'The great fiery snake ran with great rapidity all over the building. In almost every direction, up and down, crosswise and every way, I suppose, a thousand times. The thing came to the front of the building and elevated itself until it stood straight on its tail fully thirty feet in the air. I was perfectly conscious all the time, but try as I would I could not move from the spot. After the thing had remained in an upright position for, I presume, three or four minutes, there was a sudden explosion like the discharge of a cannon, and the thing disappeared entirely. With the disappearance of the strange phenomenon I felt a shock like the first one I had felt and at the same time I gained control of my limbs. I hastened to the house, told my wife what I had seen, and she thought I was crazy, but upon my insisting she consented to accompany me to investigate the matter at daylight.

"'You can imagine our surprise upon reaching the barn to find it covered with a remarkable network resembling large ropes of ice. They appeared to pass around the building in exactly the way the fiery monster had passed. It was not ice, however, but seemed to be more of a crystal, for it would not melt even when we held a flame to it, and when struck with a hatchet it simply gave a dull like sound and did not break. Upon entering the barn we were amazed as two good horses stood in their stalls immovable. They were alive, but neither could move a muscle. They seemed to be paralyzed and stood there more like statues than anything else. They were warm and breathed all right, but aside from this you could not tell they were alive. I applied the whip and they never flinched. A dog that sleeps in the barn was dead and appeared completely petrified. He was lying on the ground with his head on his paws just like he was sleeping. When I left home this afternoon everything was just as I have described it to you.'

"Mr. Weston says the house has been visited by hundreds of people and that the entire community are marveling at the strange and weird visitation. He has taken his wife and family to a neighboring farm house, and says he will not return until every evidence of the strange phenomenon has disappeared."

Source:
"Up Against It," 1893. *Atlanta Constitution* (February 27).

1894: Omen in the Oats

THE COUNTRY PEOPLE of Indiana are very much exercised . . . over the discovery of a strange and portentous marking which they find on the blades of the growing oats. On each blade they can read, plainly impressed, a letter B. Acres and acres in all parts of the county have been found to be thus curiously marked, and it is no wonder that imaginative persons can associate the presence of a letter with forebodings of evil. It is claimed that the only other times the letter was ever found on oats in this manner was just before the war of 1812 and the late civil war, and that the B stands for 'bloodshed,' which may now be looked for again. Each blade is marked, the letter, about half an inch long, being[,] as it seems, pressed into the leaf and discernible on the other side. Some say that the phenomenon occurs frequently, but none explains its origin."

Source:

"Editorial Comment," 1894. *Atlanta Constitution* (July 6).

1894: Disappearing in Midair

A SNOW-WHITE APPARITION near Frankton first made its appearance to the affrighted vision of McClelland Beagle, of Indianapolis, a track walker in the employ of the Manufacturers' Natural Gas company, and it also floated into view when Beagle was accompanied by James Haggarty. Even the horse which they were driving was alarmed. They watched the apparition until it made the circuit of John Riley's residence and then disappeared in mid air."

Source:

Untitled, 1894. *Fort Wayne* [Indiana] *Sentinel* (August 22).

1894: Shower of Frogs

A CURIOUS PHENOMENON occurred in the northern portion of the county [Delaware] Thursday afternoon [August 23]. Shortly after 3 o'clock a shower of live frogs began falling. The shower of living creatures continued for five minutes and covered a 10 acre field on the farm of Ezra Willburn. The frogs fell only on Mr. Willburn's farm, and at the time they fell the sky was cloudless. Mr. Willburn's small son was the only person who witnessed the shower, and after recovering from his surprise at such a strange occurrence he informed his father of the affair.

"The Willburns at once began catching the largest of the frogs and enjoyed a regal repast of delicious hams for supper. The neighbors were also liberally supplied. The cause of the shower is somewhat a mystery, although it is said that the frogs could have been drawn from a distant pond by a strong whirlwind and carried through the air to a point over Mr. Willburn's field."

Source:

"A Shower of Frogs," 1894. *Middletown* [New York] *Daily Argus* (August 29). Reprinted from the *Chicago Herald.*

1894: Mud Mermaids

IN A STRAIGHT-FACED account a Cincinnati newspaper began, "On the sand bar in the Ohio river near Vevay, Ind., reside two nondescript creatures horrible in appearance and habit. They are amphibious in nature and resemble in appearance huge lizards with human features." When only their upper parts were visible in the water, they looked like human swimmers and were sometimes mistaken for them.

"Of what species of animal they are no one knows, for it is impossible to get near enough to them to judge correctly," the piece went on. "The sand bar in question at low tide is covered with huge logs and stumps of trees, known in the river vernacular as snags. They have been deposited by the government snag boats engaged in keeping the channel clear. When the water is high enough to cover these snags, the creatures make their home among them. When the water recedes, they disappear into some unknown lair and wait for a rise."

Evidence indicated that the "mud mermaids," as the anonymous correspondent dubbed them, lived on fish, mussels, and other aquatic life. Sightings had started four years earlier, when a fisherman spotted the two in the water. Since he had never heard of mermaids, he was utterly perplexed and had to have the concept explained to him. The stories circulated and in 1894, when a Kentucky man, "Captain J. M. Ozier . . . who is in charge of a traveling art exhibition," heard about them while doing business in Vevay. He went to the scene hoping to see the creatures. As luck would have it, he did see one, the male entity—at only 20 feet from him—and as it swam within easy viewing distance, he drew sketches of it.

According to the piece, Ozier judged the creature to be about five feet long, probably weighing 150 pounds, and said it was yellow in color. "The body between the fore legs resembles that of a human being," it was stated. "Back of the hind legs it tapers to a point. This point in no way resembles a tail. The legs, four in number, resemble the arms and legs of the human. The fore legs are shorter than the hind pair and are used in the same manner as arms. The extremities resemble hands and are webbed and furnished with sharp claws. On the back and one-third of the way around the body appears a mass of straggling, coarse hair. The skin below the fore legs is thick and resembles elephant hide. On the arms and about the face and neck it is of a finer texture and brighter yellow color than the rest of the body."

Its facial features were strikingly human, except that its ears had sharp points and stood up like a dog's. Its face bore no signs of intelligence. It swam effortlessly and quietly. When disturbed, it would sink, not dive, below the water's surface. It apparently feared human beings and would flee when approached.

Giving the game away, the article went on to say that Mr. Ozier declared that the creature "resembles to a great extent the freak known as Zip, or the What-Is-It, which was exhibited first by P. T. Barnum." Even readers gullible enough to credit all that had gone before must have had no problem figuring out the worth of a story seeking validation for itself in a Barnum exhibit.

Source:

"Mud Mermaids," 1894. *Sandusky* [Ohio] *Register* (October 19). Reprinted from the *Cincinnati Enquirer.*

1894: Fifer's Music from Beyond

CONSIDERABLE EXCITEMENT prevails here [Frankfort] over the report that the Second ward school building is haunted, and the children are expressing fear. Residents claim that during the entire night a fifer's music can be distinctly heard emanating from the roof of the building, and some of the more superstitious claim that they have seen the form of a man walking near the edge of the roof. Other residents say that during the erection of the structure a man by the name of Entrekin, a carpenter, fell from the roof and was killed. He was a fifer in the war and is supposed to furnish the spirit now."

Source:

"This House Is Haunted," 1894. *Fort Wayne* [Indiana] *News* (December 18).

1895 and Before: A Huge Reptile

ON AUGUST 8, IN JAY County, a feared resident made one of its periodic appearances: "a hugh [*sic*] reptile of the black snake species," in the words of a press account of the next day. The creature was between 20 and 40 feet long, according to witnesses too frightened to get close enough to garner a more precise judgment.

The episode happened in a rural location just west of Muncie. A poultry dealer named William James had stopped at a farm to do business. He got out of his wagon and went to talk with two men. As they were conversing, they heard sounds of fence rails falling, and they noticed that James's horse had its ears pricked up and was looking off to one side of the road.

Moments later, the head of an immense snake poked out of the grass in that direction. The creature surveyed the situation, then crawled onto the road, trying to pass between the wagon's wheels. It was, however, too large to manage that, so it moved around it and continued on its way. The witnesses were too amazed and unsettled to do anything about it except watch.

Source:

"Indiana's Big Snake," 1895. *Fort Wayne* [Indiana] *Sentinel.*

1895: Ghost Spurns Cop

FOR THE LAST THREE nights Sergt. Dasler has been employed in searching for a ghost. This ghost has alarmed the residents of the east side for weeks past. It is the shadowy form of a woman clad in white, carrying a bouquet of flowers[,] that walks in the alley between Lewis and Madison from Hanna to Lafayette, [and] goes back[,] disappearing at some intermediate point. The people in the vicinity are greatly concerned about this ghost and some claim to have seen it vanish. Sergeant Dasler has camped on the trail but the ghost did not walk."

Source:

"An Elusive Ghost," 1895. *Fort Wayne* [Indiana] *Gazette* (July 18).

1895: "A Legion of Impalpable Forms"

IN THE MID-1870S, Gottlieb Haslinger of Big Lake hanged himself in his house. Twenty years later, when apparitions in large numbers were allegedly seen in the house and in its vicinity, local people drew a connection, without, of course, being able to prove it.

According to a local newspaper, area residents lived in fear of a "legion of . . . shadowy and impalpable forms." Mysterious lights flashed and gory heads and bloody ropes figured in the lurid tales circulating throughout the neighborhood. "Gaunt figures in white flowing robes march with stately, measured tread around the house," it was alleged in a breathless press account, "or sometimes assemble round a pile of human bones and sing a chant so weird and awful that those who have heard it have fled in frenzied terror. Skeletons dance and flit about the place. Screams are heard sometimes which are said to be invariable [*sic*] merge into deep, heartrending wails like unto those of spirits that are lost. These wails finally die away into oppressive stillness which is sometimes broken up by sharp, piercing cries of despair accompanied by groans of agony and cries of supplication."

Source:

"Ghosts Walk," 1895. *Fort Wayne* [Indiana] *Sentinel* (August 2).

1896: Masonic Temple in the Sky

THE STRANGEST PHENOMENON ever seen in Warsaw, Ind., was visible yesterday [April 21] during the big storm. It was a mirage of the Masonic Temple in Chicago. The great building hung in the sky, apparently over a wagon factory near the public square. Yesterday was a strange day in more ways than one in Warsaw. The wind was high and rain fell furiously. Hail as large as walnuts rattled on the housetops and battered the windows of buildings. A great cloud hung, after the storm had subsided, in the western sky, and on its face was pictured the Masonic Temple in complete mirage. The structure was plainly visible, and people could be seen walking about in its vicinity. Warsaw is 109 miles from Chicago, on the Pittsburg and Fort Wayne Road."

Source:

"Warsaw Reports a Mirage," 1896. *New York Times* (April 27). Reprinted from the *Chicago Times-Herald,* April 22.

1897: Airship over Indiana

THE AIRSHIP, AS IT IS claimed to be by some people, has been seen by three Fort Wayne people. . . . For several nights telescopes have been directed toward the heavens of the northwest, and at last F. Crocker and R. J. and J. L. Tretheway have been rewarded by a sight of the star, ship or whatever it may be.

"Mr. Crocker . . . was the first man to see it, and after watching the light for some time, called his wife and the Messrs. Tretheway. The two brothers were in a flat below the Crocker home, and the quartette saw the light for fully twenty minutes. When asked to describe what he saw Mr. Crocker said: 'I had taken a great deal of interest in the stories printed in newspapers about this star or airship, and was standing at the window of my flat endeavoring to get a glimpse of it. In a short time I was rewarded with a sight of it. It appeared in the west and traveled in a northwesterly direction, at a good rate of speed. It was of a yellowish color and pear shaped with the apex downward. From the sides I could see two rays of light of the same color as the main body. I am very positive that the body swayed to and fro as it proceeded on its journey. I had no glass of any kind so could not get a good sight of whatever it was. It moved rapidly and in about twenty minutes was out of sight. I called the Tretheway brothers, who are in the flat below mine, and they also had a good view of the lights.'

"R. T. Tretheway . . . said, 'Whatever the light was it seemed to me as being round with a V shaped tail to it. The color of it seemed a bright yellow and strong rays of the same colored light were constantly being shot out from the main body. The light appeared as big around as a bushel basket and traveled at a good rate of speed. I got a range on it between two chimneys that are about twenty feet apart and it took the light about fifteen minutes to pass between these two points.'

"So far as is known no other Fort Wayne people have seen the light, but many are on the lookout for it."

Source:
"Sighted Here," 1897. *Fort Wayne* [Indiana] *Gazette* (April 15).

1897: The Medium and the Airship

THE AIRSHIP HAS AGAIN been seen by Fort Wayne people. On Wednesday evening [April 14], as several members of the First Spiritual circle were leaving their society rooms after a meeting, they noticed the air ship passing over the city toward the northwest at a rapid speed. It carried two lights, a red one in front and a blue one in the rear. It is said by A. Carpenter

that the members returned to their hall and the medium went into a trance. While in that condition the medium disclosed the occupants of the ship to those gathered in the room. There were, according to the medium, two men in the car and a third living body supposed to be a dog. The spiritualists are greatly agitated over the matter and further séances may be held."

Source:
"Two Men in the Ship," 1897. *Fort Wayne* [Indiana] *Gazette* (April 16).

1897: Airship over Vincennes

IN THE MIDST OF THE excitement over airship sightings of the spring of 1897, residents of Vincennes reported seeing a mysterious aircraft pass over the city twice on the evening of April 16.

At nine o'clock, a "sphere of golden light" on the eastern horizon appeared before a number of witnesses. Those close to it said they could see the outlines of a car beneath the larger structure, though they could detect no one inside. At least one observer, Sam Judah, claimed to be within range to see the "fluttering wings" that propelled it through the air at an impressive rate of speed. Others saw the object only as a moving light heading southwestward and, after a few minutes, disappearing there.

After 10:30, the ship, or one much like it, reappeared, this time over the northern part of the city, now heading northwest. According to some, voices could be heard emanating from the ship. One man said he had glimpsed a man inside. The figure seemed to be working with the ship's machinery.

Source:
Day, Richard, and Byron R. Lewis, 1980. "Early Airship Rumors Flew in Vincennes Sky." *Valley Advance* (March 18).

1897: "A Strange Animal Which Resembles a Man"

THE FARMERS LIVING NEAR Sailor, northwest of here, are considerably aroused over the appearance in the woods of a strange animal which resembles a man. It has been reported for the last two years that a mysterious animal was inhabiting the woods, but the reports were never credited until today [April 30], when Adam Gardner and Ed Swinehart, two well-known farmers, reported that the animal was seen and that shots were fired at it. The men report that the beast walked on its hind legs and had every appearance of a man, save the body was covered with hair. The height was that of an average-sized man. When the animal saw the men approaching it jumped and started for the thick portion of the woods upon its hind legs, but afterward dropped on its hands and disappeared with rabbit-like bounds. Gardner shot at the animal and thinks he hit it, as the animal seemed lamed. A searching party is being organized to hunt for the mysterious animal."

Source:
"Resembles a Man," 1897. *Indianapolis Star* (May 1).

1897–1898: Ghost in a Fedora

A REAL LIVE GHOST has been startling the citizens near Warren, Huntington county, Ind. It is white-sheeted and wears a Fedora hat, and haunts a bridge on the Montpelier road. Traffic across the bridge is suspended at nightfall."

Source:
Untitled, 1898. *Fort Wayne* [Indiana] *News* (January 3).

1903: The Not-So-Real Airship Passes

PEOPLE LIVING IN STRATFORD, just outside Irvington, and residents in the southeastern part of Indianapolis are positive they saw an airship pass over Stratford about 5 o'clock last evening. The strange craft appeared at a high elevation sailing from the northwest, according to those who saw it, and after making almost a complete circle over Stratford, turned toward the east as if following the course of the National road.

"E. A. Perkins, president of the Indiana Federation of Labor, and Mrs. Perkins watched it through field glasses. Their attention was called to the machine by James Agnew, their next door neighbor. His chickens had begun to cackle when the machine appeared overhead. Several other people in the neighborhood watched the airship until it disappeared. All agreed that it was far too large to be a bird and shaped too much like a cigar to be a balloon. It is described as being a cigar-shaped body, but pointed in front and cut off short in the rear. Above it was a canopy protecting the center of the craft which contained the machinery.

"Two figures were seen in the center under the canopy. They could be seen by the aid of a field glass to be moving backward and forward slowly and rhythmically as if they were operating the machinery. No propeller was distinguished by those who say they watched the airship closely, but the general descriptions only vary slightly.

"The Star last night communicated by telephone with practically every village and town surrounding Indianapolis, but no one else seemed to know anything about the strange creature of the air. The Langley airship is reported safe in its morings [*sic*] and as far from flying as ever."(1)

☞

"The Stratford airship mystery has been solved.

"The strange craft [that] was seen by people in Irvington, Stratford and Tuxedo Sunday afternoon, made a landing in a cornfield opposite the home of Sam Manning, Michigan street and Emerson avenue. Fully fifty men and boys 'saw it first' and considered it their own. Finally, Leonard and Edward Manning, boys of 8 and 10 years old, took the thing home.

"Several hours were spent making repairs, and the machine will be ready for use again in a short time. It's about thirteen feet long, made of numerous pieces of tissue paper, representing all the colors of the rainbow and some others. In shape it is exactly like the submarine torpedo boat, even to the turret on the top. The total weight, including the wire framework to hold a piece of cotton saturated with alcohol, is 33 ounces."(2)

Source:
(1) "Say They Saw Real Airship," 1903. *Indianapolis Star* (September 14).
(2) "Airship Mystery Has Been Solved," 1903. *Indianapolis Star* (September 15).

1903: A Team of Ghost Horses

LAKE ERIE & WESTERN trainmen have been 'seein' things at night' in the deep cut between Peru and Denver. One of the oldest engineers, Michael Hassitt, says that on two occasions he has seen what appears to be a ghost and once he brought his train to a standstill, owing to the strain on his nerves when the apparition seemed to be upon him. It takes the form of a team of white horses attached to a wagon, in which the driver stands frantically whipping the animals over a crossing. No harm has been done, beyond scaring the trainmen."

Source:
"Trainmen See Ghosts," 1903. *Fort Wayne* [Indiana] *News* (September 21, 1903).

1904–1905: Swaying Lights on the Line

THERE IS GREAT excitement in the vicinity of Maples, a little village east of here on the Pennsylvania, because of the nightly appearance of what are believed to be 'ghost lights' along the railroad tracks at a spot about a mile east of the town. Society doings now take place in the afternoons, and lovers do their courting before sundown. There is no holding of hands after dark, for the population is keeping indoors when the shades of evening have fallen. Strange lights have been seen swaying, swinging and moving to and fro on the railroad track, and the superstitious people believe the spirit of a man who was killed there a few years ago is haunting the spot.

"Of the brave few who defy the apparition enough to go visiting at night none will go near the spot where the spirit lights flicker and cast their baleful gleams."

Source:
"Ghost Lights Seen on the Railroad Track," 1905. *Fort Wayne* [Indiana] *Journal-Gazette* (January 6).

1906: Airship Landing

JOHN WARNER, AN OLD soldier and resident of Orinoco, a suburb of which is supposed to be 'dry' on Sunday, insists that he saw an airship that night [August 26]. He says he was sitting on his back porch when he heard a noise in his barn made by his family driving horse. He went to see if the horse was sick and on returning to the house he heard a rushing noise overhead. On looking upward, he declares he saw a cigar-shaped airship, painted green and carrying green lights, which sailed gracefully down into his garden and stopped. There were four men in the ship, he says, and they informed him they were on their way to New York, from Chicago, and asked him which direction to take. He directed them as far as Seymour, when they turned on their power and sailed away."

Source:
Untitled, 1906. *Rochport* [Indiana] *Journal* (August 31).

1913: Light in the River

A MYSTERIOUS LIGHT, bobbing, winking, disappearing, flashing again and finally losing itself, was seen coming down the [Wabash] river [during a spring flood] just after dusk last night. What it was, no one living knows. The small light was first seen far up the river. Everyone on the river bank was attracted to it. Swiftly it came towards the city and people wondered. At times it would disappear, then shine dimly again. For ten minutes it came on, apparently in the middle of the vast sweep of water. It was dark enough that nothing else could be seen. When but a few hundred yards above the Brown street bridge, the little flame went out, and did not appear again. Anxious watchers strained their eyes, but there was no further sign of it."

Source:

"Mysterious Light Appears on River," 1913. *Lafayette* [Indiana] *Morning Journal* (March 26).

1927 and Before: Legend into Life

ACCORDING TO A STORY he told years later, well into his adult life, to a television station, Paul Startzman was walking near an abandoned gravel pit one day at the age of ten when he suddenly found himself facing a tiny man, no more than two feet tall. "We stood about ten yards apart and looked at each other. He had thick, dark blond hair, and his face was round and pinkish in color, like it was sunburned."

The figure was clad in a light blue garment which covered all but his ankles. He was barefoot. The little man turned and fled into the bushes, where he was lost to sight.

Not long afterward, at the same location, Startzman, in the company of a friend, allegedly had another encounter with a comparable entity. This time the figure followed them for a while.

Startzman, whose mother was an American Indian, believed these dwarfs were little people known to Indiana tribes as Paisaki (to the Miami) and Pukwudjies (to the Delaware), or "little wild men of the forest." These beings were thought to be human, not supernatural, and to live in caves or makeshift huts in the trees along the river. The Miami and the Delaware believed that the little people had been there for even longer than they themselves had been.

One Indiana legend holds that in the early nineteenth century a Methodist pastor near Marion denounced stories of little people as superstition. To underscore the point, he set out to cut down a large tree said to cover the underground entrance to the little people's residence. Suddenly, fifteen to twenty small men ran out from the base and attacked the minister. When he was knocked off his feet, they cut his throat, though fortunately not fatally. He is said to have carried the scar the rest of his life.

Source:

Swartz, Tim, 2001. "An Unnatural History of Indiana." *Strange Magazine* 21, http://www.strangemag.com/strangemag/strange21.

Iowa

1863: Men in Black

A CIVIL WAR–ERA REPORT from the southeastern part of Iowa overlaps with the long-recorded phenomenon of rock-throwing poltergeists but has its own unique characteristics. In poltergeist cases of this sort, the assailant is invisible. Perhaps the Iowa incident is closer to the phantom-attacker tradition, however, which receives explanations ranging from the psychological to the paranormal. It is broadly reminiscent of an episode from 1692 in Cape Ann, Massachusetts, when strangers who were said to be impervious to gunfire harassed the citizenry by beating on barns, throwing stones, and otherwise making nuisances of themselves.

Though the word is not used in the account, it is apparent that the victims of the curious attacks were Copperheads, as their enemies called them, or Peace Democrats, as they called themselves. Especially powerful in the Midwest, they opposed both the war (being waged under the hated Republican leadership of President Lincoln) and the abolition of slavery, fearing the latter would lead to a society of racial equality. They sought an end to the war on Confederate terms, which made them deeply unpopular among the majority of Union supporters in the North. This note is necessary to clarify certain references in the concluding paragraphs of the account that follows, which was published in at least two area newspapers:

"We learn, by a gentleman of unquestionable veracity, that great consternation prevails in Adams township, in this county, occasioned by the nightly visitations of two seeming men, at the residence of Mr. Wm. Spaulding, who lives five miles east of Blakesburg. These visitors, be they who they may, and whether in the flesh or spirits in human shape, make their appearance about seven o'clock in the evening, and remain until about five in the morning, their first appearance being on Friday [presumably October 23], a week ago.

"They seem medium sized, heavy set men *dressed in black!* [italics and exclamation point in the original]. On their first appearance, on Friday night, the family and some of the neighbors, were boiling molasses about forty rods from the house, when about seven o'clock, suddenly, clubs, cobs (they had been shelling corn during the day) and small sticks, began to fly in a shower, from a certain direction, occasionally hitting some one of the persons present, but generally falling in one small place. Once a candlestick, held by Mrs. Spaulding, was hit, nearly knocking it out of her hand. No person was then visible, but they heard something walking about with a heavy tread. About one o'clock they quit boiling, and two of the men, Harrison Wellman and J. M. Spaulding, started out in the direction from whence the missiles seemed to come, armed with clubs and brickbats, to find and chastise these strange and curious intruders. No sooner had the men started than the missiles came, larger and thicker. The fire of missiles was returned by the men, but without evidence of their hitting anybody.

"The party about the kettles returned about this time (one o'clock) to the house. After their return, the missiles seemed to strike the fence and the house with great violence. Spaulding and one of the men went out to turn out the horses, taking their guns with them. No sooner were they out of the house then [*sic*] a large club fell near them, seemingly coming from behind. One of the men wheeled, and saw a man standing near enough to be distinguished in a dark night, at whom he instantly shot. The man ran and disappeared. They turned out the horses and returned to the house.

"The next night, and for the four succeeding nights, the same state of things existed, two being seen on Tuesday night. Missiles struck the fence and house, but left no dents and marks distinguishable by daylight. In the meantime, Mr. Spaulding and his neighbors became alarmed by such strange phenomenon [*sic*], and from time to time, met at Spaulding's house to try and solve the mystery.

"On Monday night, J. W. Wellman, Wm. Hayne, Wm. Spaulding and his son, spent the night watching. Sometime in the night, one of the men looked out of the window and distinctly saw, by the light of a bright moon, a man standing before the door. After sitting awhile, they looked out again, and saw the man prostrate on a plank, lighting a dark lantern with a match. Getting their revolvers and guns ready, the party prepared to open the door, but strange to say, the four powerful men could not open it. They afterwards remained quietly in the house until 5 o'clock in the morning. One night, Mr. E. B. Day took his dog, a very sagacious animal, and tried to set her on, but she trembled, ran between her master's legs, and refused to make any more demonstrations against the ghosts.

"The above are the leading facts as related to us, of the most strange phenomenon. Mr. Spaulding at first attributed the persecution to political enmity, but certain evidence of the absence, in other places, of those he could not only suspect, at times when the visitors were seen and engaged in their operations, satisfied him that he must look for some other solution of the mystery.

"We give the facts precisely as they are related to us, merely expressing, by way of comment, our decided conviction that no Union men, in or out of the flesh, have resorted to that mode of converting Mr. Spaulding and his political friends of the error of their political ways. The story is a strange one."

Source:
"Very Strange If True!—Two Ghosts in Wapello County!," 1863. *Burlington* [Iowa] *Weekly Hawk-Eye* (October 31). Reprinted from the *Ottumwa* [Iowa] *Courier.*

1864: "In the Habiliments of the Tomb"

AN APPARITION, ROBED in white, was seen by a number of persons last night [in mid-August] about eleven o'clock, gliding noiselessly on the top of the walls at the rear of the building occupied by Kellogg & Birge. Back and forth, back and forth, went the apparition, with noiseless, almost invisible step, several times and then disappeared. By those who saw it, and the moon being quite bright[,] its every outline was distinctly visible[;] it is described as about six feet tall, and looking like a man attired in the habiliments of the tomb. It has been seen several nights in succession just as the clock struck eleven, and has excited considerable alarm in the immediate neighborhood. About twenty-five persons witnessed its appearance last night and gazed [in] speechless awe upon the picture. The interest is becoming most intense and large numbers will watch for it nightly until the mystery is explained."

Source:
"Ghost! Ghost!! Ghost!!!," 1864. *Burlington* [Iowa] *Weekly Hawk-Eye* (August 20). Reprinted from the *Keokuk* [Iowa] *Constitution.*

1864: Odd Luminous Phenomena in the Heavens

THE BEAUTIFUL PHENOMENA of Wednesday night last [August 31] has [*sic*] caused much comment among our [Davenport] citizens. We do not understand its cause, and would like to have some theory concerning it. Three distinct trains, equi-distant in appearance, tapering towards a point, like the rays of a comet, and with the light fully as bright, and standing out in bold relief on the sky at an angle of about 45 degrees, presented a subject of study for the astronomer. The phenomenon appeared about 8 o'clock p.m., and lasted about 15 minutes, at one time the sky beyond it appearing of a dull red color. A gentle breeze was blowing at the time, and the sky was somewhat overcast. About 11 o'clock the atmosphere became quite cool, and a number of meteors appeared. Who can further explain it?"

Source:

"The Meteoritic Phenomena," 1864. *Burlington* [Iowa] *Weekly Hawk-Eye* (September 10). Reprinted from the *Davenport* [Iowa] *Gazette,* September 3.

1867: "Chivalric Ghost"

IOWA IS HAVING A GHOST epidemic. The latest importation from the invisible world is described as no plebeian 'spook,' but a chivalric ghost, who, mounted on a snow-white charger, roams the forest with the speed of a Pegasus."

Source:

"All Sorts," 1867. *Titusville* [Pennsylvania] *Morning Herald* (November 20).

1874: Poltergeist at the Paint Station

THE GHASTLY GHOUL of the graveyard has broken out in Burlington and devotes his leisure hours to making things lively at the paint shops of the Burlington and Missouri River Railway. An ill-mannerly, rather boisterous but unorthodox ghost. It made its first appearance several nights ago. When the brazen hand of time proclaims the midnight hour is promptly on hand, it shies its castor into the rings, follows it and is ready for business. The lonely watchman, pacing his weary rounds, feels homesick as he notices the light in his lantern turn blue. He waits with bated breath for the manifestations which well he knows that ominous symptom prefaces. . . .

"Suddenly with a harsh clangor the doors of the shop clang open, hurled by invisible, unearthly hands, and as the discordant echoes die away in gloomy corners and shadowy nooks . . . the watchman's hair assumes the attitude of quills upon the [porcupine]. Then the ghost rattles a car door, and the faithful guardian of the shops hastens to the spot, but finds no one there, and while he looks, the ghostly spook pounds on a car-wheel with a coupling link. . . . It kicks at windows and slams the doors and dances unearthly double-shuffles on the tops of the cars. . . . One night several persons watched; they heard the noises but did not make any thorough inquiry or investigation into the causes thereof."

Source:

"A Railroad Ghost," 1874. *Burlington* [Iowa] *Weekly Hawk-Eye* (March 19).

1875: Trotting Ghost

A GHOST IS HAUNTING Summerset. It trots up and down the railroad track between the depot and the junction, clanking chains, and waving a gory hand in the haunted air. The Summersetters go to bed at dusk, sleep with their heads under the covers and don't get up till after sun rise."

Source:

Untitled, 1875. *Burlington* [Iowa] *Weekly Hawk-Eye* (February 11).

1880: Captured Serpent Alleged

ONE OF THE WITNESSES to the serpent described below, Dr. C. L. Horner, reportedly believed the snake to be an anaconda, though such snakes are not native to North America and could not long survive in its climate. According to conventional herpetology, anacondas grow no larger than 33 feet. The account reported:

"Mr. E. G. Lobeck, of Slayton, Iowa, is in the city [Des Moines] and from him we learn the particulars of the capture of another mammoth serpent near Slayton, the mate to the one captured some time ago. . . . The monster was first seen by Dr. C. L. Horner, stretched out on a sandbar of the Des Moines river, about six miles from Slayton. Mr. Lobeck thinks he can say without exaggeration that the serpent was fully forty feet long, and at its thickest was nine inches in diameter. It is of a darkish brown color with golden stripes. It was captured by tracking it to its cave, which is the same one in which the other snake was captured, and fencing it in by driving stakes in front of the hole. When the monster realized his position he made a herculean move for liberty, and in so doing he caved in the earth for eight feet around him. . . . The spot which is shown for its habitation is one of the wildest sections in the west, and is just such a place where one would look for a monster of this kind."

Source:
"Another Big Snake," 1880. *Burlington* [Iowa] *Hawk Eye* (September 16). Reprinted from the *Des Moines Register.*

1883: "A Literal Shower of Birds"

A REMARKABLE PHENOMENON occurred at Independence during a heavy thunderstorm on the night of the 1st inst. last [May 1]. Many were aroused by a pelting against the windows, and supposed it to be hail; others thought it was caused by bats. But the next morning thousands of birds were found all over the city, some dead and some alive. Wherever a door had been open the place would be full of them. It was a literal shower of birds, and how and whence came they? In size the birds were a trifle larger than snow birds, and their color much like that of a quail. No such bird was ever seen there. One theory is that they were drawn into the vortex of a Southern cyclone and carried as far as Independence, where they were dropped."

Source:
"A Queer Bird Story," 1883. *New York Times* (May 12). Reprinted from the *Davenport* [Iowa] *Gazette,* May 9.

1884: Wild and Hairy Woman

THE NEWS OF A QUEER specimen of the female sex in the woods near Gordon's Ferry, a place about twelve miles north of here [Dubuque], reached here this morning [July 17]. When first discovered by hunters she was standing like a statue in a clear space with her back towards them. They, wishing to get a square look at the strange apparition, got around in front of her, but they no sooner had done so than, with an unearthly scream, she darted off through the woods and ran about 300 yards, when she stopped, got behind a large tree, and, with a wild glare, viewed the visitors at a distance. After another ineffectual attempt to discover her identity she was lost to their sight. They explored the woods for over two hours, but were unable to discover any trace. The hunters describe the strange creature as follows: She was apparently about 20 years of age, with lithe and sinewy form, a receding forehead, and eyes which shone with an unnatural luster. Her hair, about three feet long and black as jet, hung in disheveled locks over her shoulders and back. She was in an almost nude condition, and was minus shoes and stockings."

Source:
"A Wild Woman," 1884. *Daily Nevada State Journal* (August 21).

1885: Bizarre Haunting

A GHOST STORY OF AN unusual style comes from a Swedish settlement on the Rock Island Road, northwest of this city [Davenport]. There are any number of people willing to vouch for the truth of the story, and it is certain that part of it is true beyond all question. A man named Richardson lived on a small farm near the village of Tiskaville in a humble but contented way. He was industrious and well liked by his neighbors.

"On Wednesday of last week [July 15] his youngest child, a daughter, ran screaming to a neighbor's before daylight, saying that a huge man, all covered with fire, had come into the house and carried off all but herself. It was supposed the house was on fire, and aid was quickly afforded, but the house was found to be all right. Nothing at all was disturbed, but no one was about. The horse and buggy remained in the stable. The clothing left off by members of the family on going to bed was found where it had been left. The vicinity was thoroughly searched, but without avail. No train had stopped, and no water was near. It seemed as if the ground had opened and swallowed the family up.

"A neighbor's family moved into the house to take care of the things and the child and were nearly scared to death the first night. They assert that suddenly the house was filled with a strange white light and the voice of Richardson was heard calling his daughter. She responded, and instantly the light disappeared and a great shower of small stones fell upon the roof. The same scene has been enacted nightly since and the whole community is aroused. The child does not appear to be in the least alarmed at the voices."

Source:
"Iowa's Remarkable Ghost Story," 1885. *New York Times* (July 23).

1895: "Spectral Lantern"

THERE SEEMS TO BE no doubt of the existence of a specter in the Tyrone bottoms, west of Albia, and engineers and firemen on the Burlington declare they have seen it walking the track and swinging a lantern. One man declared it was the form of a woman whose husband was killed on the railroad some time ago. The moans of the woman can be heard for several rods, but she has not yet succeeded in stopping any trains with her spectral lantern. Many railroad men look upon the ghost as a hoodoo, and predict an accident at that point sooner or later."

Source:
"Ghost with a Lantern," 1895. *Fort Wayne* [Indiana] *News* (April 9).

1897: Airships over Western Iowa

A QUEER AIRSHIP was seen three weeks ago by trainmen at Sioux City, Ia. Twice after that the same men had glimpses of the machine. Now several residents of the western part of the city claim to have been added to the list. The witnesses do not all insist that the aerial mystery is an airship, but they positively deny that it can be a balloon. It is evidently under perfect control, travels in a bee line at a uniform distance from the earth and carries a row of lights at each side of the color of incandescent lamps."

Source:
"Iowa's Airship," 1897. *Delphos* [Ohio] *Daily Herald* (April 3).

1897: Airships over Eastern Iowa

THE AIRSHIP, WHICH has frequently made its appearance in neighboring states, has been seen by hundreds of people in eastern Iowa last night. Reports from all along the line of the Burlington, Cedar Rapids & Northern railroad say that the mysterious aerial craft has been seen by every operator and station agent between West Liberty and Cedar Rapids. They all describe it about the same way. A bright glaring headlight, revealing a glistening steel hull, dim winglike projections on either side and a hissing sound as it glides through the air. Excitement exists at all the above mentioned points, and people crowd the streets of the towns and cities in the hope of catching sight of the object."

Source:
"Airship in Iowa," 1897. *Oshkosh* [Wisconsin] *Daily Northwestern* (April 9).

1897: Scaring the Horses

IN WOLF CREEK TOWNSHIP, just across the Big Sioux River from the Nebraska border, wealthy farmer Richard Butler, in a horse-drawn wagon on his way home on the evening of April 14, noticed a light in a field along the road. As he tried to make out what it was, he made out a dark object surrounding the illumination. He was able to determine that the lights were coming through windows in the object.

As his eyes adjusted to the dark and light, he saw that it was long and narrow, shaped like a corset box, and 30 to 35 feet long. It was 6 or 7 feet high and just as wide.

At that moment Butler's horses saw the object and bolted for the opposite side of the road. The wagon tumbled into the ditch, and Butler went flying. By the time he pulled himself together and out of the ditch, the mysterious machine was moving southward, ascending at a 45-degree angle.

Source:
"Farmer Describes Airship," 1897. *Arizona Republican* (April 18).

1899: "Lady in Black"

LAPORTE HAS A REAL live ghost, and a black one at that. . . . 'The lady in black' . . . appears in various parts of the city at unseemly hours of the night and when addressed mysteriously disappears."

Source:
"Items from Exchanges," 1899. *Nashua* [Iowa] *Reporter* (September 28).

1901: Coal House Ghost

INTEREST IN THE COAL house ghost is daily increasing. Saturday night [February 23] a large party, probably 35 people, went in the Central depot [in Mason City] in carriages, backs and afoot, and patiently waited[,] some of them till three in the morning[,] for a sight of the spirit. A dozen or so were standing just back of the depot building when at 2:30 they were surprised and startled to see the figure of a man step from what seemed to be the blank side of the coal house and stand perfectly still for a mo-

As his eyes adjusted to the dark and light, he saw that it was long and narrow, shaped like a corset box, and 30 to 35 feet long. It was 6 or 7 feet high and just as wide.

ment. It seemed to have a peculiar quantity of disseminating light and at times the light was so dazzling that the spectators were blinded.

"Finally one of the party got his nerve to the proper condition, and pulling a revolver pointed blank at the figure. Immediately there was a slight atmospheric disturbance and a sound like the rustling of leaves and the ghost disappeared again. Again he fired and again the figure disappeared and again the party felt the atmospheric disturbance, like the rush of air that follows in the wake of a fast moving express train[,] and again there was silence. When the spirit made his appearance this time, according to the stories of the spectators, he held both hands to his head, and moving slowly off began to ascend in the air, and when directly over one of the largest coal buckets[,] dropped noiselessly into it and disappeared from view. Immediately the crowd, who had been standing motionless watching the strange phenomenon, rushed to the spot, but the coal house was empty and there was no sign of a ghost.

"They went away and not again that evening was anything seen of his majesty although several allege that at different times they heard a sound as of coal rattling down the sides of the house. These are the stories as told by the people who were there. And in connection with all the talk has arisen the question, is this the ghost of John Widdy? John Widdy was a railroad man on the Iowa Central. On Dec. 3, 1884, he was caught between the bumpers of two freight cars which he was endeavoring to couple, and his head was crushed to a jelly, causing instant death. This was directly in front of the coal house. Widdy was a man about five feet eleven inches in height and answers perfectly to the description of the ghost given by those who claim to have seen it."

Source:
    "A Real Ghost," 1901. *Nashua* [Iowa] *Reporter* (February 28).

1907: Ghost in a Mine Shaft
WITH THE REPORT that a ghost is haunting shaft No. 13 in the mines at Oskaloosa the . . . miners are in a fever of fear. Numerous miners have reported seeing the wraith. It floats along beside the men[,] and at one time when a party of men hunted it down and got it into a corner[,] the figure suddenly vanished into thin air. Some say it resembles a soldier in a uniform and a slouch hat. The entire camp is stirred up by the weird apparition."

Source:
    "Say Mine Is Haunted," 1907. *Nashua* [Iowa] *Reporter* (March 21).

Kansas

1873: Serpent in the Sky

ON THE MORNING OF June 26, according to an article in the *Fort Scott Monitor*, residents of the town looked up at the sun, halfway up the sky, to see "the form of a huge serpent, apparently perfect in form," surrounding the disc. The paper claimed it was reporting the testimony of two credible individuals who were willing to sign affidavits attesting to their incredible experience. The serpent was visible, they said, for several moments.

Sources:

"Singular Phenomenon," 1873. *Fort Scott* [Kansas] *Monitor* (June 27).

1875: Flying Serpent Preserved

A FEW WEEKS AGO we referred to a lady living in the southern part of the city [Leavenworth] having seen a flying snake in her peregrination through that delightful portion of the metropolis. Yesterday we were met by a friend, who inquired in an excited manner, if we had ever seen a snake that had wings, and 'flew through the air with the greatest of ease'? From his statements we learn that while two boys named Remington and Jenkins, the former from this city, and the latter a Platte Countian, were hunting in the woods, a serpent was seen approaching them, about four feet above the earth. Jenkins took off his hat, and, throwing it over the snake, succeeded in capturing it. It is over one foot long, spotted, and has wings about the size of a man's hand. The boys have the serpent preserved in alcohol."

Source:

"A Flying Snake," 1875. *The Two Republics* [Mexico City] (September 15). Reprinted from the *Leavenworth* [Kansas] *Times*.

1882: Shower of Pebbles

A CURIOUS PHENOMENON was noticed yesterday afternoon. While the rain was falling briskly, a shower of pebbles, ranging from the size of a pea to that of a hickory nut, occurred. It was very brief, but extended over almost the entire city. The pebbles were round, and had been worn by the water. They were probably picked up by the wind."

Source:

"Additional Local," 1882. *Atchison* [Kansas] *Globe* (March 27).

1882: Rainbow Cloud

A CURIOUS PHENOMENON appeared in the heavens in a southeast direction this morning at 11 o'clock, being a small cloud showing all the colors of the rainbow, the sky around it being perfectly clear."

Source:

Untitled, 1882. *Atchison* [Kansas] *Globe* (May 21).

1882: "An Etching on the Darkness"

WHILE WALKING IN WEST Atchison last evening, a reporter sat down in front of an open store to rest, and heard a story which impressed him very much because of the earnest manner in which it was told. There were three men in the group, and the reporter learned by their conversation that one was a freight engineer on the Central Branch, the other his fireman, and the other a shop hand. The engineer said, in brief, that a few nights ago, while running on the West end, and while approaching the bridge where Brit Craft and his fireman were killed, that he distinctly heard an engine whistle ahead of him. Thinking there must be some mistake, he leaned out of the cab window to look ahead. At this moment he was within a hundred yards of the bridge, and while he looked a phantom engine came around the curve.

"The sight was so strange that he was powerless, and did not shut off steam, and the next moment the strange apparition disappeared through the bridge. Before he had time to look again his own engine was on the bridge, and across it. His recollection of the phantom is that it was drawn in white lines, an etching on the darkness, and he declares that Brit Craft was leaning out of the cab window looking ahead. The fireman was engaged in throwing in coal, and he noticed this because he saw him through the boiler head and steam dome.

"The engineer who told the story apologized a great deal for being so ridiculous, but he declared with great earnestness that what he said was true. The fireman who accompanied him was at the moment engaged in packing down coal, but he agrees that the engineer was white as death, immediately after the bridge was passed, saying to him in a tremor: 'I have seen the ghost of Brit Craft's engine!'"

Source:
  "Additional Local," 1882. *Atchison* [Kansas] *Globe* (August 29).

1886: Ghost of an Injured Child

A NUMBER OF BOYS living on South Seventh street [in Atchison] claim that the ghost of a little child, with the top of its head cut off, can be seen on a tree west of Fred Mills' residence every night after ten o'clock. A large number of people went out last night to see it, but if any of them succeeded, they were scared so badly that they are keeping quiet to-day."

Source:
  Untitled, 1886. *Atchison* [Kansas] *Daily Globe* (July 14).

1887: Ghost of the Narrows

A LATE GHOST STORY gained so much circulation at Rushville that the railroads were unable to keep their watchmen at the Narrows. A number of men were killed at the Narrows, and a ghost is said to walk there every night. At this particular point the railroads are cut through a steep bluff, and a watchman is stationed there. The ghost frightened all the men off until Ed. Canfield appeared, who said he would brave the ghost for $5 a night. This amount was paid for a while, but yesterday a man was found who agreed to work for less money, and Canfield was discharged."

Source:
  Untitled, 1887. *Atchison* [Kansas] *Daily Globe* (December 21).

1891: Personal Precipitation

THE STORY OF A wonderful phenomenon comes from Rossville, nineteen miles west of Topeka, on the Union Pacific. For nineteen days, it is said, rain fell incessantly on the orchard belonging to H. Klein, a prominent Rossville resident. This orchard is in the town and is bounded on the east by Mr. Klein's residence, on the other three sides by lines of fences. The rain did not fall outside of Mr. Klein's premises, but for nineteen days there was no intermission in the fall, and it was only stopped by a cold snap."

Source:

"Mr. Klein's Private Rain," 1891. *Woodland* [California] *Daily Democrat* (December 30). Reprinted from the *Chicago Inter-Ocean*.

1892: Mud Fall

A FEATURE OF THE SERIES of recent storms in the west was a shower of mud along the line of the Union Pacific railroad in the vicinity of Onago, Kan. The south and east sides of houses were plastered with mud, and a Union Pacific train was so covered that its headlight was invisible. The phenomenon has not been explained."

Source:

Untitled, 1892. *Stevens Point* [Wisconsin] *Journal* (April 23).

1894: War in the Heavens

WHEN H. W. J. SMITH, a longtime resident of Dickinson County, and a neighboring farmer, B. W. Blue, stepped outside following a visit at the home of Andrew Thompson, who lived 3 miles from Manchester, they looked up at the clear sky and saw "something like a large luminous ball" toward the northeast, about 30 degrees above the eastern horizon. They had no more opened their mouths to speculate that it was a meteor or comet than the object shot toward the west about 3 degrees, then returned to its original position. Then, in Smith's words:

"It opened as a casket with a hinge, presenting on its right a cross—most beautiful, golden, corrugated and furbished. At the left of this was a living man clad in citizen's style, with a plain crown on his head. His form was symmetrical, his countenance bright and permissive—a perfect son of man.

"The casket closed, and away it went to the eastern horizon like a meteor. There it oscillated as if for a time to be emptied and refilled, re-

On his head was a military hat, the crown blended with the man's hair. On each side of the man's head was a horn, and a cross was erect behind him. He stepped out and forth and began action, never stopping to rest or turning his back on the enemy.

turning on the same path to its original place. It opened, presenting a portly man, with sword and scabbard on his thigh, a cross on his breast and on his head a crown of many glittering jewels, like stars. He looked beautiful, but was partly hidden by an obtrusive rider on a black or dark horse.

"These were hidden or overshadowed by a haughty woman in costly royal attire, who seemed to rule over both. Then these were eclipsed by the coming of a military leader with sword in right hand, elevated ready to strike, the scabbard cast away, a cross on his right breast and a square and compass on his left. On his head was a military hat, the crown blended with the man's hair. On each side of the man's head was a horn, and a cross was erect behind him. He stepped out and forth and began action, never stopping to rest or turning his back on the enemy. He retreated eastward to within about five degrees of the horizon, then began to advance with heavy martial tread, like one tramping the wine press and wielding his sword.

"About 11:40 p.m. as we stood watching the phenomenon, blood was seen to stream forth from the casket and spread far and wide, apparently two-hundred miles in extent. Mr. Blue, who is a veteran of the [Civil] war, said it was like the blood of the battlefield, only a deeper red. The warrior seemed at times to be in blood to his knees and above.

"At 12:15 I retired, but Mr. Blue remained watching until 2 a.m. and says the warrior was yet parading the skies and was joined by another, who advanced to meet him from the east. The casket vanished after this warrior stepped out. Myself [sic] and Mr. Blue saw the first and second scenes. Others saw part of the second. Mr. Thompson and Mr. Blue saw all the second."

Source:

"Battle in the Heavens," 1894. *Woodland* [California] *Daily Democrat* (October 3). Reprinted from *Evangelical Visitor*.

1896: Unknown Aerial Object

IT WAS THE PRIVILEGE OF A chosen few of Bronson's people to behold a prophet of the future last Friday evening [December 1]. We might call it a special dispensation in consideration of the fact that people elsewhere were not so fortunate. The way it occurred was somewhat as follows:

"On last Friday evening one of our townspeople glanced . . . up to the heavens and beheld a large object borne on the air high above the haunts of man. Authorities differ as to height from one to five miles and in size from that of a hat to a threshing machine. This object came sailing out of the west about five o'clock with a swift, steady motion. When directly over town it stopped for five minutes to take a picture or a census of the town. . . . What this object was we cannot say. . . . The consensus of opinion calls it a flying machine, perhaps the one lately operating in California."

Source:

"A Flying Machine Seen," 1896. *Bronson* [Kansas] *Record* (December 17).

1897: Monstrous Reptilian Creature

MR. B. P. WALKER, POSTMASTER at the town of Logan, is authority for the story that the farmers who live in the valley of Crystal Creek, in Phillips county, are greatly excited there over the appearance of an enormous reptile which reputable men, who declare that they have seen it, say is not less than fifty feet long. It has the sinuous body of a snake, but its tongue is not forked, and on its head are two short horns. Its color is green, with dirty white spots. It feeds upon small animals and fowls, and has an enormous appetite.

"One morning a farmer lost sixty chickens, and he followed the trail of the reptile until it disappeared in the creek. Another farmer lost forty young pigs in forty-eight hours, and a German testifies that the monster killed his plow horse, which was feeding near the creek, by a single blow of its enormous tail. It has been shot at several times, but its hide is proof against bullets. When in anger it lifts its head ten feet in the air, protrudes its tongue three feet or more, and utters a whine like a puppy crying for its mother. The farmers are afraid of it, but will make a united effort to kill it. It hides in a swamp and water, and, it is supposed, comes from the deeper water of the Solomon river, some miles distant."

Source:

"A Kind of a Santer in Kansas," 1897. *Statesville* [North Carolina] *Semi-Weekly Landmark* (May 18). Reprinted from the *Topeka Dispatch*.

1897: Railroad-Bridge Specter

LAWRENCE HAS A GHOST STORY. An old abandoned Northwestern railroad bridge spanning the Kaw river at the northeast of the city is getting rickety, and is not considered very safe even for foot passengers. But the superstitious among these do not cross over any more at night, for an apparition has been seen there many times of late.

"Six weeks ago Lizzie Madden, a Negro woman, was crossing the old bridge about 11 o'-clock at night, when a tie broke and she went through head first into the mud twenty feet below. She was found dead head downward the next morning.

"It is thought to be Lizzie Madden's ghost that appears there at night. A few nights ago George McCann was crossing over from his work at the canning factory and, having heard of the appearance, he was loaded for it. When he came to the place where the Madden woman fell through, the same white figure that had frightened so many suddenly came into view. McCann shot at the object twice and it did not run, but quickly vanished.

"As this object appears only on dark nights, it is thought that some one has rigged up a 'ghost' that is white on one side and black on the other. This theory explains the sudden appearance and disappearance of the object."

Source:

"Lizzie Madden's Ghost," 1897. *Arizona Republican* (December 4).

1899: Picture in a Horse's Eye

THERE ARE MANY CURIOSITIES in Kansas.

For instance, in Solomon, Kansas, is a horse with the picture of a woman in his eye. It is said that the phenomenon was caused by electrical shock as follows. About a year ago while the horse was on the farm of J. P. Sullivan of Solomon, lightning struck a tree close to where the animal was standing. Mrs. Sullivan was between the horse and the tree, and the electrical shock produced the phenomenon which the horse now carries in his right eye."

Source:

Untitled, 1900. *Naugatuck* [Connecticut] *Daily News* (June 18).

1919: Multiple Balls of Light

AROUND 6:30 P.M. ON October 8, 1919, at a busy downtown intersection in Salina, witnesses observed the sudden appearance of what was described as "a ball of fire as large as a washtub floating low in the air." When it hit the northwest corner of a business establishment about halfway to the top, it ripped out a brick, smashed a second-story window, then exploded with a pistol sound.

In the explosion's wake, baseball-sized balls of fire filled the air. The *Monthly Weather Review* reported that they "floated away in all directions. Some of these balls followed trolley and electric-light wires in a snaky sort of manner and some simply floated off through the air independently of any objects near by."

Though this sounds very much like a manifestation of ball lightning, the account mentions nothing about the sorts of storm conditions generally believed to generate the phenomenon.

Source:

"Ball Lightning at Salina, Kans.," 1919. *Monthly Weather Review* 47: 728. Cited in William R. Corliss, ed., 2001, *Remarkable Luminous Phenomena in Nature: A Catalog of Geophysical Anomalies.* Glen Arm, MD: Sourcebook Project, 131.

Kentucky

Circa 1845: Knitting Needles from the Sky

ONE NIGHT, IT IS CLAIMED, knitting needles fell out of the air and onto the property of F. W. Curry of Harrodsburg. The fall took place during a rain-and-wind storm and came down over an acre's worth of space. The needles were of various sizes. Some were whole, some broken, and some stuck into the ground, while others fell onto their sides. Some residents took the needles and used them for needlework.

"The most substantial citizens in the county will substantiate the fact," a newspaper stated. "No explanation has ever been given. . . . Mrs. John Thomas has some of the needles to this day."

Source:

"A Shower of Knitting-Needles," 1876. *Atlanta Constitution* (April 8). Reprinted from the *Harrodsburg* [Kentucky] *Observer and Reporter*.

1857: "Dreadful Reptile"

ABOUT THREE WEEKS AGO [apparently in September] five men went to gather whortleberries in the mountainous part of Harlan County, Kentucky, and in their travels came to a small branch at the foot of a steep ridge, w[h]ere they discovered a smooth beaten path, or rather slide that led from the branch up the ridge. Curiosity tempted them to know its meaning, and they followed the trail to the top of the ridge, w[h]ere to their astonishment, they found about an acre of ground perfectly smooth and destitute of vegetation, near the center of which they discovered a small sink or cave, large enough to admit a salt barrel.

"They concluded to drop in a few stones, and presently their ears were saluted with a loud, rumbling sound, accompanied with a rattling noise; and an enormous serpent made his appearance, blowing and spreading his head, and his forked tongue protruded. The men were struck with wonder and afright, and suddenly the atmosphere was filled with a smell so nauseating that three out of the five were taken very sick, the other two, discovering the condition of their companions, dragged them away from the abode of death. About ten feet of the snake had to their judgment, made its appearance, when they hurried home and told what they had seen to their neighbors.

"The next day were mounted some ten of the hardy mountaineers, armed with rifles, determined to destroy the monster. On approaching within one hundred yards of the dwelling of his snakeship, their horses suddenly became restive, and neither kindness nor force could make them go any nearer. The men dismounted and hitching their horses proceeded on foot with files cocked to the mouth of the cave. They hurled in three or four large stones, and fell back some fifteen steps when the same noise was heard as before, and out came the dreadful reptile, ready, as his looks indicated, to crush the intruders.

"About the same length of the snake had appeared from the hole, when eight or ten bullets went through his head, and, as the monster died, he kept crawling out until twenty feet of that huge boa lay motionless on the ground. It was a rattlesnake, with twenty-eight rattles—the first was four inches in diameter, the rest decreasing in size to the last. With difficulty the men dragged him home and his skin can now be seen by the curious in Harlan county."

Source:

"A Virginia [*sic*] Snake Story," 1857. *Gettysburg* [Pennsylvania] *Compiler* (October 19).

1866: "Some Demoniacal Personage"

IN A LETTER DATELINED February 17, Bracken County, and published in a local newspaper, Nathaniel G. Squires recounted an utterly fantastic story that he insisted was a true experience:

"The people of this neighborhood are in the greatest state of excitement in consequence of a remarkable visitation or apparition, of some demoniacal personage in our midst.

"On Monday night last [February 13], after myself [*sic*] and family had retired to rest, we were suddenly aroused by a great outcry from the Negro quarters—which are immediately to the rear of the house—in which prayers vied for supremacy with blasphemies, men, women and children screaming 'fire' and 'murder' at the top of their voices, all conspiring to create a scene worthy of a pandemonium.

"Terribly startled, my wife and I sprang from our bed. The room was illuminated as brightly as by a flood of sunlight, though the light was of a bluish cast. Our first and most reasonable conclusion was that the Negro cabins were being consumed by fire. We rushed to the windows and beheld a sight that fairly curdled the blood in our veins with horror, and filled our hearts with the utmost terror. My daughters, shrieking loudly, came flying into my room, hysterical with fear. This is what we beheld:

"Standing to the right of the upper cabin, near the fence that separates the negroe's [*sic*] garden from the house-yard, was a creature of gigantic stature, and the most horrifying appearance. It was nearly as high as the comb of the cabin, and had a monstrous head not similar [dissimilar?] in shape to that of an ape; two

short very white horns appeared above each eye, and its arms were long, covered with shaggy hair of an ashen hue, and terminated with huge paws, not unlike those of a cat, and armed with huge and hooked claws. Its breast was as broad as that of a large sized ox, its legs resembled the front legs of a horse, only the hoofs were cloven. It had a long tail armed with a dart shaped horn, which it was continually switching about. Its eyes glowed like two living coals of fire, while its nostrils and mouth were emitting sheets of blue-colored flame, with a hissing sound, like the hissing of a serpent only a thousand fold louder. Its general color, save its arms, was of a dull dingy brown. The air was powerfully impregnated with a smell of burning sulphur.

"The poor Negroes were evidently laboring under the extremest terror, and two of them, an old woman and a lad, were actually driven to insanity by their fears, and have not recovered their reason up to this writing. I do not know how long this monster, demon or devil, was visible after we reached the window—possibly some three seconds. When it vanished it was enveloped in a spiral column of flame that reached nearly to the top of the locust trees adjacent, and which hid its horrid form completely from view. The extinction of the flame was instantaneous, and with its disappearance we were relieved of the presence of this remarkable visitor.

"If ours had been the only family visited by this unearthly creature, I should have kept silent, and perhaps tortured my mind into the belief that it was a hallucination. But precisely the same apparition made its appearance at my neighbor's, Mrs. Wm. Dole, appearing there in precisely the same shape in which it presented itself to us, save the head, which appeared to those who witnessed it at Mrs. Dole's to resemble that of a horse. At Mr. Adam Fuqua's[,] another neighbor, its head was that of a vulture. On Tuesday night it appeared at Mr. Jesse Bond's, there wearing the head of an elephant. At these places it made the same appearance as at my house—excepting only the changing of the head—and disappeared in the same manner. These parties are all reliable ladies and gentlemen, and at my request have made oath to what they witnessed.

"What it is, what its object, what its mission, is something that passes my poor comprehension. What I have written is simple, unadorned truth. You are at liberty to use this in any manner you may esteem."

Aside from other obviously implausible claims, it is hard to believe that Squires could have taken in so much detail of the monster's appearance if the sighting—under the most traumatic of circumstances—lasted "possibly some three seconds." Nonetheless, it would not be the last extraordinary claim about a shape-shifting entity from Bracken County (see 1868 item below).

Source:

"A Queer Story," 1866. *Titusville* [Pennsylvania] *Morning Herald* (March 15). Reprinted from the *Mount Sterling* [Kentucky] *Sentinel.*

1868: Centaur on the Road

A "PROMINENT TOBACCO MERCHANT," according to a Kentucky-based correspondent of a Cincinnati newspaper, was traveling through southern Bracken County along Willow Creek, 2 miles from Brooksville, on the evening of October 10, when suddenly an "object" appeared on the road in front of him. He gave the following testimony to the reporter—or, at any rate, so it is alleged:

"The object was about six feet in height, and walked upright. The face was at times that of a man, very pale, with curls of flame falling over his shoulders; eyes of sulphurous blue, changing constantly in size, one moment large as a tin cup, and then gradually decreasing in size until it was almost invisible. Its arms were those of a man, and hands deadly pale. In one hand, it held a torch, and in the other a sword that seemed to be about four feet in length. Its lower extremity was that of a horse, with legs well proportioned and hoofs as those of a horse. Its tail, which was about three feet in length, was of flame. Its breath was a solid sheet of fire, which vibrated with the heavings of its breast, like the pendulum of a clock. It was certainly the most frightful object I ever beheld.

"It walked off to the side of the road, and then vanished. When it disappeared I immediately put spurs to my horse and galloped by the spot where I had seen it. When I arrived at the summit of the hill, about two hundred yards off, I looked back and saw the object in the same spot where I had first beheld it. I stopped my horse and watched it for a moment; it walked over to the left side of the road, and mounted a rail fence that stood there [and] commenced running toward me. I did not stay to see the remainder of the drama."

The merchant rushed to town, where a party composed of "lawyers, doctors, preachers and tradesman, armed with guns and pistols" assembled and headed off to the encounter site. There the creature was waiting for them—or it wasn't. It turned out that not everybody could see it. Those who did see it saw it running at a high rate of speed, back and forth. At one point it passed the crowd, and several fired at it without apparent effect. At 11 p.m. it was gone.

It was back the next night and for some nights afterward. Some people claimed to have seen it a year earlier. "The county is astir," the correspondent wrote, "and the people are eager to get rid of their unwelcome visitant." There was at least one previously published account of an encounter with a somewhat comparable entity in the county (see 1866 item above).

Source:

"Devil Loose in Kentucky," 1868. *Atlanta Constitution* (November 4). Reprinted from the *Cincinnati Enquirer.*

1876: The Kentucky Phenomenon

IN MARCH 1876, A widely distributed press account reported: "A shower of flesh has recently fallen over quite an extended range of country in Bath county, Kentucky, and is attracting much attention among men of science. A quantity of the matter having been transmitted to Prof. [J. Lawrence] Smith, of Louisville, for analysis, that gentleman reports as follows:

"'In my mind this matter gives every indication of being the dried spawn of the batrachian reptiles, doubtless that of the frog. They have been transported from the ponds and swampy grounds by currents of wind, and have ultimately fallen on the spot where they were found. This is no isolated occurrence, I having come across the mention of several in the course of my reading. The only one I can now fix the date of, is recorded by Merschanbrock, as occurring in Iceland in 1675. The matter is described by him as being glutinous and fatty, which softened when held in the hand, and emitted an unpleasant smell when exposed to the action of fire. The ovum of the egg of the batrachian reptiles is a round mass of transparent jelly, in the center of which appears a small black globule. In the present case the passage through the air would have dried up more or less of these gelatinous matters, so that the interior as I found it, is still soft and gelatinous.'"

Smith's theory, however, found few takers. Mount Sterling attorney J. M. Bent, who had investigated the site and sent the samples to Professor Smith, told a *Courier-Journal* writer that "I never knew spawn to be at all red, and the flesh that fell contained a little of that color, though in the main white. . . . The large amount of the flesh also causes me to believe it was not spawn. I cannot well conceive how so much spawn could be taken up by winds and then deposited" (quoted in Smith and Mangiacopra 2002, 71).

As the *Times* and other New York newspapers ridiculed Kentuckians for their supposed credulity and superstition, the *Courier-Journal* retorted that the phenomenon was not supernatural, merely a curiosity that a Kentucky scientist, Prof. Smith, had satisfactorily explained. It interviewed him again, and he was sticking to his theory of frog spawn. He added, "I should state that, of the few small pieces I saw, and which were thick enough not to be entirely dried up, it was only the exterior that was reddish. When cut into, the interior, the soft gelatinous matter, was colorless."

But *Scientific American Supplement* had another "proper explanation," as Leopold Brandeis termed it in an article in that publication (cited by Fort 1941, 45–46). It was only nostoc, a fresh-water algae with gelatinous qualities.

Other scientists thought this was a stretch. A. Mead Edwards, head of the Newark Scientific Association, queried Brandeis and soon afterward looked at a sample in the possession of a colleague identified only as Dr. Hamilton. Hamilton had declared the sample to be "lung tissue." After his own examination, Edwards agreed. He located other samples from other sources and detected, he said, muscular fibers. Though puzzled, he tended to agree with the hypothesis—surely a wild one—that the stuff was nothing more than buzzard vomit, notwithstanding either the great quantity that had fallen or the improbability of buzzards being able to throw up such a mass of stomach content and still maintain flight.

All that is certain about this strange occurrence is that some kind of biological material fell out of the sky. What it was, and how it got there, remain a mystery for which no fully convincing solution has yet been proposed.

Sources:

Craig, Berry, 2003. "Meat Shower Part of History." *Cincinnati Enquirer* (April 28).

Fort, Charles, 1941. *The Books of Charles Fort.* New York: Henry Holt, 45–46.

Smith, Dwight G., and Gary S. Mangiacopra, 2002. "Southern Falls—Flesh and Stones: A Reexamination of Charles Fort's Two Classic 19th Century Falls from the Skies." *The Anomalist* 10: 64–88.

"A Wonderful Phenomenon," 1876. *Freeborn County* [Minnesota] *Standard* (March 16).

1879: Child Flying

CAPT. BEN S. DRAKE, OF this city [Lexington] . . . while passing a vacant lot at the corner of Spring and High streets, [in September,] saw what he thought was a man lying on the grass. He approached it, when the object vanished. One week later the same operation was repeated. People began to talk about it. Night before last a crowd of about fifty visited the spot, and saw what they are willing to swear was a child, which, at their approach, rose into the air and vanished. Excitement now rose to fever heat, and last night, between the hours of 9 and 12, not less than 800 people, white and black, rich and poor, visited the spot to see the strange apparition, but it did not make its appearance. When the sun rose this morning there were, by actual count, forty-nine people standing around the lot, having remained on watch all night. An old house, which the superstitious claimed was haunted, formerly stood upon the lot."

Source:
"A Ghost Story," 1879. *Sedalia* [Missouri] *Daily Democrat* (October 30). Reprinted from the *Cincinnati Commercial.*

1880: Leaper and Flyer

ON JULY 28, 1880, according to a story of uncertain provenance, a strange being appeared in the city of Louisville. He was seen by a number of witnesses, who described him as being tall and thin, with a long nose, pointed ears, and long fingers. Clad in a shiny uniform, he wore a cape. On his chest a blue light flashed on and off. He could also leap great distances with ease, hopping from the ground to a rooftop of a house or barn in no more than a second or two. Another account has him wearing a helmet and claims that the blue light spouted a blue flame. When accosting female witnesses, he tore at their clothes but apparently did not otherwise molest them.

The following day, the *Louisville Courier-Journal* reported a story no less strange. Early the previous evening, two men looked out over the Ohio River, where they observed a flying machine of some sort. In the middle of it sat a man who seemed to be operating the device with his hands and feet. He and it were at some considerable altitude. The machine would descend, and then the operator would move faster, causing it to ascend again. At one point he changed direction from southwest to southeast. According to a later *Courier-Journal* story, a family in Madisonville, Kentucky, southwest of Louisville, saw a "circular form [which] changed to an oval. . . . There seemed to be a ball at each end of the thing." This sighting occurred an hour or two after the Louisville incident.

The phantom-leaper story is strongly reminiscent of a British tradition, little known elsewhere, of an entity known as "Springheel Jack" or "Springheeled Jack." The stories were first reported in London newspapers in the late 1830s. The figure was said to wear a cloak, to have a lamp on his chest, to emit blue flames, and to be able to leap great distances (thus the nickname). Generally similar alleged sightings were reported well into the twentieth century (the last known one was in 1996, in the English county of Cornwall), in at least two cases in explicitly UFO contexts. The first suggestion that Springheel Jack was an extraterrestrial was published in 1954.

Sources:
Dash, Mike, 1996. "Spring-Heeled Jack: To Victorian Bugaboo from Suburban Ghost." In Steve Moore, ed. *Fortean Studies,* vol. 3, 7–125. London: John Brown Publishing.
Edwards, Don, 2000. "Weirdness Used to Be a Tall Tale, Not a Movie." *Lexington* [Kentucky] *Herald-Leader* (October 10).
"A Flying Machine," 1880. *Louisville Courier-Journal* (July 29).
"The Flying Machine," 1880. *Louisville Courier-Journal* (August 6).
Rife, Phil, 1997. "Springheel Jack Invades America." *Fate* 50, 11 (November): 16–18.

1886: Scenes from a Haunted House

THE FEMALE INMATES OF a house on Green street [in Louisville], between Sixth and Seventh, had a dreadful experience about 1 o'clock yesterday morning [July 7]. About that time, in the middle of their midnight revels, all the lights in the house were suddenly extinguished. The women were tossed about by unseen hands, missiles of every kind began to fly about the rooms, unearthly yells were heard in every part of the house and for some time the whole house was a veritable pandemonium. The women fled in fright, declaring that a ghost had taken possession of the house. Some time after the noises had ceased one of the women returned and lighted the gas throughout the house. The furniture in three rooms was found smashed to pieces, the bed clothing cut to shreds and general havoc and confusion everywhere. A curious crowd soon filled the place, and the premises were searched, but no trace of the mysterious visitors could be found.

"The women firmly believe that the work was of preternatural agency, and claim that the house has been haunted for a number of years. There is a story current in the vicinity that a double murder was committed in the house years ago, and that the bodies of murdered men lie buried in an abandoned well beneath the house."

Source:
"A Ghost in Possession," 1886. *Elyria* [Ohio] *Daily Telephone* (July 8).

1893: Light in Winter

WILLIAM O'CONNOR, WHO resides at Kirksville, in this county [Madison], is very much puzzled and excited over an experience he had with a ghastly fire phenomenon, three miles from this city [Richmond], recently. He professes to be a skeptic on the subject of spiritualism, and relates a remarkable and interesting story of his late adventure.

"In company with two young men and two young ladies, Mr. O'Connor left Richmond, on Monday night, Jan. 16, in a sleigh drawn by two fleet horses. They were on their way to a party at Kirksville. The night was cold, and a light snow was falling, but it was not too dark to see objects twenty-five yards away. When they reached a point about three miles southwest of this city[,] one of the young men saw a brilliant light over on a hillside, about one hundred yards from the roadside, and called attention of the party to it. The crowd thought it was the reflection of a lamp burning in a house nearby, but as the latter was 300 yards away and the light on the hillside exceedingly bright, they were unwilling to accept this. They stopped to investigate the strange scene. At this time the light began to slowly circle round, rising a great distance from the ground, then falling to the earth again in large circles.

"Mr. O'Connor describes the appearance of the light as electric in brilliancy, and says there was no human near the spot; had there been he would have been visible, as the light was so bright. The curiosity of the young men arose to a high pitch at this time, and two of them alighted and waded through deep snowdrifts, and stood for a quarter of an hour watching the strange phenomenon.

"Being chilled, the party resumed their journey. They drove to the farmhouse where the party was to be given, and witnessed the same peculiar spectacle which they had come in contact with an hour before. At the farm gate they again watched the light another quarter of an hour, which continued to approach them, circling as it came.

"Mr. O'Connor and companions are unable to fathom the mystery, and others who have witnessed the same sight at or near this spot are likewise puzzled to know what it is."

Source:
"Ghostly Lights," 1893. *Marion* [Ohio] *Daily Star* (February 6).

1893: Colored Discs

AN ASSOCIATED PRESS (AP) dispatch out of Cincinnati dated August 26 took note of what it called a "solar phenomenon" allegedly experienced by persons living in and around Leslie, Kentucky, in Cumberland County, the previous day. Half an hour after sunrise, residents who happened to be outside noticed that the sun had a "very peculiar color."

Soon thereafter, thousands of round discs, judged to be the size of wagon wheels, filled the sky between the earth and the sun. All of them were moving, and as each approached the ground, it would change shape. Some became triangle-shaped, others assumed a square appearance, "and others taking on various proportions and odd forms." Each disc, too, had its own color—red, green, black, or another shade—but on reaching the earth, each turned a dense purple. According to AP, "It appeared to

All of them were moving, and as each approached the ground, it would change shape. Some became triangle-shaped, others assumed a square appearance, and others taking on various proportions and odd forms.

those witnessing the phenomenon that the sun would swallow up some of the disks, and then again they would appear to sink into the solid ground and then disappear."

All of this took place in total silence, except for the shouts and prayers of the terrified observers, most of whom believed that Judgment Day had come. The display ended after an hour.

Source:
"Judgment Day," 1893. *Arizona Republican* (August 27).

1896: Pink Creature

RESIDENTS OF HIGHTOWER, a tiny town near the larger Morgan, reported encounters with a peculiar animal in the spring and early summer. It looked like no animal they had ever seen before. As tall as a month-old calf, it was of a pinkish color, entirely devoid of hair. It could run as fast as a fox.

It first appeared on William Jones's farm, then on his neighbors', but its home base seemed to be the Jones property. A newspaper stated, "Women . . . are discussing the probability of its being a supernatural vision, appearing in advance of some dread calamity, but men are out in posses with firearms, searching the wooded sections, with the purpose of solving the mystery."

Source:
"Strange Animal," 1896. *Arizona Republican* (June 23).

1897: "A Huge Oblong Shape"

FOR THE PAST FEW DAYS reports have been flying over the country that a mysterious airship was passing over the land, and was apparently headed toward Kentucky. Strange were the stories told of this mysterious visitor. Wherever it went it spread terror and confusion among the stock and awe and superstition among men. A glance at this prodigy coursing through the heavens bright with red and green lights, caused cats and dogs to flee to places of safety, cows to bellow in their barns and horses to kick their stalls to splinters. In some of the small towns over which the airship passed it was supposed to be an omen of wrath to come. Revivals were held, and the inhabitants prepared themselves for the judgment day. It was last seen on Saturday [April 10] at Winemas, a little town in Indiana. Now Augustus Rodgers of near Louisville claims it passed over Louisville last night.

"Nothing had been heard of it for several days when Rodgers, who lives two miles south of the city, came out of his cabin an hour before daylight to attend to his stock. There immediately above him at a distance of 400 feet[,] according to his story, was a terrifying and yet a beautiful sight. A huge, oblong shape, says Rodgers, apparently about 40 feet by 15 feet, brilliantly lighted, for it was yet dark, and flying through the air at a speed of 100 miles an hour, met his eyes. The vision, Rodgers says, was in the form rather of a barge than a ship, with massive proportions and solidarity. Rodgers called to his wife, who came out, and together they watched the strange sight as it disappeared to the southeast. Before it vanished uncouth and enormous shadows flickered from all parts of the ship, and both Rodgers and his wife saw a form, like that of a man, standing at the front of the ship and directing its course.

"Exactly ten minutes later John S. McCullough, who lives a short distance back of Churchill Downs, was driving down a lonely road toward town. It was still very dark, and McCullough had great difficulty in seeing the road before him, but it was suddenly illumined by a great light. . . . He pulled up and tied his horse and stepped onto the road. As yet he could see nothing, but the woods and road were as bright as day. A moment later he heard a whirring sound, and the airship was over his head. He says as the ship flashed out of sight a small black object leaped from off it and struck a few feet from McCullough. He picked himself up and went to it and found it to be a half-burnt coal.

"Rodgers and McCullough are both reputable citizens and as they are willing to swear to what they saw[,] their story is attracting attention. Altogether the whole affair is most incomprehensible and has an uncanny look."

Source:

"Airship Passes in the Night," 1897. *Louisville Evening Post* (April 13).

Louisiana

1866: A Strike from Above

TWO GENTLEMEN, A few days ago, while riding along the road a short distance from this place [Shreveport], witnessed a curious occurrence during the day time. A rain was coming up, preceded by a slight sprinkling, when at a short distance ahead they saw a large ball of fire descend slowly from the clouds and affix itself to the trunk of a tall, dead pine, at the height of a few feet from the ground. Both called to each other simultaneously to notice the strange object, which to use their own words, 'blazed up where it stood like a candle.' It so continued for a few seconds, when it suddenly exploded with a tremendous detonation, tearing the tree into a thousand splinters and setting fire to the portion of the stump that remained. A considerable area was filled with falling foliage, and fragments scattered in every direction. Immediately upon the explosion, a streak of fire was seen shooting off horizontally from the tree, following the surface of the ground, passing within fifty or sixty feet of them, and of the character of a stream of lightning, as it is often seen descending from the clouds when it strikes. The sight was terrific in the extreme. The air became strongly impregnated with a 'pungent, sulphurous odor.'"

Source:

"A Singular Phenomenon," 1866. *Brooklyn* [New York] *Eagle* (July 7). Reprinted from the *Shreveport Southwester.*

1890: Dancing Star

[The phenomenon] is to be seen early after dusk in the South [near Vernon]. It has many motions, and will dart upward, run around, fall down, dance, and move about over a considerable space, and is never still but a few seconds at a time. Sometimes it seems to shoot over a space of fifty yards, and occasionally sparks of fire are given off."

Source:

Untitled, 1890. *Lafayette* [Louisiana] *Advertiser* (March 15). Quoting the *Vernon* [Louisiana] *Patriot.*

1891: "Strange Feathered Monster"

One day in January, when hunter August Heiss of Liverpool Parish sighted a strange bird, he opened up with two well-aimed shotgun blasts. The bird fell out of the sky but was still alive when he came upon it. He bludgeoned it to death with an old fence rail.

Neither he nor anyone he spoke with had seen anything like it. Measuring 9 feet from wing tip to wing tip, it was snow white except at the wing's extremities, which were jet black.

Characterizing the bird as a "strange feathered monster," press accounts noted, "The bill from the back of the head measured two feet, with an extensible sack or pouch of gutta-percha-like substance hanging to the under jaw and attached to the bill, very much as a dip net is to its staff. The bird had evidently traveled a long distance and could scarcely fly, and Mr. Heiss is of the opinion that the fowl was one of those spoken of by African travelers. Enough feathers were taken from the bird to make a large pillow. They are of an excellent quality, white and soft as down."

Source:
Untitled, 1891. *Fort Wayne* [Indiana] *Sentinel* (January 24).

1892: Shrieks, Laughter, and Screams in the Night

A GHOST OF EXCEPTIONALLY lively and jocose nature is agitating the vicinity of what is known [as] the old Cross Roads near Lake Charles, La., by its midnight shrieks of laughter and bursts of hilarity. The Cross Roads is a small settlement lying four miles south of this town, where two old county roads formerly met. The people say that every night between 12 o'clock and 1 they are suddenly awakened by a great noise of shouting and shrieks as of many voices. Then a single voice, apparently a man's, takes up the task of making the night sleepless. The laughter seems that of a person in a great glee or intoxicated, but invariably ends in a loud scream of anger or pain.

"After this nothing more is heard until the same hour on the next night. Every effort has been made by the people to penetrate the mystery of this occurrence, but up to date no solution has been found. One peculiarity is that the laughter seems to be in each house and to be as distinctly heard in one as another. It is apparently the noise of a person seated on the roof or shouting through the ceiling. Various surmises are afloat as to the identity of the returned spirit, but though addressed by the names of all whom it could possibly be, no response is to be won from it. It is generally believed, however, to be the ghost of a peddler who, in a drunken frolic, was killed by a prominent planter for some silly jest. This mystery has been puzzling the people for something over a week now."

Source:
"A Lively Ghost," 1892. *Gettysburg* [Pennsylvania] *Compiler* (May 24).

1897: "A Machine of Unique Invention"

ON APRIL 21, THE *NEW ORLEANS Picayune* related that a few hours before, just after midnight, an employee who had stepped outside the newspaper office spotted a passing airship, one of many being reported in the United States that month and the subject of much speculation and derision. He ran back up the steps and excitedly recounted his sighting of a mysterious craft, 50 to 60 feet long, bearing a powerful searchlight and heading toward the northwest. His colleagues went out to see for themselves, but observing nothing, they subjected the unhappy man to jokes and ridicule. Still, the newspaper noted, "His general reputation for truth and veracity is such . . . that his friends cannot help but believe that he saw something, and as the airship was seen in the vicinity of Natchitoches yesterday, it is not unlikely that it may have come from this direction."

The article went on: "A young man connected with the Picayune claimed to have seen the airship here on April 1, but all he saw was the searchlight, as it was dark at the time. It was generally treated as an April fool joke, but the subsequent excitement caused by the alleged appearance of the strange craft in other cities leads to the belief that there may be some truth in it."

On April 17, Stephenville farmer C. L. Mcilhany had reported speaking with the crew of a landed airship. The two identified themselves as S. E. Tillman and A. E. Dolbear, as reported by the *Dallas Morning News* of April 19. The two airmen reportedly said they were experimenting with an aerial device financed by New York investors who preferred to remain anonymous. The people of New Orleans apparently followed this story with interest. The *Picayune* account of April 21 said: "The two successful navigators of the airship, Dolbear and Tilman, as they gave their names to the people of Stephenville, Tex., were supposed to have been in the city last night, but it was in all likelihood a hoax, as cards sent to their rooms failed to get any response."

The *Picayune* article also supplied more information about the sighting in Natchitoches: "Last night at about 1:30 o'clock [in the early morning of April 20], as a gay crowd were returning from

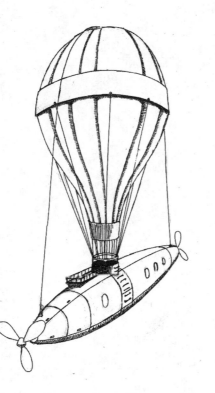

He ran back up the steps and excitedly recounted his sighting of a mysterious craft, 50 to 60 feet long, bearing a powerful searchlight and heading toward the northwest.

a reception given by Company I at the Armory Hall, they were treated to a strange sight, coming from the western heavens. That spectacle consisted of a massive airship, the first story being of balloon shape, and the under car being of conical or cigar form. When first seen it seemed but a faintly animated spark, which, as it came nearer, gradually grew brighter. After a few moments its entire form was clearly perceptible and developed a structure [rest of line illegible] in semi-darkness. Its destination appeared to be northerly, and it gave indications, from its course, of coming from Texas. It was evidently a machine of unique invention, as its movement was of an undulatory character, and bore striking resemblance to the movement of some huge bird. It was visible, though about a thousand feet high, for nearly half an hour, and was a rare curiosity to those who had the good fortune to witness the phenomenon. The remarkable fact in connection with its visitation was that, as its course neared the city, the light that illuminated it became suddenly low, and did not again rise until the city was left in darkness."

Source:

"The Airship: One Man Saw It in New Orleans Last Night," 1897. *New Orleans Picayune* (April 21).

Maine

1853: Balls of Purple Fire

A SEVERE SNOW STORM, accompanied by lightning, occurred in portions of Maine on the 13th ult. [February]. A correspondent . . . writing from Bass Harbor, Mt. Desert, describes the storm, as it appeared there, as awful and sublime. The lightning was of a purple color, and sometimes appeared like balls of fire, coming in through windows and doors, and down the chimneys, while the houses trembled and shook to their very foundations.

"Mrs. E. Holden was near a window, winding up a clock; a ball of fire came in through the window and struck her hand, which benumbed her hand and arm. Captain Maurice Rich had his light extinguished, and his wife was injured. He got his wife on to a bed, and found a match; at that instant another flash came and ignited the match, and threw him several feet backwards.

"A great many persons were slightly injured. Some were struck in the feet, some in the eye; while others were electrized [sic], some powerfully and some slightly. But what was very singular, not a person was killed or seriously injured, or a building damaged; but a cluster of trees, within a few rods of two dwelling houses, was not thus fortunate. The electric fluid came down among them—taking them out by the roots, with stones and earth, and throwing all in every direction. Some were left hanging by their roots from the tops of the adjacent standing trees, roots up, tops down.

"The lightning, after entering the earth to the depth of several feet, and for a space of some eight or ten feet in diameter, divided into four different directions. One of the courses which it took led through the open land, making a chasm to the depth of several feet, and continued its march unobstructed by the solid frozen ground, or any other substance, to the distance of 370 feet; lifting, overturning, and throwing out chunks of frozen earth, some of which were 10 or 11 feet long by four feet wide; and hurling at a distance rocks, stones and roots. The power here displayed was truly awful, and had it fallen on a building, it would have thrown it, with its inmates, into ten thousand fragments."

Source:

"Singular Phenomenon," 1853. *Gettysburg* [Pennsylvania] *Republican Compiler* (March 14). Reprinted from the *Gardiner* [Maine] *Fountain*.

1854: "The Appearance of a Miniature Human Being"

ON THE MORNING of Jan. 2d, while engaged in chopping wood a short distance from my home in Waldoboro, I was startled by the most terrific scream that ever greeted my ears; it seemed to proceed from the woods near by. I immediately commenced searching round for the cause of this unearthly noise, but after a half hour's search I resumed my labors, but had scarcely struck a blow with my axe when the sharp shriek burst out upon the air. Looking up quickly, I discovered an object about ten rods from me, standing between two trees, which had the appearance of a miniature human being. I advanced towards it, but the little creature fled as I neared it. I gave chase, and after a short run succeeded in catching it.

He is of the male species, about eighteen inches in height, and his limbs are in perfect proportion. With the exception of his face, hands, and feet, he is covered with a hair of jet black hue.

"The little fellow turned a most imploring look upon me, and then uttered a sharp shrill shriek, resembling the whistle of an engine. I took him to my house and tried to induce him to eat some meat, but failed in the attempt; I then offered him some water, of which he drank a small quantity. I next gave him some dried beach [*sic*] nuts, which he cracked and ate readily. He is of the male species, about eighteen inches in height, and his limbs are in perfect proportion. With the exception of his face, hands, and feet, he is covered with a hair of jet black hue. Whoever may wish to see this strange specimen of human nature, can gratify their curiosity by calling at my house in the eastern part of Waldoboro, near the Towbridge tavern. I gave these

facts to the public, to see if there is any one who can account for this wonderful phenomenon.

J. W. McHENRI."

Source:
"A Wild Man in Waldoboro," 1854. *Hornellsville* [New York] *Tribune* (January 25). Reprinted from the *Thomaston* [Maine] *Journal.*

1870: "By Some Unknown Power"

A SINGULAR PHENOMENON occurred recently in the field of John Gould, Jr., in Lisbon. A loud noise was heard in the vicinity, on the same day the shock of earthquake was experienced at Richmond [Maine]. People rushed out of doors and looked around to discover the cause of the noise, but nothing unusual was to be seen. Since then, it has been found that a large mass of earth has been lifted from its place, in Mr. Gould's field, by some unknown power. The earth removed is nearly in the form of a parallelogram on the surface. It is about twelve feet long and four feet wide, is fully a foot thick, or to the depth of the frost. It is as regular, and the corners as well defined[,] as though cut by a saw, and was thrown out, apparently, by some tremendous power exerted on all parts alike, and it was deposited 'right side up' half its width from the place it formerly occupied. The ground one side of the hole is puffed up about six inches, to the rising land, about a rod distant."

Source:
Untitled, 1870. *Morning Oregonian* (March 31). Reprinted from the *Lewiston* [Maine] *Journal.*

1882: Path of an Immense Serpent

IN EARLY 1882, ACCORDING to an account published in a Maine newspaper, a man named Charles S. Hunnewell and two friends, Joseph and Frederick Harryman, embarked on a fishing expedition to Chain of Lakes in the east-central part of the state. They went a little north of the upper part of the lakes to try their luck in a stream emptying into the larger body, where they saw an incredible sight.

The first to see it was Hunnewell. It looked like the trail left by some huge, snakelike creature that had come down the stream, then crawled out of it to cross a meadow, then another stream. It appeared that the animal had continued some distance over a bog until it reached the woods. There it apparently turned in a wide circle and retraced its path to the original stream.

"It then passed down the stream some distance and entered a large floating bog on the other side," the Calais newspaper alleged. "In this bog the trail ended in overflowing water, from which no indications exist that the serpent returned."

This route would have taken the creature along a winding path for between a quarter and a third of a mile. The path itself was more than 4 feet wide and, depending, apparently, on what it had crawled over, 2.5 to 3 feet deep. Figuring from the distance between curves in the trail, the three men estimated that it was at least ninety feet long. Logs and brush seemed to have been pushed out of the thing's passageway as if it were clearing obstacles as it moved.

The following day, Hunnewell went to a nearby logging camp, where he spoke with brothers Sewell and Hiram Quimby. Though deeply skeptical, the latter agreed to accompany Hunnewell to the site to see for himself. Quimby was suitably astonished. Soon afterward, his brother and other fellow loggers examined the extraordinary track. That weekend, many people from a nearby town flocked to the scene. Hiram Quimby judged that the creature must have weighed 60,000 pounds. The Quimby brothers, the paper attested, were "well known in Calais as reliable and truthful men."

The newspaper account also refers to a "Mr. Hall," who evidently had made an earlier claim, presumed known to readers, to having actually seen the serpent. Unfortunately, the earlier story seems to be lost to history.

Source:

"Maine's Land Serpent: Its Weight Estimated at 60,000 Pounds," 1882. *St. John* [New Brunswick] *Daily Sun* (February 9). Reprinted from the *Calais* [Maine] *Times.*

1885: Captured Sea Creature

THE SEA SERPENT HAS just been captured, or, rather, the creature that has started all the sea serpent stories on the New England coast has just been towed into this harbor [Portland]. Capt. Cobb, of the schooner Dreadnaught, which arrived this morning, says: 'Were about five miles from Halfway Rock when we saw a break in the water ahead, and one man sang out, "there's a sea serpent!" I looked, and there was something that did look like a big snake dead ahead. We made sail but couldn't overtake him. Then we lowered boats and gave chase. It was a five mile job of it before we came up with him.'

"'Yes,' interrupted another man, 'and all we could see was that big, snakelike head lifted out of the water three or four seas ahead. We came up and struck at him with a harpoon. It took him right inside, and would have fastened a right whale, but it didn't make an impression on him. It turned the harpoon, but the critter kept right ahead. You might fire cannons at him all day and it wouldn't do a bit of good if you didn't hit his head.'

"'Then we struck at him again,' said Captain Cobb, 'and sent a boat alongside. He turned over on his back, making a great swell. Then we struck him in the neck, and, after making a big fight[,] he yielded, and we towed him alongside the Dreadnaught. He will weigh from 1,200 [pounds] up. . . .

"'He's got some kind of armor that may be called his shell. His neck would not draw in the turtle fashion. His head is a cross between a catfish and a snake, and he's about as bad looking as he can be. A doctor came on board this morning, and he says he's a serpent. I've no doubt he's what they've been calling a sea serpent along the coast. If we hadn't captured him we'd have reported that we saw and chased a sea serpent. I've traveled all over the world and seen all kinds of turtles—deep sea turtles among the rest—and if he's a turtle, it must be a new breed, that's all.'

"Captain Cobb has brought the Dreadnaught up to the Custom House wharf and Dr. Thomas Hill is to be appealed to and asked to name the monster. In the mean time, the snake or turtle, as the case may be, promises to get well of his wounds and manifests a strong desire to smash things."

Source:

"A 1,200-Pound Sea Monster," 1885. *Elyria* [Ohio] *Weekly Republican* (July 30).

1891: "Emanating from Some Heavenly Body"

REPORTS COME FROM here [Winthrop] regarding the wonderful aerial phenomena on Monday [February 23] over a wide section. It was evidently most noticeable in the middle and northern part of the State. The nature of the disturbance has unsettled the minds of the people. Nothing just like it is remembered in Maine before. It was a very brilliant light emanating from some heavenly body, resembling in size the full-orbed moon, followed by a terrific explosion, the crash shaking the earth in a terrible manner. Preceding this the sky was a kaleidoscope of rapidly changing colors. Some think it was a meteor, others an earthquake, and others a wonderful electrical phenomenon."

Source:

"That Meteor," 1891. *Bangor* [Maine] *Daily Whig and Courier* (February 25).

1896: Haunted Town

IN JUNE AN ARTICLE in the *San Francisco Examiner* observed, "It is doubtful if any town in all New England has more ghosts and haunted houses than Easton, Me. It may well be called a haunted town. Although the citizens are sensible and well to do farmers, with all sincerity and almost to a man they vouch for the truth of many tales of remarkable and startling supernatural visitations."

One house in particular laid claim to dramatic manifestations. From time to time, people had tried to live in it, but in each case they were driven out, complaining of supernatural harassment. Finally, one day (apparently) in 1896 an unnamed young man, described as "the scion of one of the first families," decided to stay in the house overnight to investigate. He left the light on when he climbed into bed, but that did not prevent the ghost of a huge dog from coming through the closed door and staring menacingly at him before vanishing.

The young man next saw a white-clad woman with uncombed, wild black hair falling into her face. The ghost was weeping and wringing its hands. The clanking sound of a heavy chain filled the house, along with unnerving groaning noises. The frightened investigator repaired to the dining room and stayed there the rest of the night. "The house has never been entered since," the newspaper stated.

The same female ghost—either that, or one exactly like it—allegedly appeared in daylight in another Easton home. A man "of undoubted veracity" encountered it when he looked up from his chair and saw the figure approaching. He was naturally startled, but he assumed the woman was a normal human being, albeit one of rather alarming appearance: "a remarkably white face, luminous dark eyes, and masses of dark hair hung in wild disorder," as the newspaper put it. The gentleman stood up and was approaching her when she stared unsettlingly into his eyes, seeming to "penetrate his very soul." Then she disappeared. It was his one and only experience of the entity.

Source:

"Maine's Haunted Town," 1896. *Stevens Point* [Wisconsin] *Gazette* (June 10).

Maryland

1849: Musical Question

THE GOOD PEOPLE OF Piney grove . . . have been thrown into a considerable excitement, owing to strange sounds resembling the finest music, similar to that of the accordion, which follows a young lady, about sixteen years of age, who resides in the family of Miss Teaky Green. The sound is distinct, and, it is said, responds promptly to any question. The young lady affects to be, if not really, very much alarmed at this strange visitation. We have seen several respectable persons who have visited the house, who vouch for the truth of this story. The young lady, to all appearances, say they, has no agency in producing the sounds, but we suspect that she is endowed with the singular power of ventriloquism, which she is exerting as a hoax or a trick, to frighten the family in which she resides. The music is said to be soft and lovely, beyond description."

Source:
"A Musical Ghost," 1849. *Adams* [Pennsylvania] *Sentinel and General Advertiser* (September 3). Reprinted from the *Rockville* [Maryland] *Journal.*

1860: Light upon the Rail

AN ENGINEER EMPLOYED on the Northern Central Railway in Maryland informs the Baltimore Republican that on Friday night last [May 4], while running a freight train over the road, and when about ten miles from Baltimore, he was suddenly startled by seeing directly before him what he supposed to be the front light of another engine coming towards him—not over a hundred yards distant.

"He immediately blew the whistle to put down the brakes, and finding that he was about to run upon it, he reversed the engine and did everything in his power to stop, but finding it impossible, he gave the alarm to the fireman and rushed to the rear of the engine for the purpose of jumping off to save his life, when, upon turning again to take another look ahead, he was completely astounded by the discovery that the light had disappeared, but where to or how, no one was able to say. The story was substantiated by all those who were on the train, who state that they all saw the light directly ahead, very distinctly, and rather larger than the usual front lights.

"After running a short distance the train was stopped, and those engaged upon it instituted a search for the cause of this mysterious light, but all their efforts were in vain, as there was no light of any kind to be seen in the whole neighborhood. The Republican says it will be remembered that a man was killed at this very spot almost a week since, and it is the supposition of those who saw it, that the light was caused by some supernatural agency, as they are of the opinion that it was entirely too large to have been the reflection from a hand lantern. As we are not inclined to believe in 'ghosts,' we have some doubts about this, but it was certainly a singular phenomenon of some kind."

Source:
"Supernatural Light on a Railroad," 1860. *Western Journal of Commerce* (May 10).

1874: Spewing Ball Lightning

"Just before sunset last evening [July 4] a tremendous thunder-storm broke over the city [Baltimore], accompanied by a tornado and a remarkable meteorological phenomenon. In the western suburbs a whirling cloud descended until it seemed to rest upon the earth, and a darkness black as midnight enveloped everything within its scope. The cloud was circular and moved as if impelled by a cyclone that circled around its outer surface. While it continued, objects twenty feet distant were imperceptible, and a smell like burning pervaded the atmosphere. This was caused by the incessant flashes of lightning which, on the line of railway, assumed the form of moving balls of fire, following each other in quick succession, and casting a lurid, blue glare through the darkness. At one time, within a space of 300 feet, five of these balls of fire were counted at the same moment."

Source:
"Terrific Storm in Baltimore," 1874. *New York Times* (July 6).

1879: "A Substance . . . That Looked Like Snow"

"An aerolite passed over the farm of Dr. P. O. Cherbonnier, of Talbot, on Friday of last week [June 24]. The Doctor was from home, but his wife and daughter saw it very distinctly, although the sun was shining. It went in a Southern direction, appeared to revolve like a rocket, and then exploded, giving a loud report, and disappeared behind the woods. Now comes the strangest part of the story. About 1 o'clock P.M.—five hours after the aerolite passed over it—a substance commenced to fall on the farm that looked like snow. The growing corn and grass and shrubbery were covered with a powder resembling lime or magnesia. A quantity of it was collected and kept until the Doctor's return. He exhibited a specimen of the powder at this office yesterday morning. Some of it had ad-

hesed to the corn-blades, and some was loose. He will send it to the academy of science at Baltimore for analysis."

Source:
"Aerolite," 1879. *Denton* [Maryland] *Journal* (July 5). Reprinted from the *Easton* [Pennsylvania] *Star*.

1879: Anomalous Illumination

"About seven o'clock or a quarter before on Monday evening last [November 10] this whole section of country was suddenly illuminated by a bright red light, nothing like lightning, which lighted up the whole atmosphere as light as mid-day. Everybody that saw it agreed that it was a strange and unusual phenomenon. It was as though ten thousand gas jets had been suddenly turned up."

Source:
Untitled, 1879. *Denton* [Maryland] *Journal* (November 15). Reprinted from the *Milford* [Maryland] *News*.

1881: His Soul Goes Marching through the Clouds

In 1881 the nation was traumatized by the shooting of President James A. Garfield on July 2. He lingered until September 19 before dying. In the wake of his passing, a series of miraculous appearances were reported:

"A little girl saw a whole platoon of angels marching and countermarching to and fro in the clouds after nightfall, their white robes glistening with a mystic light. Her father and neighbors were summoned and saw the same sight, which, however, soon vanished from their profane gaze. In another locality a farmer saw a week afterwards bands of soldiers of great size, equipped in dazzling uniforms, their musket steels quivering and shimmering in the pale, weird light that seemed to be everywhere, marching with military precision up and down

There is no doubt of the fact that there were many who thought they saw Garfield in the clouds. In Talbot county the illusion was seen by many. A farmer living near Clara's Point, on going out into his yard after dark[,] saw, as he described it to his neighbors, angels and soldiers marching side by side in the clouds, wheeling and going through every evolution with military precision and absolutely life-like and natural.

unseen avenues and presenting arms at the sound of unheard commands. It is also reported that many people many miles away, situated at the lower end of the peninsula, saw the extraordinary phenomena at the same time. A few go so far as to say, in spite of the ridicule of their associates, that they distinctly saw in the midst of the soldiers, and conspicuous by his size and commanding presence, the hero President himself, pale but with his every feature distinctly and vividly portrayed.

"There is no doubt of the fact that there were many who thought they saw Garfield in the clouds. In Talbot county the illusion was seen by many. A farmer living near Clara's Point, on going out into his yard after dark[,] saw, as he described it to his neighbors, angels and soldiers marching side by side in the clouds, wheeling and going through every evolution with military precision and absolutely life-like and natural.

"These illusions were due doubtless to the strong impressions [occasioned] by the long continued sufferings and death of the President and the peculiar auroral or meteoritic displays which so often give a weird-like appearances [*sic*] to the ever changing panorama of the heavens."

Source:
"Excited Imaginations," 1881. *Oshkosh* [Wisconsin] *Daily Northwestern* (October 12).

1882 and Before: A Monster Crossing the Road

IN ELK NECK, NEAR the small town of Cecil, a number of individuals claimed to have seen a giant snake. The sightings started in the 1870s, but a sighting in July 1882 got particular attention, possibly because the witness allegedly had a particular clear view of the animal. The report ended speculation, at least locally, that the observers were in error and were really seeing something less extraordinary.

"A gentleman whose veracity is beyond question and whose head is too level to mistake a shadow for the substance" was passing down the Elk Neck road, near Mill Creek, when he spotted what he first thought was a log on the road about 25 yards in front of him. Then he realized it was moving, and he recognized it for what it was: a huge, dark-colored, snakelike reptile.

"How much of the forward part of the reptile had already disappeared in the bushes," the local paper reported, "it was impossible to tell, but the portion in view extended entirely across the road, and there was still a portion in the undergrowth on the side from which it was passing. The road at this point is about 18 feet in width, and it is estimated that his snakeship is at the lowest calculation 10 feet greater in length, making him in all 28 feet long." It had the thickness of a stovepipe.

As the witness's cart approached it, the snake picked up its speed and vanished into the woods along the highway.

Source:
"The Snake of the Eastern Shore," 1882. *Denton* [Maryland] *Journal* (July 29).

1883: Yellow Rain

A QUEER PHENOMENON visited this section last week. The ground was covered, after two or three slight showers of rain, with some kind of dust resembling sulphur or the pollen of flowers. The water in ditches was covered with it, and on the roofs of houses where the shingles make a shoulder to catch it, and on porch floors where the boards meet and in the cracks of the pavements it gathered in quantities. The peculiar shower, whatever they [*sic*] were, seems to have spread over the [Chesapeake] bay and was noticed in Baltimore and elsewhere on the Western Shore."

Source:
"A Shower of Sulphur," 1883. *Denton* [Maryland] *Journal* (May 19).

1883: Red, White, and Blue Lights

THE PEOPLE OF DORCHESTER county are excited over recent supernatural appearances in the midst of them. Hundreds of people are said to have seen three bright lights—red, white and blue—coming down the Chesapeake, following the track pursued by vessels that pass thro' Fishing Creek and Fishing Bay. They almost constantly change their relative positions, and are sometimes near the earth and then again high up in the air. They cast a bright lurid light upon the water, and nobody can account for them."

Source:
"Mysterious Lights," 1883. *Denton* [Maryland] *Journal* (September 22).

1883: The Red, White, and Blue

BETWEEN 5 AND 6 O'CLOCK yesterday morning the employees on the early train on the Baltimore and Hanover Railroad between this place [Hanover] and Emory Grove, witnessed a most beautiful and startling phenomenon in the eastern heavens, while the train was passing between Smith's Station and Valley Junction. The sky that morning was fairly aglow with crimson and golden fires, when suddenly, to their great astonishment, an immense American Flag, composed of the National Colors, stood out in bold relief high in the heavens, continuing in view for a considerable length of time. It gradually faded from view, and was replaced by a white flag or streamer, with a black band extending across the center. What this vision or omen portends, is a matter of conjecture, and it has excited considerable interest among the superstitious, who declare that all sorts of troubles are about to come upon the country, while others assert that it is a sign or forerunner of war at a not very distant day."

Source:
"An Omen in the Skies," 1883. *Frederick* [Maryland] *Weekly News* (December 27). Reprinted from the *Hanover* [Maryland] *Spectator.*

1884: Suddenly, Hordes of Crickets

AT 7 O'CLOCK MONDAY evening [September 6] a sudden and satisfactory shower of rain visited this city. It continued only a few moments, scarcely moistening the pavements and almost as soon as ended[;] the streets were perfectly dry. A very singular phenomenon in connection with the rain was the appearance of myriads of crickets which came immediately after the shower ceased and from that on to midnight could be counted by the scores, any where. They infested dwelling houses, stores, hotels[,] etc., and appeared especially partial to ice cream saloons. The sidewalks and bed of the streets were covered with them for a time. About a million of them raided *The News. . . .* The crickets disappeared during the night. The sudden appearance created much speculation."

Source:
"A Phenomenon," 1884. *Frederick* [Maryland] *Weekly News* (September 11).

1885: Ball Lightning on the Track

AS THE 4 O'CLOCK TRAIN from Baltimore was a few miles out of Washington yesterday the passengers were amazed to see a ball of fire as big as a man's hand rolling along the line of the rails before and behind. This was a little before the thunder storm began. The conductor of the train says that he saw the ball of fire following the track, and that when it apparently reached the middle of the track it exploded with the noise of a cannon shot, distinctly jarring the cars, and that afterwards there was a distinct smell of sulphur in the cars. The passengers who also witnessed or felt the phenomenon, thought somebody had attempted to blow them up with torpedoes. This phenomenon has often occurred on the wide stretch of the Kansas plains, but an old railroad man is authority for the statement that he never experienced it in this section before."

Source:
"A Strange Phenomenon," 1885. *Elyria* [Ohio] *Republican* (July 30).

1889: Luminous Cylinder

A BRILLIANT LUMINOUS cylindrical body" passed over Oella (now part of Catonsville), at an altitude of several hundred feet, on the evening of February 14. Heading toward the Patapsco River, in a northwest direction, the cylinder suddenly effected a curve and descended. It landed about a mile from the small town. It was visible for a few seconds, brilliantly lighting the ground around it. "Just at the time of its appearance," a newspaper said, "a number of villagers were returning from work, and the uncommon sight created considerable alarm among them."

Source:
Untitled, 1889. *Hagerstown* [Maryland] *Herald and Torch Light* (February 21).

1889: Silent Spook

THERE IS A SMALL unoccupied log house on Main street, just west of St. Paul's Protestant Episcopal church, in Sharpsburg. Last Friday night [February 15] two young men named Spong and Delaney[,] hearing, as they supposed, an unaccountable noise in the building, concluded to investigate the cause of it. They procured a lantern and entered the house, Delaney in advance with the light and Spong bringing up the rear. Suddenly Delaney uttered a cry of fright, fell back into Spong's arms in a fainting condition and was carried out into the street, and it was some time before he entirely recovered consciousness. Then he explained that he saw a figure in white at the head of the stairs, that his cry was to it, and when no answer was returned he was overcome with fright. Spong also got a rather unsatisfactory view of the apparition. In a short time a number of persons explored the house without discovering any clue to the nocturnal visitor that had shocked Delaney's nerves so sensibly. The 'ghost' has not made a second appearance."

Source:
"A Sharpsburg Ghost," 1889. *Hagerstown* [Maryland] *Herald and Torch Light* (February 21).

1894: Creature with No Head

BURRSVILLE NOW OWNS A real ghost. The ghost in question was seen some nights ago by two reputable citizens, and was shaped like a colt, calf, dog, or some other animal, but had no head. One of the men caught it, but allowed it to resume its course after a short tussle."

Source:
"Burrsville," 1894. *Denton* [Maryland] *Journal* (January 27).

1894: Strange Stars

ABOUT TWELVE O'CLOCK on Saturday night [July 28] a strange phenomenon was witnessed in the western sky, which lasted only for a few minutes. The phenomenon consisted of a number of shooting stars which seemed to go backward and forth and in some instances almost coming down to the house tops. At this season of the year it is not uncommon to witness stars shooting around in nearly every direction, but such an unusual display as that of Saturday night is not often seen."

Source:
"Meteoritic Display," 1894. *Frederick* [Maryland] *News* (July 30).

1895: Market House Spook

A PATRICK STREET MAN, who claims to be an authority on spooks, says that he saw the foot-prints of a ghost just outside the Market House this morning. He also said that the brownie in Tom Pope's store window nodded to him while he was going up on the same side of the street. There has for some time past existed a suspicion in the minds of quite a number of market-goers that an apparition of some sort sits on the front porch of Dave Rice for a short time previous to the opening of a Saturday morning market, and just before daybreak it will skip around the building and then suddenly vanish. The fat woman said that she remembered hearing the people talk about the Market House spook long ago, when she and Billy Albaugh used to play together, but it had since disappeared and not been spoken of until this morning."

Source:
Untitled, 1895. *Frederick* [Maryland] *News* (July 6).

1898: Phantom Woman Minus Head

"TRAINMEN ON THE Baltimore and Ohio and West Virginia Central railroads employed near Cumberland, Md., were recently frightened by the shape of a headless woman that makes her appearance at Greenwade's siding, near Twenty-first bridge, between Cumberland and Keyser, W. Va. Freight trains are side tracked there, and when the trainmen are waiting a headless woman emerges from an old culvert or bridge and walks up and down the track. Whenever any of the men attempt to follow her, she disappears. One railroad man was so badly frightened that he left the service of the road. Others say that if the headless woman keeps up her antics they, too, will quit. . . . The men declare that the ghost can be seen almost nightly."

Source:
"Headless Woman's Ghost," 1898. *Marion* [Ohio] *Daily Star* (January 8).

1901: 500 Frogs

"A SHOWER OF SMALL frogs and toads fell in the vicinity of Red Hills, several miles south of Frederick, Md., on Saturday [June 8]. The phenomenon lasted but a few minutes, but it is estimated that nearly 500 small frogs fell to the ground. The curious sight was witnessed by several people. Several years ago a similar shower was witnessed in the same vicinity."

Source:
Untitled, 1901. *New Oxford* [Pennsylvania] *Item* (June 14).

1915: "Luminous Ball of Large Proportions"

"LAST SATURDAY EVENING [September 4] at about 8 o'clock the attention of a number of the Hobbs villagers was attracted by a phenomenon in the direction of the depot and in the form of a luminous ball of large proportions."

Source:
"Hobbs," 1915. *Denton* [Maryland] *Journal* (November 15).

Massachusetts

1819: Rot from the Stars

ON AUGUST 13, 1819, between 8 and 9 p.m., a fireball "of a brilliant white light resembling burnished silver" descended at an oddly slow rate of speed and came down in the yard of an Amherst house occupied by Erastus Dewey. Unaware of the event, Dewey did not leave his house until the next morning, when he made a bizarre discovery, a mass of strange-looking material some 20 feet from his front door. Soon afterward, Professor Rufus Graves arrived on the scene and examined the stuff for himself.

It was a "circular form," he would write in the *American Journal of Science,* "resembling a sauce or salad dish bottom upwards, about eight inches in diameter and one in thickness, of a bright buff color, with a fine nap upon it similar to that on milled cloth. . . . On removing the villous coat, a buff colored pulpy substance of the consistence of good soft soap, of an offensive, suffocating smell appeared; and on a near approach to it, or when immediately over it, the smell became almost insupportable, producing nausea and dizziness. A few minutes exposure to the atmosphere changed the buff into a livid color resembling venous blood." Graves placed the material in a half-pint tumbler. At first, the paper said, "it was observed to attract moisture very rapidly from the air. . . . It soon began to liquefy and form a mucilaginous substance of the consistence, color, and feeling of starch when prepared for domestic use."

Then, however, it began to evaporate. Within two or three days, only a dark residue was left on the glass. When the professor rubbed it between his fingers, he saw it transform into a fine, odorless ash. Subsequently, Graves got into a bitter dispute with Amherst College chemist Edward Hitchcock, who insisted that the material was a "species of gelatinous fungus, which I had sometimes met with on rotten wood in damp places, during dog days." The witnesses had been entirely mistaken in linking the fungus to the meteoritic fall. Graves, however, emphatically countered that there could be "no reasonable doubt that the substance found was the residuum of the meteoritic body."

The incident is hardly an isolated case. It is simply one episode in the long, controversial history of a phenomenon known in Welsh as *pwdre ser,* which translates as "rot from the stars" or "star jelly." Once found credible by scientists, it is now rejected as the outlandish product of witness error. Rightly or wrongly, witnesses over the centuries have reported meteorites that leave blobs of jelly in their wake. Reports like these continue even now, as do attendant debates about their meaning.

Sources:
Burke, John G., 1986. *Cosmic Debris: Meteorites in History.* Berkeley: University of California Press, 80–81.
Corliss, William R., ed., 1977. *Handbook of Unusual Natural Phenomena.* Glen Arm, MD: Sourcebook Project, 497–502.
Fort, Charles, 1941. *The Books of Charles Fort.* New York: Henry Holt, 41–44.

1837: "Monster . . . from Nowhere"

THE OLD INHABITANTS of Gloucester . . . [speak] . . . of a . . . redhot monster [that] appeared . . . to four men who were digging in the night on Farmer Page's land for Captain Kidd's hidden treasures. The creature was seen and sworn to by all four of these rogues, and their description of it is to be recommended to novelists, poets and playwriters [sic] who wish to produce a sensation. The 'burning beast' was like this:

"'It was a large animal, staring eyes as big as pewter bowls. The eyes looked like balls of fire. When it breathed as it went by, flames came out of its mouth and nostrils, scorching the brush in its path. It was as big as a cow, with dark wings on each side like a bat's. It had spiral horns like a ram's, as big around as a stovepipe. Its feet were formed like a duck's and measured a foot and a half across. The body was covered with scales as big as clam shells, which made a rattling noise as the beast moved along. The scales flopped up and down. The thing had lights on its sides like those shining through a tin lantern. Before I saw it I felt its presence and I smelled something that was like burned wool as it went by. I had a feeling of suffocation when it came near me. The monster seemed to come from nowhere and to go away in the same manner.'"

Source:
"Jolly Old Ghost," 1896. *Coshocton* [Ohio] *Democratic Standard* (February 14).

1841: Sky Squid

DURING THE HAIL storm of Wednesday afternoon [July 7], a Mr. John Seaver witnessed a very singular phenomenon whilst riding over the Tremont Road. Among the hail-stones which fell, was an animal, ten inches in length, and four inches and a half in circumference, known to . . . the fisherman as the *Squid*. It is supposed to have fallen from a great height, from the fact that the hail-stones which fell with it were ascertained to be salt. The animal has been preserved in spirits."

Source:
"A Singular Phenomenon," 1841. *Adams* [Pennsylvania] *Sentinel* (July 12).

1849: Sea Serpent More Than 100 Feet Long

A CORRESPONDENT OF THE Boston Journal, indorsed [sic] by the Editor as a gentleman

He appeared to be from one hundred to one hundred and fifty feet in length, about as thick round as an oil cask, with large and very bright eyes, and sixteen projections or humps upon the surface of his body, below the head.

well qualified to judge of such matters, states that, while cruising in Nahant Bay, on Saturday the 25th ult. [August], they saw floating upon the surface of the water, at the distance of about one quarter of a mile, a huge monster, supposed to be the far-famed sea serpent, previously observed in the same vicinity, on several occasions, by different individuals. He appeared to be from one hundred to one hundred and fifty feet in length, about as thick round as an oil cask, with large and very bright eyes, and sixteen projections or humps upon the surface of his body, below the head. He appeared to take no notice of the craft on board of which the writer was, and the progress of which was so much impeded by a heavy swell, as to be unable to approach any nearer—her speed through the water being exceeded by that of his snakeship. It was so near the shore at the time referred to, that, with the aid of a good spyglass, it might, in the opinion of the correspondent, have been seen from Nahant head."

Source:
Untitled, 1849. *Alton* [Illinois] *Telegraph and Democratic Review* (September 14).

1850: Black Rainbow

A *BLACK* RAINBOW was seen by the citizens of New Bedford, Tuesday evening [February 19] about 8 o'clock. Its direction was from the Northwest to the Southeast; it was visible about 20 minutes" (italics in original).

Source:
"Singular Phenomenon," 1850. *Adams* [Pennsylvania] *Sentinel and General Advertiser* (February 25).

1855: This Train Is Bound for Plymouth

It WAS IN THE MIDDLE of the summer night, between 3 and 4 o'clock, swore persons who lived near the railroad track a mile from Plymouth, that the distinct whistle could be heard. It sounded exactly like an approaching train, but none was scheduled, and none could be seen.

Four local men who vowed to solve the mystery showed up at the site at a designated hour. They had just made themselves comfortable when they heard, off to the north, the railway whistle, and then, as the Plymouth paper put it in the purplest of prose, "the distant clatter of wheels . . . louder, nearer, nearer—the click of the rails in their chairs." The account went on to say that "the rush of steam was as plain in their ears as if the lantern glared before them—the shriek of a demon whistle close at hand made them leap from the track as the train thundered down the grade—the hot breath of the panting steed was in their very faces as it passed—as the unearthly scream ceased, they heard the brakemen screwing up their brakes, the tinkle of a bell and a sound of meeting cars, as if the invisible specter monster of the road had reached his journey's end."

Some weeks later, one of the witnesses, a man with interest in the Spiritualist movement then popular in America, visited a spirit circle in Boston. He would report that no one knew him except one other sitter, but he had never confided his experience to this individual. Nonetheless, an alleged spirit communicating through automatic writing announced that it had to leave the circle. Asked why, it replied, "To run the train." What train, specifically? "The Old Colony train," it said. This train ran through Plymouth, the witness/sitter knew, though no one spoke. The spirit went on to say, "In life I was an engineer upon that road. At stated intervals, a train, unseen of mortal eyes, takes the spirits of pious dead to the Pilgrim home they consecrated by their lives. Mr. —, who is with you, will say whether he has ever heard or knows aught of that train."

Unimpressed, one newspaper editor wrote, "It is a little singular . . . that spirits who can enter rooms with closed doors, and who are ever present at the call of mediums, should require a train of cars to transport them to Plymouth!"

Source:
"The Dead Man's Train," 1855. *Hornellsville* [New York] *Tribune* (December 27).

1864: The Blob

A LARGE METEOR WAS SEEN to fall near the shore of Parker's pond, in Hubbardston, on the night of the 9th ult. [October]. On visiting the spot the next morning a mass was found of a gelatinous, light colored, semi-transparent substance, described by some parties to be as large as a hogshead. A gentleman who visited the spot three days afterwards, after a large quantity had been carried away, and much more trampled into the earth, and dissolved and evaporated, says that he could at that time have gathered two bushels of the debris. A specimen of the mass was presented to the Natural History Society at its regular meeting last evening. Although tightly corked in a bottle, it had diminished considerably in bulk, and was partially dissolved. It was a light straw color, and had a strong odor of sulphureted hydrogen, with a sulphurous taste. A chemical analysis will be made by Professor Bushee, which, with more information in regard to the appearance of the meteor when first discovered, will probably be obtained and presented at the next meeting."

Nothing, however, came of the incident, and there was no follow-up to the story.

Source:
Untitled, 1864. *Janesville* [Wisconsin] *Daily Gazette* (November 16). Reprinted from the *Worcester* [Massachusetts] *Spy*.

Circa 1869: Specter Crew

THE BURIAL OF John Winters recalled to old-time fishermen a tradition of a modern Flying Dutchman with its ghostly crew that was believed to roam the seas in pursuit of a ship that had sent them to the bottom, relates a correspondent from Gloucester. Winters was the last survivor of the crew of the Gloucester schooner, Charles Haskell, which in a storm in March, 1869, ran down and sank a Salem schooner and its entire crew on Georges fishing banks. He died at the Fishermen's Snug Harbor in his eighty-second year, repeating almost to the last the tale of the ghost ship supposed to have pursued the Haskell throughout its career as a fisherman.

"Once off Eastern point, at the entrance of Gloucester harbor, Winters said, a schooner ran down the wind, hove alongside the Haskell, and its phantom crew climbed the rigging, declaring themselves the ghosts of the Salem fishermen.

"Winters and others of the Haskell's crew refused to fish in the ship again and a new crew was taken on. These returned with a similar story of ghostly visitations at sea, took their dunnage bags and quit. Another and still a fourth crew were shipped, but each came to port with a renewal of the story of a ship shrouded in white and a specter crew, and the Haskell was hauled up, unable to get men. It finished its seagoing as a sand freighter, and the Salem ship was not heard of again."

Source:
"Pursued by Ghostly Ship," 1921. *Middlesboro* [Kentucky] *Daily News* (February 18).

1870: Red Light, Invisible Train

THE ENGINEER OF the freight train on the Boston and Lowell Railroad, which leaves Boston about 3 o'clock in the morning, has on several occasions discovered a red light swinging at a furious rate at the Woburn Station, where the train stops for water. The light would sometimes be in front and sometimes in the rear of the train. When the engineer would stop his train and send some one to learn why the signal to stop was made[,] the messenger would be greatly surprised to see the light vanish. Investigation has proved that no person was there with a lantern, and the brakeman and conductor concur also in having beheld the phenomenon, which, so far as is known, is without visible cause.

"Some laborers living on the line of the above station state that a few mornings since[,] they were coming down the road in a hand car, when they suddenly heard the approach of an engine and train, and knowing that no train was due in the vicinity at that hour, they became greatly frightened, and, jumping out of the car, threw it off the track to await the train which they thought was coming at a rapid pace upon them, but which, it is needless to say, did not come.

"The superstitious regard the affair as a fore-warning of some disaster, while the spiritualists have the ready theory that it is the spirit of a man who was killed there about two years since."

Source:
"A Railroad Haunted by the Spirit of a Victim," 1870. *New York Times* (January 31). Reprinted from an unnamed Boston newspaper.

1875: Carrying the Serpent

GEORGE M. BALL, A young man of eighteen years, employed on the farm of Mr. Perry, in Westfield, has been accustomed to drink while milking, night and morning, a quantity of warm milk. A few mornings since, failing to take his accustomed draught, something came up in his throat, choking him, and he fell over senseless. A son of Mr. Perry was surprised to see a snake's head protruding from Ball's mouth, but on attempting to seize it, the serpent retreated down his throat. A powerful emetic was administered, and in a short time the young man vomited up a 'hooked adder' two feet and eight inches long, and about as thick as two fingers of a man's hand. It lived only five minutes. Ball has probably carried the serpent for at least twelve years, as he was accustomed to drink from a small brook when a boy. Since parting from his tenant his health has greatly improved, and his appetite is a little more reasonable."

Source:
"Snake Story," 1875. *Elyria* [Ohio] *Constitution* (December 2). Reprinted from the *Springfield* [Massachusetts] *Union.*

1883: "Hydra-Headed Monster"

A FEW DAYS SINCE[,] George, son of John Reichel, an honest and truthful lad, found in the vicinity of the land lighthouse [in Boston] a snake five feet in length, a hydra-headed monster, which had four heads at one extremity and

three at the other. The monster was killed, and is now corked up in a bottle of alcohol."

Source:
Untitled, 1883. *Grand Traverse* [Michigan] *Herald* (October 4). Reprinted from the *Boston Post.*

1885: "Bright Line of Light"

THE FIRE IN WATERTOWN Monday evening [March 9] was attended by a phenomenon which attracted the attention of residents of all the surrounding country. The fire illumined the sky above it to a height, as seen from Cambridge, of 30 degrees or thereabout, which was not unusual. Above, however, in a direct line from the apparent center of the fire and, within 10 of the zenith, glowed a bright line of light. The phenomenon continued until the fire disappeared, and therefore may be supposed to have been connected with it; but in Boston, Lynn, Hyde Park, and other cities and towns where the existence of the fire was unknown[,] people were utterly unable to account for it. The prevalent theory was that a great comet had suddenly appeared in the earth's atmosphere."

Source:
Untitled, 1885. *New York Times* (March 11).

1886: Bug-Eyed

FOR SOME TIME QUEER stories have been told of unusual proceedings at the Worcester Steel Works at night. Complaints were made of assaults committed and of the general conduct of the men who insisted on going in and out at will, until at length, to check these alleged outbreaks, the managers requested police protection. Patrolman Dealey was detailed to go on duty there after 9 o'clock at night and remain until early in the morning.

"To-day [July 22] Michael Gleason, one of the employees, told a Times correspondent a queer yarn about an experience that he had recently. He said that one night he was walking through the mill, which was well lighted by electricity. When he passed the boiler house he saw a strange man standing inside with his hand on the throttle of an engine which had not been fired up. The man looked at Gleason for a moment. Gleason spoke to him jokingly and asked him if he was going to start up. The stranger's countenance did not change and his eyes seemed to jump from their sockets. The man was unknown to Gleason, who had worked in the mill a long time and knew everybody.

"Turning quickly[,] Gleason ran to the other end of the mill, very much frightened. Large drops of perspiration stood out upon his face, and suddenly he swooned away. Gleason said that he was cared for by his companion, and when he recovered he told him what he had seen. He described the man's appearance minutely, even to the striped jacket which he wore, but no one knew him. Finally some one recalled the fact that it was a perfect description of an engineer who was killed two years ago at the very spot where the strange apparition was seen by Gleason. Many of the workmen are firm in the belief that it was the ghost of the dead engineer."

Source:
"A Ghost at the Throttle," 1886. *New York Times* (July 23).

1887: Frogs from Rain

AFTER THE FIRST SHOWER the paved streets and sidewalks [of Boston] were covered with small frogs about the size of a large, black cricket. During the second shower I watched the brick sidewalk in front of my house very carefully to see whether they came hopping along from behind the pavements, or suddenly appeared in the center of same. Presently a little dark bunch appeared in the center of the brick sidewalk, then another, until there were four close together. At first they did not move, but I kept my eye on them, and in one or two minutes they began to hop, and proved to be frogs.

"After the shower the streets, as far as examined in the direction from whence the wind and rain came, were found to be covered with the frogs, while a few rods in another direction not a frog could be found. The same phenomenon was noticed in the streets of Brookline, which was exactly in the direction of the wind at that time. To say they came up through the pavements, or hopped along from some vacant lot, is to say an absolute impossibility does occur. As soon as they recovered from the effects of the fall they showed their natural instincts by trying to get away from the streets and the passing people."

Source:
"Raining Frogs," 1887. *Atchison* [Kansas] *Daily Globe* (August 10).

1892: Oceanquake and Luminosity

A CURIOUS PHENOMENON is reported by Pilot Sullivan, who brought the steamship Trinaeria into port yesterday. . . . On Monday last [July 25] the pilot boat was cruising for incoming ships in the vicinity of Georges Shoal. About 7 o'clock in the evening the crew were suddenly startled by a noise which seemed to come from directly underneath. First it had a grating sound, and then it seemed as if the metal sheathing of the hull was exposed to a submarine hailstorm. The sharp, rattling noise continued for about six seconds. The vibrations were steady, and, according to the pilot, were undoubtedly caused by an earthquake. For a moment the crew thought the keel had grated on the bottom, but they were reassured a moment later by the lead, which showed plenty of water underneath the vessel.

"Before the crew had time to cease wondering at the subaqueous disturbance their attention was demanded by a brilliant and startling display overhead. The heavens from zenith to horizon were suddenly streaked by a rift of light, serpentine in shape and clearly defined. The rippling coil of light was dazzlingly brilliant, and it was several minutes before there was any perceptible diminution to its luster. For nearly an hour the heavens were illuminated by the gigantic scroll, which looked like a tracing in phosphorous, and then it gradually faded away, and finally disappeared. Throughout the time, the sea remained perfectly calm.

"On the same date and at the same hour the same phenomenon was observed from the steamship Trinaeria, which was then sixty miles distant from the pilot boat. Capt. Thompson of that vessel says the light glanced across the sky as swiftly as a streak of lightning. It looked to him like a huge serpent. The earthquake shock was not felt on board the Trinaeria, but heavy tide rips were observed. A marked unsteadiness in the barometer was noted. At the beginning of the aerial display it fell from 29.39 to 29.29, rapidly rising again as the light faded from the skies."

Source:
"An Earthquake at Sea," 1892. *New York Times* (July 30).

1893: Floating Hand

A STORY IS TOLD . . . and given in all sincerity, which is to say the least very uncanny. It is because the young women earnestly request it, that their names are not given to add strength to the tale. They live on the Prescott corporation, or rather did live there until a few days ago, occupying one large room in which were two beds. A few nights ago they were out together for a visit[,] returning home shortly before ten o'clock and immediately retiring. One of the young women said they all fell asleep, she being the last one to pass to the Land of Nod. She says she was awakened sometime during the night by the pressure of a hand upon her. She awoke with full possession of her senses as if the cause had been a great one. She looked toward the foot of the bed and there pressing downwards was a ghostly hand. She says she screamed and awoke her bedfellow who in turn saw the hand, and screamed in fright.

"The hand floated in the air over the bed, and in terror the two young women watched it. One got out of bed, crossed to the other bed and awoke the other roommates. All four watched

the ghostly hand as it seemed to gently rock up and down. Suddenly the bell in the High street church belfry struck twelve o'clock, and in a flash of light which illumined the entire room the hand disappeared. The young women lighted the lamp and all night they sat up not daring to go to sleep, or to remain in the dark. When daybreak came they hurried to a clergyman to whom they told their story. He advised them, if they were timid[,] to leave the room and not to sleep there again. Like a sensible man that he was, he did not add to their timidity by advising them not to look on the incident as an ill-omen. He simply told them to leave the room if they did not want to stay there. This they did. . . .

"It may be incidentally remarked that the young women saw it was not the effect of electric illumination on the street, neither had they been eating Thanksgiving mince pie, nor swapping highly improbable ghost stories during their visit."

Source:
"Latest Ghost Story," 1893. *Lowell* [Massachusetts] *Daily Sun* (December 6).

1894: Ghost in a Boat

THIS CITY [WALTHAM] has a ghost story. It is said to have frightened gangs of workmen employed by the Gov. Gore estate on Grove street. A night crew was at work there recently pumping out a small pond. The morning after the next night the foreman found that his gang had not been at work. He hired another crew and put the men at the pumps that night. Again in the morning he found the gang had skipped. He began to think something was wrong.

"The foreman sent for the men. They said they had seen a ghost. He laughed at the men, but they were firm in their belief, and no amount of coaxing on the part of the foreman could induce them to return to work on the job. Since then it is said no night gang has been employed. One of the men who were frightened thus describes his experience:

"'There was a gang of men from Silver Lake at work on the night in question. Just at midnight I heard an exclamation of terror from one of my companions. Looking up, I saw a sight I will never forget. On the brink of the pond is a small boat house, and from the door of it we saw a boat with a man in it[,] whom [*sic*] we knew had been dead over a year, [and we saw the man] emerge and come slowly toward us. The ghost, if it were [*sic*] one, was holding a red light in his hand. We waited no more but left for home as quickly as our legs could carry us.'

"This was repeated the following nights, according to the story of the man, and no night gang has been employed since. The story is vouched for by two reputable men, who say they would not stay there over night if the estate were given to them."

Source:
"A Ghost Story," 1894. *Lowell* [Massachusetts] *Daily Sun* (January 9).

and finally the light gave a turnaround, swerved directly down into the village of Wilmington, and was seen no more.

"There are various opinions about the lights on the two mountains ten or a dozen miles apart on the same night. Local 'philosophers' are 'probing the matter,' and it is possible that outside lore will be required to fathom the mystery. We do not care to render any verdict this week.

"Interest in the case was heightened when on Sunday evening at 10 o'clock the same mysterious light, or some other, in a circle apparently of ten feet, in the shape of a horse shoe, with arms protruding from the inside, passed slowly by F. A. Gleason's house, a few feet from the ground, and went off over Charles Sawyer's into the mountain beyond. This light so nearby was of whitish yellow hue. Mr. Gleason and family were the only ones who saw this. If these strange lights have been seen in other towns we would like to know."

Source:
Untitled, 1900. *North Adams* [Massachusetts] *Transcript* (October 6).

1900: Like a Horseshoe with Arms

LAST WEEK MONDAY NIGHT [September 24] at 12 o'clock James Dinwiddle and George Buckley saw a bright light on Mt. Pleasant that cannot be accounted for to date. The light would go up and down and then contour around in a circle several rods, and then come back to the place of starting.

"What adds interest to the phenomenon is the fact that the same night at 2 o'clock in the morning J. Rooney happened to look out of his window and saw a similar light on Haystack mountain, in Wilmington. Mr. Rooney says it darted around so much that he thought it would bear watching so he called up his family and hired men. The spectacle, they all said, was wonderful. The light would go far up toward the heavens and then come straight down and diverge to the right or left for a few rods and then go up again. This was repeated time and again during the two hours the party watched it,

1909: Zigzagging Light

NORTHBRIDGE HAS A GHOST, or a ghost has Northbridge, residents of the heretofore peaceful village are uncertain which. For several times, at about the same hour, a mysterious light, varying in size from a small bulb to that of a bushel measure, has appeared and performed queer antics on the high ledge near Wayside.

"First treated as a joke, continued nightly repetitions have caused the phenomenon to become a serious reality and the village and its neighborhood are discussing the affair, while scores of citizens are seriously frightened. At least three families are packing up their household goods with the announced intention of moving out of town.

"One night fully 200 persons assembled in the vicinity of the ledge, but when the light appeared many women screamed and hurried home, evincing no desire to continue the investigation. Half a dozen armed men had the temerity to go to the top of the ledge, and, in close formation, shoulder to shoulder, tried to catch up with the light. Like a will-o'-the-wisp the light zigzagged along the ledge, climbing up the trunk of a tall pine tree, from which point it was visible for a considerable distance, descended rapidly within a few yards of the watchers and, mounting the crest of the hill, disappeared in the nearby pond."

Source:

"Town Hunts Ghost," 1909. *Trenton* [New Jersey] *Evening Times* (May 25).

Michigan

1858: Stomach Snake

THE FOLLOWING SINGULAR story is told of a man named Beach, who had swallowed a snake, in Michigan:

"For the past seventeen years the sufferer has been satisfied that there was a living animal of some kind in his stomach. If he drank liquor the animal would seem to become drunk. This he judged from the fact that it remained perfectly quiet until the effects of the spirits wore off. At times when he partook of food offensive to the animal, it would become agitated and roll about with a motion which could be felt by placing the hand upon the stomach. Having tried many physicians without being relieved, Beach was induced to apply to a German doctor, who recommended the process of starving the intruder out.

"This advice was adopted, and the patient succeeded in inducing the animal to come up into his throat, but for fear of strangulation he swallowed vinegar and drove it back. For four months means were tried to relieve the man's stomach of its unwelcome guest, and finally, on Friday of last week, he passed an entire snake, measuring just three feet in length. It was somewhat decomposed, and had evidently lost four or five inches of its tail.

"As to its original size, our correspondent cannot determine. Its head measured crosswise just one inch and a quarter. Its teeth were about one-eight[h] of an inch long. From the formation of the head the correspondent thinks the reptile is of the common water snake species. The man is now doing well, and is in good spirits in consequence of being relieved of his hideous tormenter. Our correspondent, who is well known to us, and in whose assurances we can place the utmost confidence, is knowing to all the facts we have stated above."

Source:
"Starving of a Snake out of a Man's Stomach," 1858. *Gettysburg* [Pennsylvania] *Compiler* (November 8). Reprinted from the *Sandusky* [Ohio] *Register*.

1868: Light as Bright as Day

A GENTLEMAN EMPLOYED at the Detroit and Milwaukee depot [in Detroit] states that about 12:45 o'clock yesterday morning [January 31] he witnessed a most curious phenomenon, which found no parallel in his own experience. He was passing upward Woodward-avenue, when he suddenly discovered that he was surrounded by a very brilliant light. The moon was down, the sky was clouded, and there were no street lamps lighted, so that the night was unusually dark. His first thought was that there was an explosion in the street lamp, under which he was just passing, but looking up he discovered that this could not be. He then supposed it to be the aurora borealis, but there was none of the well-known characteristics of this illumination of the heavens. The thought of meteors next occurred to him, but he saw no meteoric bodies, and the appearance differed altogether from what he had before witnessed when large meteors had fallen in close proximity to him. The whole city was illuminated as light apparently as day.

"He was looking up the street and saw both buildings and trees, and then turned and looked down, seeing the whole length of the avenue to the river and the Canada shore beyond. The light lasted from a quarter to half a minute. He described it as very peculiar. It was not a flash, nor yet a steady light, but seemed to come in waves, and he could think of no comparison except the flapping of the wings of a bird. This phenomenon was also witnessed by another person. A gentleman employed at the Michigan Central was at the time passing along Third-street, near Howard, and his description of the appearance agrees with that given above. Another gentleman living in the eastern part of the state also witnessed it. The same phenomenon was also seen at Ypsilanti by a gentleman who became somewhat nervous thereat."

Source:
"Curious Phenomenon Observed in Michigan," 1868. *New York Times* (February 3). Reprinted from the *Detroit Post* (February 1).

1870: "Monster Resembling a Snake"

A VERY PRONOUNCED fish story comes from Barry county, Michigan, where, we are 'told,' a monster resembling a snake is residing in Harwood lake. It is represented as being about twenty feet long and from five to seven inches in diameter. It has been seen a number of times, once, last summer, by a Swede, who supposed it to be the devil. The other day his snake-ship was seen by a boy, who was fishing at the time and had quite a string of fish lying on the bank. The boy says the snake came out of the water, and with one gulp swallowed fish, string and all, and then came for him, and he, not particularly wishing for such an associate, made tracks for a more congenial companion. 'It is said' that the teeth of this serpent are three or four inches long."

Source:
Untitled, 1870. *Titusville* [Pennsylvania] *Morning Herald* (June 13).

1878: Ghost in the Woodpile

QUITE A SENSATION has been created on the Chicago & Lake Huron Railroad, near Olivet, in consequence of a singular apparition which has just made its appearance in that vicinity. . . . A spook came out of the wood-pile a few nights since and stopped the eastward-bound train. It is described as a human form robed in snowy white, and appeared on the track a few rods in advance of the engine. The engineer blew the whistle, but the mysterious form refused to yield the track. The train was stopped, and a party went ahead to reconnoiter, when the strange personage retreated, and when they retreated would follow them. To all questions that were asked it gave no response. They ordered it off the track, and it refused to budge, when they fired several bullets through its heart, but, instead of crying out, it danced a hornpipe on the rails, and seemed to delight at their discomforture. Finally the engineer mounted the engine and pulled the throttle, and just as the exasperated engine was about to make mince-meat of the stranger, it disappeared in the air.

"The news soon spread in the vicinity, and the next night farmers and trackmen went to the spot, and, behold! the strange figure confronted them. They set dogs on it, who seemed to be grappling with an object, but no blood was found. The men, armed to the teeth, boldly went forward, but it retreated, and when they receded would follow to a given point. A party outflanked it, and came upon it to solve the mystery, but it vanished heavenward.

"Not being satisfied, the party went to the spot the next day, when to their horror, an old man robed in black came out of the wood-pile and took his wonted position on the track as if to dispute their passage, his long silvery locks and snowy-white beard floating in the breeze. The dogs were called into requisition against it, but to no avail. He carried a death-like smile throughout, and retreated at their advance as before, and, on being surrounded, again vanished heavenward.

"We heard from the scene a couple of days ago, and the country thereabouts is all excitement. The question is: Has some old man been murdered and buried there, and this was his apparition? Certainly it is flesh and blood. Scores of visitors from other counties have been there and took observations, and all have gone away mystified. At last accounts no light had been gained as to the strange apparition."

Source:
"A Mysterious Apparition," 1878. *Elyria* [Ohio] *Republican* (May 2).

1884: Walled Lake Reptile

GREAT EXCITEMENT prevails at Walled Lake, occasioned by the presence in the lake of a huge reptile, said to be the largest one ever seen in that vicinity. This monster was first seen by Jas. Monroe, who lives near the lake. He says, and his friends entertain no doubt as to the truthfulness of his assertion, that this snake measures, according to his calculations, no less than 30 feet in length, and is seemingly about 10 inches in diameter. Hundreds of curious people are daily seen on the shores surrounding the lake watching for the frequent appearance of this huge reptile."

Source:
"News of the Week," 1884. *Grand Traverse* [Michigan] *Herald* (June 19).

1884: Fall of Chalk

AN EXTRAORDINARY PHENOMENON is reported as having occurred in the Straits of Mackinaw, and the inhabitants along that turbulent water are trembling yet. On the 13th [of June] a light warm shower fell at Wangoshance, a point of land extending out into Lake Michigan. Immediately after the rain great quantities of a dry, chalk-like substance fell, turning the lake to a milky hue and covering the ground to the depth of an inch. Some of the people tasted it and were taken violently ill, and the substance was afterwards pronounced a species of lye, but not so heavy as potash. A good many theories are advanced as [to] the cause of this extraordinary feat of nature, some claiming it comes from

the limestone pits, and others that it is the ashes of forest fires packed by the rain."

Source:
"Northern News," 1884. *Grand Traverse* [Michigan] *Herald* (June 26). Reprinted from the *Detroit Journal*.

1884: "Rainstorm of Stones"

MONDAY AFTERNOON [November 1] a rainstorm of stones commenced falling in Castleton township, Barry county, and continued at intervals up to Friday. They began falling on the farm of Sylvester Osborne, and so thickly that men engaged in husking corn upon Osborne's place were compelled to suspend work. Charles Osborne was hit by several of the falling missiles, but not seriously injured. The people were greatly excited, and many have visited the spot and witnessed the phenomenon. The stones are of a dark, volcanic nature, and are not said to fall with great velocity."

Source:
"News of the Week," 1884. *Grand Traverse* [Michigan] *Herald* (November 6).

1885: Cemetery Light

CONSIDERABLE EXCITEMENT has been caused lately at Au Sable by the report that for several weeks past a strange blue light or ball of fire has been seen every night in the village cemetery, rising from the grave of Fulton, a victim of the late Burner horror. Those who claim to have seen it say that it rises to the height of 6 or 8 feet and then falls back to the grave like a burning star. One night recently some 500 persons assembled at the cemetery to see the phenomenon, but it did not appear. The superstitious think it an evil spirit . . . while others say it is escaping gas."

Source:
Untitled, 1885. *Grand Traverse* [Michigan] *Herald* (October 22).

1892: "Singular Whirring Noise"

A PARTY OF YOUNG MEN were boating in Lake Huron just off the west end of Round Island one summer day in 1892 when a "huge snake" swam into view. It was heading in a southeasterly direction toward the mainland, not showing any obvious hostility or even much interest in the viewers. Still, they were so frightened that they immediately returned to nearby Mackinac Island, where they were vacationing.

According to a *Chicago Tribune* story, one of the witnesses, J. Frederick Stevenson, made a formal statement that evening. He swore that the creature looked "black and oily" and, except for its extraordinary size, looked like a snake. What was not snakelike was its "singular whirring noise." The noise sounded like something machinery would make. The *Tribune* noted, "Mr. Stevenson is very cool and deliberate—almost reticent—in his statements and seemed not at all soliticous for hearers or anxious for credence."

A party of young men watching the boaters through field glasses allegedly corroborated the sighting, which so frightened one that she fainted.

It should be cautioned that the *Tribune* of the latter nineteenth century occasionally carried unverifiable tales of lake monsters and sea serpents. An attempt by cryptozoologist Gary S. Mangiacopra to locate other records of the Round Island incident found no evidence that they existed.

Source:
Mangiacopra, Gary S., 1979. "Water Monsters of the Midwestern Lakes." *Pursuit* 12, 2 (Spring): 50–56.

1894: Superior Serpent

THE MASTERS AND CREWS of the steamer Geo. F. Williams and schooner H. A. Hawgood have a great sea serpent story to tell just now. They state that when about half way between Copper Harbor and Whitefish Point, coming down Lake Superior, with the tow line dragging in the water, as the steamer was not being pushed, a jerk at the line was noticed by the crews of both boats. Captain Ellis, of the steamer, went to discover what was the matter, but could see nothing. Captain James Owen, of the Hawgood, however, states that he and his crew saw a great snake-like monster cross the bows of the schooner, and feel certain that it was this that caused the shock. They further aver that the thing turned almost in the direction of the schooner and passed it close by, so that Captain Owen, the man at the wheel, and one of the seamen could see the form distinctly, although the color could not be made out on account of the darkness. The monster, they say, moved through the water with an undulating motion, and that its back, or rather the middle of its body, was sometimes arched out of the water as high as eight feet. This is certainly a first class fresh water serpent story."

Source:
"A Lake Superior Sea Serpent," 1894. *Sandusky [Ohio] Register* (September 10). Reprinted from the *Cleveland Leader*.

1895: Swamp Fire

ONE EVENING IN late September, near Springfield, a farmer noticed a ball of fire sailing out from a swamp. It passed along a field until it got to the road on which the witness was traveling on horseback. The ball was close enough that the farmer thought he could approach it, but as he tried to do that, it shot off at an accelerating rate of speed, sparks flying behind it all the while.

It disappeared into the woods, but any relief he felt was short-lived. Soon afterward, it reappeared right in front of the horse's legs. Once again it streaked silently toward the woods.

The shaken farmer abandoned his trip and returned home. He theorized that he had seen the ghost of a man who some years earlier had committed suicide in the swamp.

Source:
"Thinks He Chased a Ghost," 1895. *Fort Wayne [Indiana] Sentinel* (September 27).

1897: Lights on the Water

A SENSATIONAL GHOST story [comes] . . . from Boughner and Mills lakes, in Mills township, Ogemaw county. Weird lights are seen on the water every night, and the sound of groaning and weeping is very audible. A party set out the other night to investigate. Upon their approach the lights, which seemed to float on the surface of the water, immediately disappeared and the groaning ceased.

"Last summer a young woman was drowned in one of the lakes and three years ago a dead man was found on the shores of the lake. It was discovered he had been murdered for his money.

"These lakes are near the village of Shearer, a number of whose inhabitants are said to have left town because of this ghost scare."

Source:

"Weird Lights," 1897. *Detroit Evening News* (March 29).

The thing swooped down from the sky and a half a dozen farmers immediately surrounded it. While they were examining the strange craft a creature nine and a half feet in height clambered over the side and grew eloquent in an unknown tongue.

1897: Space Giant

REYNOLDS, IT IS CLAIMED, not only had the pleasure of looking at the airship, but several people had the rare good fortune to become acquainted with the navigator. The thing swooped down from the sky and a half a dozen farmers immediately surrounded it. While they were examining the strange craft a creature nine and a half feet in height clambered over the side and grew eloquent in an unknown tongue. One of the farmers hospitably extended his hand, but in the country the visitor comes from this seems to be considered an affront. The big fellow swung one of his legs and the farmer retired in disorder with a broken hip. Then the unknown sprang into his aerial craft, turned on some strange power and the whole thing darted away. There is no still in the vicinity of Reynolds that is known to the revenue authorities and a sharp lookout is being kept for moonshiners."

Source:

Untitled, 1897. *Grand Rapids* [Michigan] *Evening Press* (April 16).

Minnesota

1868: Monster in the Mississippi

On June 29 the *St. Paul Dispatch* reported: "Great excitement was created near the levee last evening by the discovery of the presence of a huge water monster about the size and the shape of a large alligator, in the river. Incredible as it may seem that a tropical animal, if it is one, should have strayed so far from the lagoons of the lower Mississippi, it is vouched for by so many reliable witnesses that we are compelled to believe that the monster was seen as aforesaid.

"He was first discovered just after the Northern Belle came in, swimming in the wake of that boat. He was there seen by two reliable Germans, Wm. Schmid and Henry Protius, who caught a fair and plain sight of the terrible animal. They state that he was not of the largest size of alligators, being only some twenty feet long, but was a good sized animal, nevertheless, and enough to inspire one with fear. He sank once after their attention was called to him, and reappeared again. By this time the boat was quite near the Transfer House, and several others saw the monster. They set up a yell, of fright, we presume, which alarmed the Kraken, and he sank, remaining under some time.

"He was next seen near dusk, by Wm. Carey, who was standing on the railroad track, but saw him rise out of the water a few yards from shore. The monster lay there blowing and uttering a sort of gurgling or bellowing noise for several minutes. Mr. Carey says he was dark colored, had large angular scales, and a head longer than broad, with a thick neck and long jaws, amply armed with teeth. He switched a pretty liberal supply of tail about, which had a ridge on the back several inches high. The body he could not see, the monster keeping that under water.

"Several boys, whose names we could not learn, or we would have secured their statement, saw him at the same time. They were setting [sic] on a log quite a distance out in the boom fishing, when they perceived the monster. They at once fled with a yell of terror, leaving their lines. The monster must have got alarmed at this, for he at once sank and swam under water for some distance, as could be seen by his tail. It made waves like a steamboat.

"Mr. Carey and the boys alarmed the men and others bathing near there, and they came out of the water very quick. Some [of] the Germans living near there got their rifles, but could not get a shot at the monster. He has not been seen to-day, but could turn up again. Some think he is an alligator that has escaped out of one of the numerous menageries that are traveling on the river. Our idea is that it is not an alligator, but some water monster that has strayed from the lower river. It is well for bathers to be on the lookout."

A follow-up story appeared in the press the next day from Donald McDonald, a West St. Paul fisherman and prominent resident. As quoted by Mark Hall (2003), he said: "I was out in my boat last evening [June 29] about 10 o'-clock, examining my trot lines, which are stretched from Raspberry Island to the West St. Paul shore, about a quarter of a mile above the depot. The moon was shining quite brightly, enough to see objects quite distinctly. While quietly pulling my boat along the water, by the line, I noticed the water agitated and heard

The monster lay there blowing and uttering a sort of gurgling or bellowing noise for several minutes. Mr. Carey says he was dark colored, had large angular scales, and a head longer than broad, with a thick neck and long jaws, amply armed with teeth.

something splashing in it like persons swimming and diving. Looking that way, I saw a large object lying on the surface of the water that made me tremble with fear. It appeared to be 30 or 40 feet long, and was two or three feet thick, perhaps more. It[s] color was dark, but whether it was covered by hair or scales I could not tell well, although but a few feet from it. It lifted its head repeatedly out of the water, and looked like a huge serpent, which I believe it is. I see you think it may be an alligator. Sir, I have seen alligators thousands of times. This is not one. I was a sailor many years but never saw anything like this. I watched it for some time. It seemed to be eating fish. After a while, it moved off, and swam exactly like a snake."

Ferryman Pierre Prideau told of his own sighting the previous evening: "I had been over to Prince's Mill on Sunday evening, and was returning in my skiff, near dark, with a man I was ferrying over, name unknown to me, when the man said to me, 'Look there, what is that[?]' I turned round, for my back was to it, and saw a hideous sort of fish or snake, I could not tell which. It seemed to be about 30 or 40 feet long, and several feet thick. It was much like a tremendous water turtle, only it had no shell, and its head stuck out on a long neck, as thick as a barrel, and it had long jaws like a gar fish, only several feet long. Its tail was many feet long, and it wagged to and fro in the water like a fish. It was covered with large scales. It was lying near the shore when we saw it, near the point, partly out of the water. We looked at it a moment or two, when the man I was rowing with got frightened and took the oar and made a great splashing in the water, which alarmed the thing, for it raised its great head, and looked at us a moment so fiercely that we about dropped dead with fright, when it all at once sank out of sight, and we rowed ashore as fast as we could."

Sources:
  Hall, Mark, 2003. "The Beast of Carver's Cave."
    Wonders 8, 4 (December): 108–121.
  "A Water Monster," 1868. *St. Paul* [Minnesota]
    Dispatch (June 29).

1872: "My Body Lies Buried"

THE WINTER OF '72 WAS a severe one in western Minnesota, settlers in that section will well remember. Snow fell to a considerable depth, and those days in Murray county there were no roads, no fences, no anything—but a broad, bleak prairie. On one cold, bleak morning in January, John Weston and two neighbors each yoked his oxen to his sled, and they started in company across the prairie for wood, distance of some fifteen miles. The neighbors, after loading their sleds[,] started for home, leaving Weston to follow, which he did for the first half of the journey, about two miles behind. A blizzard now overtook them, and they wandered aimlessly about on the prairie, hoping to reach their firesides, but in vain. One of the neighbors, however, reached home late at night, and the other was found still alive in the morning. But poor Weston! Where was he? His team had been found frozen stiff and half buried beneath the snow the next day, but Mr. Weston could not be found.

"For weeks the search was kept up by the neighbors for the missing man, but without avail. Nothing was known of him till about the first day of February. Mr. W. W. Cosper, a neighbor living about one mile and a half from the Weston claim, was attending to his cattle at the barn. The morning was mild and foggy. Mr. Cosper saw a man approaching and was both surprised and delighted to see that he was his old friend Weston.

"'Good morning, Mr. Weston. Where in the world have you been keeping yourself? We all thought you were buried under the snow.'

"'Mr. Cosper, I was overtaken by the storm and lost. My body lies buried six feet beneath the snow on the northeast corner of my farm. Farewell!'

"Mr. Cosper stood for a few minutes struck dumb with awe. He went into the house and told his wife and family what had occurred. From there he went with his wonderful story to his neighbors. He told them that he had never seen a ghost in his life until that day. But as sure as his name was Cosper he had seen the spirit of John Weston and had conversed with it.

"All his friends were, of course, surprised. But knowing the truthful reputation and unblemished character of the narrator, his religious zeal, his intelligence and freedom from superstitious beliefs, they could come to but one conclusion: he had been made the victim of some huge practical joke.

"The truth or fallacy of the story could be proven. A number of the neighbors with spades, shovels[,] etc., repaired to the spot designated by Mr. Cosper. The excavation began; fruitless at first, and many began to jeer Mr. Cosper for the deceits of his imagination when W. M. Davis exclaimed:

"'Here he is!' and the lifeless body, cold and stiff, of John Weston was raised from its snowy tomb.

"Most of the actors in this truly wonderful occurance [*sic*] are still living, many of them in the little village of Fulda. Any and all of them were willing to subscribe and affix their affidavits to the truthfulness of the story. Mr. W. W. Cosper is an American, fairly educated, a member of the Methodist church of Fulda, and superintendent of the Fulda Sunday school, a quiet, unassuming, kind-hearted man, loved by his brethren and respected by all."

Source:
  "A Veritable Ghost," 1883. *Grand Traverse*
    [Michigan] *Herald* (May 10).

1872: "Awful Apparition of the Dead Man"

IN MARCH, DURING A BLINDING snowstorm, at Randall Station on the St. Paul and Pacific Road's main line, a terrible disaster occurred, and several persons were killed. One was a section foreman who was known to be particularly dedicated to his job, to the extent that he had turned down offers to work elsewhere, far from the remote, windswept prairie landscape on which his station was located.

A longtime railroad employee named Connelly, with a reputation for seriousness and sobriety, replaced the dead man at the station. At first things went well, but in due course they took a turn for the worse—much worse. Connelly would claim that the ghostly form started appearing at his bedside, making furious motions and seeming to try to say something to him, though no sound emanated from the apparitional throat. Soon the ghost was throwing him violently out of bed, leaving handprints and fingernail marks on his body. Then the harassment began to occur during the daylight, while Connelly was on the job.

He kept silent about all of this for fear of being laughed at, though he did apply—to no avail—for transfer or at least the chance to build a house elsewhere in the section. Then, one day in the fall, according to a St. Paul newspaper:

"One evening, after the labors of the day were closed, and as Mr. Connelly and the men under his charge were eating their supper, the door of the house opened noiselessly, and in the doorway, and in the full gaze of all who were present, stood the awful apparition of the dead man. The shadow remained long enough to make a number of demonstrations of a revengeful character, and then disappeared, apparently melting into space. An awful feeling of terror fell upon that small party of men, and for a time they were speechless, gazing into each other's faces with eyes distended with horror. They were not men easily frightened, and some of them had looked death in the face without flinching. But this unearthly supernatural visitation, which was recognized at once by all as the spirit of the man they had all well known while living, was more than they could stand."

It was only after that encounter that Connelly finally spoke about what he had been through. And it was not over. Tools disappeared mysteriously. The engineer saw the apparition at night on a number of occasions, in each case working on the track with a crowbar and motioning as if giving directions to a gang of fellow section workers. Sometimes it would stand on the tracks, its arms extended as if in warning. "The engineer says at such times the engine acts as if plowing its way through drifting snow," it was reported, "and although he pulls his engine 'wide open' the speed of the engine is sensibly decreased until it reaches a certain point, when it will plunge ahead as though relieved from some obstruction."

Source:
"After Death," 1872. *Titusville* [Pennsylvania] *Morning Herald* (November 11). Reprinted from the *St. Paul* [Minnesota] *Pioneer,* October 25.

1875: "Some Monster as Yet New to Naturalists"

ACCORDING TO AN 1875 ISSUE of the *Faribault* [Minnesota] *Republican,* "Shieldsville has a sensation in the shape of an aquatic monster that inhabits the lake [Lake Mazaska], whether sea serpent, devil-fish, alligator, or sturgeon, or some monster as-yet-new to the naturalists, is to be determined. About two years ago an old man living in Shieldsville reported having seen some large fish or animal swimming in the lake, but as . . . he was wont to get [drunk] little credence was attached to his statement. About three weeks ago, however, Mr. Dennis McAvoy, one of the proprietors of the Shieldsville Mills, while riding with his wife near the shore of the lake, saw an object in the water which resembled a basswood log in color, but which was plainly seen to be making its way through the water. Mr. McAvoy subsequently saw the same object from the window of his house, when it was swimming in the lake, at about thirty rods distance. The portion of the body then above water was about six feet in length, but due allowance being made for the distance, must have been considerably larger. The heart and tail were submerged. The color was of a reddish brown. Mr. McAvoy's miller, and three or four others[,] have also seen the same animal, at different times lately. The turtles have had to bear the blame of the spoliation, but the people now begin to think the 'What-is-it' has taken them. As Barnum will be along soon, there will be a good opportunity for him to prospect in the Shieldsville lake for a curiosity for his museum."

Sources:

Hall, Mark, 2003. "The Beast of Carver's Cave." *Wonders* 8, 4 (December): 108–121.

"What Is It?," 1875. *Faribault* [Minnesota] *Republican* (August 4).

1881: Strange Flying Phenomenon

PEOPLE IN WISCONSIN and Minnesota are talking with awe about the ball of fire followed by a fan-shaped tail of flame which passed over La Crosse [Wisconsin] and Minneapolis, just above the housetops on Monday night [August 1], and wondering what its character and errand were. Head and tail together seemed to be about forty feet long, and the apparition moved with great velocity in a horizontal line and without [within?] easy range of a rifle ball. It passed over La Crosse at 8:30 o'clock, and fifteen minutes later dawned upon the astonished gaze of Minneapolis, 130 miles distant by airline."

Source:

Untitled, 1881. *Fort Wayne* [Indiana] *Daily Gazette* (August 5).

1883 and Before: Shooting at the Serpent

AT GENEVA, "LINCOLN Holmes shot at the serpent with a charge of shot. He has been seen here in the lake [Geneva Lake] at different times for 28 years. Twenty-one years ago E. Ellingson shot at him with a charge of shot. He is about 18 feet long."

Source:

"Local Correspondence: Geneva," 1883. *Freeborn County* [Minnesota] *Standard* (May 3).

1887: "Countless Millions of Bugs"

ST. PAUL WAS LAST NIGHT treated to a phenomenon in the form of clouds of what are variously called Green Bay, Sunday and day bugs. About 10 o'clock a breeze sprang up from the south, and with it came countless millions of bugs, which swarmed at every light, often becoming so thick around many street lamps as to almost obscure the light. Around the electric light masts they seemed to congregate in greater numbers than elsewhere, and in the vicinity of Bridge square, Seven Corners and at the park, at the head of Third street, the streets were literally covered with the pests.

"Along the Wabash street side of the Second National Bank the sidewalk was covered to a depth of over a foot; around the market house, at whatever point an electric light was located, the sidewalk was covered with them. The Merchants' hotel received a liberal share of the bugs, the steps leading to the veranda being completely hid from sight, and it is estimated that more than a wagonload of the bugs could have been taken from the front of the building.

"In Rice park was witnessed a curious sight. The trees near electric lights were covered with bugs, giving the trees the appearance of being moving masses of life, while the electric light wires were strung with insects. It is probable that after striking the wire they were unable to get away on account of the current. At 2 o'clock in the morning the streets in the vicinity of Bridge Square, which had been cleaned, were again covered with them, and they still continued to come."

Source:

"A Shower of Bugs," 1887. *Marion* [Ohio] *Daily Star* (July 15).

1889: Dirty Snow

A PECULIAR PHENOMENON occurred at Aitken the other afternoon. At 4:45 o'clock it became so dark that lights were required in business houses. The air was filled with snow that was as black and dirty as though it had been trampled on. The dirt was very fine, something like emery, and contained particles that had a metallic luster. This dirty snow fell to the depth of half an inch, and the atmosphere at the time presented a peculiar greenish tinge. Solid chunks of ice and sand were reported to have been picked up in various places. Several parties saved small vials of the sand or dirt as a curiosity. At twelve o'clock it was still falling fast, but the snow became as white as usual."

Source:
"Snow and Dirt," 1889. *Freeborn County* [Minnesota] *Standard* (April 18).

1892: Miracles at St. Michael's

FILLMORE COUNTY IS IN the southeastern part of Minnesota. Canton village is in the southwestern part of Fillmore county. In Canton village is a small Catholic chapel called St. Michael's; that chapel has a small window, and in that window there is, or recently was, a phenomenon which many people considered a miracle.

"It was a picture of a woman holding an infant child. Neither the woman nor the child looked at all like the standard pictures of the Madonna and infant Jesus, as the child represented a very chubby and broad jawed baby of the healthy western type, while the woman bore the likeness of a stout Swedish or Norwegian nursegirl. Nevertheless there they were. There was no way of accounting for their being there, and imagination did the rest. It was considered a miracle, there was a rush of visitors, 'miraculous cures' were reported, and now Canton is celebrated.

"The story is very curious. The circular window had been without glass for some time when Father Jones ordered a new one from the only store in the village and had it put in place. The sun was shining full on the glass when the glazier left it, and a few minutes later a little girl on the opposite side of the street cried out, 'Oh, see the pretty picture.'

"Passersby followed the line of her pointing finger and saw the image in the glass. A crowd collected, and Father Jones, the village priest, was sent for. He saw it and so did all the rest, but Catholic priests are as a rule very wary of giving encouragement to such things, and he decided that it was in the glass originally. He had that glass taken out and another put in, but the picture came again, not so boldly outlined as before, but still plain enough to be recognized. Father Jones then went up inside the spire and looked through the glass, but on that side it was perfectly clear and free from lines of any kind.

"Bishop Cotter, of that diocese, was then consulted and pronounced it a delusion, and as thousands of people were crowding the little village to see the wonder[,] he ordered the glass taken out. Meanwhile an invalid suffering from an incurable disease entered the church to pray and went away completely cured, as he alleged. This started a second pilgrimage. A boy with curvature of the spine entered the church and went away straight and sound. Then all the cripples and invalids of that section thronged to Canton, and at last accounts the little village of four or five hundred people was literally overcrowded with the lame, sick, blind and maimed.

"It is, to say the least, a puzzling affair. It is a standing rule in the Catholic church that no event is to be considered miraculous until every possible natural explanation is disproved and until every reasonable doubt destroyed. One explanation, and the one generally accepted by skeptics, is that some talented photographer or manipulator of lenses has been amusing himself by transferring an image from some distant point of vantage, the process being the same as that used in producing the stage 'ghosts,' which excited such wonder some twenty years ago. But what about the 'cures' of which over a score are now reported? It's a very pretty problem as it stands."

Source:
"The Cures at Canton," 1892. *Trenton* [New Jersey] *Times* (December 23).

1897: Martians over Minnesota

GOOD MORNING, HAVE you seen the airship? If not you are unlucky. It was sighted by at least a thousand people last night who kept the Pioneer Press telephone in Minneapolis hot from 8 o'clock till midnight. Flying along in his long, low, rakish craft, the inhabitant from Mars executed more maneuvers in the heavens over toward Minnetonka last night than could be the substance of ten thousand Fourth of July rockets. It is idle to ask why Mr. Mars does not deign to light. The plain fact remains that all's not well in the sky of Minnesota.

"Last night about 8 o'clock R. G. Adams . . . who is in business at Lake street and Nicollet avenue, caught sight of the strange craft flying low in the heavens in the direction of Minnetonka. He called his father and mother to witness the phenomenon and, with the aid of a field glass, was able to see the outlines of the ship. 'It was about 8 o'clock,' said Mr. Adams last night, 'when I saw the thing. Through the glass I saw an object that appeared to be about 18 or 20 feet long. It was shaped like a cigar and in the middle and on top of it was a square light. This light was alternately white, green and red, as the navigator cut through the sky. He was going at a high rate of speed when I saw him. He would dip and shoot down for, say[,] a half-mile with a green light, and then mount with the speed of a rocket showing a white light. As he floated he changed his light to red or green. The light looked as big as a plate through the glass, but to the naked eye appeared about the size of an orange. I could distinctly see the vague outlines of the craft. The lights were clearly defined in the middle and on top of the thing in a square box that looked about as large as a locomotive headlight.'

"Mr. Adams was found at Lake street and Nicollet avenue with an excited crowd of men and boys[,] a few of whom had seen the flying machine. Among those who saw the ship were B. K. Melville and L. S. Davis, to whom the light looked about the size of an orange. They say it changed color as described by Adams and observed it hovering over Minnetonka waters in an uncertain way. The strange apparition did not leave the Minneapolis horizon until nearly midnight, and was seen from several other places in the city. The best test of the veracity of all these witnesses is that their descriptions agree.

"Next the light appeared over Lake Minnetonka. Druggist Newell of Excelsior 'phoned that he had seen it, giving the following description: 'It seems to be coming towards Excelsior from the direction of Hotel St. Louis. We can see a green and a white light. Sometimes it takes a shoot down and we can see the green light plainly. Then it rises slowly up. It dodges around and sometimes seems to go almost in a circle.'"

Source:

"Does He Hail from Mars?," 1897. *St. Paul Pioneer Press* (April 12).

1897: Insect Shower

ONE SUMMER NIGHT, in an otherwise clear sky, a single cumulus cloud appeared over Hutchinson. One of the witnesses, John Zeleny, later produced an account for *Science* of this curious occurrence. The cloud, he wrote, was a third of a mile long and about one-fourth that length in thickness. It was also luminous, though Zeleny, a trained observer, could think of no reason it should be so. The luminosity consisted of a "uniform, steady, vivid, whitish light."

The cloud rose from the eastern horizon and floated till it was above the town. At that point a mass of insects descended from the sky and covered the ground, as many as 50 to 100 per square foot. "These insects proved to be a species of hemiptera and were nonluminous," according to Zeleny. "They had apparently been induced to take wing by the bright object in the sky."

Source:

Zeleny, John, 1932. "Rumbling Clouds and Luminous Clouds." *Science* 75: 80–81.

1898 and After: Rune Stone

MORE THAN A CENTURY after it was unearthed on a farm in west-central Minnesota, the Kensington rune stone—often referred to by its initials, KRS—continues to stir debate and drive speculation about who besides Amerindians may have been in North America before 1492.

The story's origins, as most accounts have it, date to a discovery made on November 8, 1898. A Swedish-American farmer named Olof Ohman, who lived 3 miles northeast of Kensington in Douglas County, was using a winch to remove a poplar or aspen tree on a small hill surrounded by swamp. As the tree was pried from the soil, the roots came into view. They were clasped around a gray, tombstone-shaped rock, 2.5 feet high, 3 to 6 inches thick, 15 inches wide, and weighing 200 pounds, according to subsequent measurements. Ohman's ten-year-old son Edward, observing markings on one edge, pointed them out to his father. On closer examination, the writing was seen to continue from the edge to the broader face, the side that had faced downward when exhumed. The stone had lain so close to the surface that the top of it had nearly protruded from it.

Ohman showed the stone to a neighbor, Nils Flaten. Flaten also saw the tree, which in an affidavit a decade later he remembered as being 8 to 10 inches in diameter. Ohman said he had no idea what the characters on the stone represented. For a short period of time he stored the object near his property. As word of its existence spread locally, so did rumors that the writings were a coded reference to a hidden treasure left by white or Indian robbers. Hopeful seekers dug up various sites that they thought might hold the alleged treasure, but all such efforts proved futile. Soon the local bank was displaying the stone in its front window, bringing curiosity-seekers and fresh speculation. At least some of Kensington's residents, mostly Scandinavian Americans, now recognized the characters as runes, a form of writing used by medieval Northern Europeans and familiar to Kensington-area immigrants from history books and other sources.

Despite the discovery's potential importance—if true, it would radically alter scholarly and popular understanding of the European discovery of America—no newspaper took note of it, and the outside world remained unaware of it until two months after the KRS had come to light. On January 1, 1899, John P. Hedberg, a real-estate agent in Kensington, wrote Swan J. Turnblad, the Swedish American publisher of the Minneapolis-based *Svenska Amerikanska Posten,* to report the recovery of a stone with apparent "old Greek letters" on it. Hedberg enclosed a sketch of the characters, asserting that it was an exact copy. (It was not, a consideration that would figure in subsequent discussions of the KRS's provenance.) Turnblad turned the letter and drawing over to the University of Minnesota, and in short order it came to the attention of Professor Olaus J. Breda, a specialist in Scandinavian languages.

On January 14, the first printed account of the KRS appeared in the university's weekly *Ariel,* which quoted Breda's rough translation. According to Breda, "The language of the inscription presents a queer mixture of Swedish and Norwegian (Danish) words, the spelling of some words being such as to give the word a flavor of the old language" (quoted in Blegen 1968, 20). In his judgment, the writing most likely did not date back to the fourteenth century, as the carved message alleged, but to a modern source.

The generally agreed-upon translation is as follows: "8 Swedes and 22 Norwegians on an exploration journey from Vinland westward. We had our camp by 2 rocky islets one day's journey north of this stone. We were out fishing one day. When we came home we found 10 men red with blood and dead. AVM save us from evil. We have 10 men by the sea to look after our ships, 14 days' journey from this island. Year 1362."

In February and March, newspapers in Minneapolis and Chicago carried articles about the KRS. By this time, Northwestern University Professor George O. Curme, an authority on German philology, had examined both a copy of the inscription and the KRS itself. In his estimation, the language was suspiciously modern and therefore "ungenuine." He had shared the inscription with a Swedish linguistic authority, Professor Adolf Noreen of Uppsala University. Noreen rejected it as spurious. Interviewed by the *Minneapolis Journal* (February 22), Breda stated bluntly that "the whole thing was a hoax, or the result of an effort on the part of some one in part familiar with runic inscriptions to amuse himself" (quoted in Blegen 1968, 20). Even so, Breda urged that two runic experts, Ludvig F. A. Wimmer (Denmark) and Sophus Bugge (Norway), be shown the inscription. They were, and they, too, weighed in with a negative assessment. So did Bugge's Christiana University colleagues Gustav Storm and Oluf Rygh. With Bugge, they wired the *Minneapolis Tribune,* which reported that "the so-called rune stone is a crude fraud, perpetrated by a Swede with the aid of a chisel and a meager knowledge of runic letters and English" (quoted in Wahlgren 1958, 9).

Significantly, in the context of future claims that champions of the KRS would make, three small-town, nonacademic Minnesotans, two of them living not far from Kensington, provided generally accurate translations in letters to newspapers. These translations were printed within days of the appearance of facsimiles of the KRS inscription in press accounts of the discovery. Clearly, knowledge of runic writing was hardly unknown in late nineteenth-century Minnesota.

Within two months of the KRS' becoming known outside west-central Minnesota, there seemed no good reason for anyone to take it seriously. Archaeological hoaxes—of everything from "petrified men" to inscribed plates—were ubiquitous all through the previous century. They reflected a pranksterish, and perhaps uniquely American, sense of humor—and, beyond that, a frontier resentment of the learned—which expressed itself in efforts to fool both the gullible and, better, the professors. If that was the point of the KRS hoax, if that is what it was, it had, to every available appearance, failed. The KRS was returned to Ohman, who stored it in a shed.

That, however, would not be the end of the story. Owing to the promotional genius and near-fanatical devotion of a Norwegian American writer, the KRS would come under renewed scrutiny and survive to figure in theories and controversies into the early twenty-first century. In the summer of 1907, Hjalmar Rued Holand of Ephraim, Wisconsin, visited Kensington and met Ohman. Holand would give various accounts of how this happened. The first was that mere chance had brought him there, since he had not heard of the KRS during the burst of publicity in the winter of 1899. In fact, Holand was a reader of the Chicago newspaper *Skandinaven,* which in late February and early March 1907 had afforded the KRS extensive coverage. Moreover, on March 15, the paper carried a long letter from Holand. It was not about the KRS, but it was about a rune stone, this one a proposed monument to honor Norwegian explorer Leif Ericson. The letter was published only five days after *Skandinaven* ran an extended article on the KRS.

In any event, when Holand met Ohman, Ohman gave him the stone, apparently under the mistaken impression that Holand would give it in turn to the Norwegian Society of Minneapolis. The next year, 1908, Holand wrote his first book on alleged Scandinavian explorations of early America. Written in Norwegian, the book devoted a chapter to the KRS, which it depicted as genuine. Holand insisted there, and elsewhere to the end of his days, that the supposedly simple, uneducated folk of Kensington did not possess sufficient knowledge to fake runic writing; he ignored specific evidence to the contrary, in the form of the already-mentioned instant translations by other small-town Minnesota immigrants. To critics who had argued that the word *opdagelsefard* ("voyage of discovery") existed neither as a word nor as a concept in the Scandinavia of the fourteenth century, Holand countered that the word may have been spoken orally before it was captured in print.

Beyond that, he theorized that Paul Knutson, a baron who served under King Magnus of Norway, had led the expedition into the North American interior. In 1354, according to an undisputed document, the king ordered

Knutson to sail to Greenland to restore the flagging Christian faith there. Nothing more is known; in fact, the sole evidence of the voyage is the order itself. Though he had no proof that any such thing ever happened, Holand speculated that Knutson and his men had sailed from Greenland to Vinland (in Newfoundland, on the northeastern coast of North America), known to the Norse since A.D. 987. From Vinland, he said, they passed through to Hudson Strait and Hudson Bay. An expedition then went onward to what is now west-central Minnesota, where some of its members fell victim to hostile natives.

In 1908, the Norwegian Society of Minneapolis appointed a three-man committee to investigate the KRS and the circumstances of its discovery. Chairman Knut O. Hoegh did most of the work, traveling to Kensington as well as to Elbow Lake, 20 miles from Ohman's farm, where another alleged Scandinavian artifact had surfaced. Little is known about the Elbow Lake stone, except that, according to a historian of the KRS controversy, it "seemingly had an inscription of some 150 to 200 characters within a double circle eight and six inches in diameter" (Blegen 1968, 56–57). Though he was not educated in the subject, Hoegh thought that the markings looked something like runes. The stone has long since disappeared, as have affidavits Hoegh collected the following year from Kensington residents who were at Ohman's farm around the time that the KRS was unearthed.

Holand spoke to the Chicago Historical Society on February 3, 1910, and brought the KRS with him. University of Illinois philologist George T. Flom was there and took issue with Holand on a number of points. Determined to get to the bottom of the affair, Flom launched his own inquiries, which took him to Kensington that spring. Besides interviewing Ohman and other principals, he transcribed the inscription of the actual KRS, now housed in St. Paul at the Minnesota Historical Society. In an address to the Illinois State Historical Society on May 6, he blasted pro-KRS claims. The stone, he said, looked as if it had been "shaped and chiseled in recent times." Furthermore, the narrative had a "thoroughly modern character."

Particularly suspicious were details that would not have figured in any authentic period account, such as the reference to comrades "red with blood and dead" and the improbably precise numbers of the slain. And how, in the midst of imminent threat, would the survivors calmly wait for one of their band to carve an account on stone? Most important, however, the runes themselves gave every indication of having been put there by someone who knew "Dalecarlian runes," from the Dalecarlia (Dales) region of west-central Sweden—precisely where Ohman had grown up (he had emigrated in 1879). Flom added that both Ohman and his close friend Sven Fogelblad (who had died in 1897), a minister in Sweden and a schoolteacher in America, were familiar with runic writing (Blegen 1968; Wahlgren 1958).

A committee of the University of Illinois Philological Society reviewed Flom's conclusions. Its members, seven scholars conversant in Old Norse, identified the language as deriving from a modern Swedish dialect. Dismissing the KRS as a hoax, they speculated that it had taken its inspiration from popular speculation about a Scandinavian presence in pre-Columbian America.

Another committee, however, this one representing the Minnesota Historical Society museum, endorsed the KRS—notwithstanding a body of negative evidence it encountered and either did not publish or downplayed. Heavily influenced by Holand, it was led in name by clergyman Edward C. Mitchell but in practice by geologist/archaeologist Newton H. Winchell, who, in common with the other members, had no expertise in Scandinavian languages. Committee members did, however, consult with authorities in this area. One of them, Chester N. Gould of the University of Chicago, critiqued Holand's methodology in a lengthy memorandum. The use of the anachronistic *opdagelse* alone was enough to discredit the KRS, he said. But the committee largely ignored this and other skeptical commentary. Even so, Winchell, who visited Kensington on three occasions, was hardly uncritical. He acknowledged that on "first inspection" the inscription looked to be of recent vintage. He also noted, though he would not publish the fact, that he had found a stone "exactly" like the one on which the runes had been carved not far from the site—certainly consistent with, though not proof of, a locally executed hoax. Though Holand then and later portrayed Ohman as unsophisticated, Winchell's interviews with the farmer revealed a more complex character who, though his English was halting, was "a more intellectual man than I had supposed." Ohman also impressed Winchell as honest and candid.

Jeweler Samuel Olson, who had known Ohman for nearly three decades, told Winchell that if the KRS was a hoax (as he deemed possible), the hoaxers were Ohman, Fogelblad, and Ohman's neighbor Andrew Anderson. Olson added that when he knew Ohman in another area town, Brandon, Ohman had actually created runic characters. Another neighbor, Joseph Hotvedt, who had been farming near Kensington before Ohman moved there, remarked that Ohman had both spoken of runes and possessed books on the subject. (Subsequent investigators were able to verify this latter allegation.) In later years, prominent Norwegian American Rasmus B. Anderson, author of an 1874 book on possible Scandinavian explorations of America before Columbus, met Andrew Anderson—who now lived in western North Dakota—and discussed the KRS affair with him. Fogelblad and Andrew Anderson were so close that in the summer of 1897 Fogelblad had died in Andrew Anderson's Kensington house. Andrew told Rasmus that Ohman was widely read; moreover, he—Andrew Anderson—and Fogelblad had been "well versed" in runes. Though he did not confess that the three had manufactured the KRS, Rasmus Anderson would write that he "gave me some significant winks." In response, the pro-authenticity Winchell produced letters from both Anderson and Ohman denying any knowledge of a hoax.

Still, Professor Johan A. Holvik of Concordia College in Moorhead, Minnesota, received a letter from a man he regarded as reliable that contained information on Ohman, Fogelblad, and their possible motives in forging the KRS. According to Henry H. Hendrickson, postmaster in Hoffman (just northwest of Kensington), one day in 1890 Ohman boasted in Hendrickson's

hearing that (in the informant's paraphrase) "he Would like to Figure out Something that Would Bother the Brains of the Learned." Hendrickson claimed that Fogelblad had conceived the hoax and Ohman had carried it out. If true, this confirms suspicions that the KRS was planted beneath the tree before 1898 and left there for "discovery" later.

Further confirmation of Hendrickson's allegations came many years later, when Mrs. Arthur Nelson of Seattle visited the Minnesota Historical Society museum on July 13, 1955. She told curator F. Sanford Cutler a curious story, passed on from her grandfather Moses D. Fredenberg, operator of a tool shop in Alexandria (a larger town and commercial center northeast of Kensington). The KRS, Fredenberg had said, was the creation of "two Swedes" who had come to his shop from another town seeking chisels to manufacture the alleged artifact. Mrs. Nelson's cousin supplied further details to Cutler, saying that Fredenberg had spoken of two young Scandinavian Americans who boasted they were "going to have some fun." Fredenberg referred to them as pranksters who would go to any length to fool people. He did not recall their names, but one may reasonably infer that they were Ohman and Fogelblad. In an affidavit signed on May 9, 1965, Elbow Lake farmer Victor Setterlund swore that Fogelblad, a friend of his father's, had confided to the elder Setterlund that he once created a rune stone, then buried it. The younger Setterlund further asserted that a man named Nils Underdahl told him that while visiting the Ohman farm, Underdahl had actually seen Fogelblad carving the stone (Blegen 1968, 78).

As Minnesota historian and KRS critic Theodore C. Blegen conceded, "The statements lack precision, and the episodes to which they refer took place long ago. The principals are all dead" (ibid., 79). Though hardly more than suggestive anecdotes, however, they are consistent with other lines of evidence indicating that the KRS was carved probably a short time—no more than a few years—and planted, by Ohman and Fogelblad, beneath a tree before its "discovery." Subsequent developments would name yet another active participant.

Concerned that the museum committee was ignoring the body of negative evidence, Warren Upham, secretary and librarian of the Minnesota Historical Society, cautioned Winchell that no fewer than ten scholars had declared the KRS a forgery. They did so, he wrote, "based on the language and runes of the inscriptions, order and usage in sentences, which they consider too modern in word forms, inflections, order and usage of sentences, etc" (Wahlgren 1958, 101–102). In February 1910, replying to an inquiry from Upham, Norwegian scholar Olaus Breda reaffirmed his 1899 skeptical appraisal. How odd, he thought, that something that had lain for five centuries in Minnesota's harsh climate should have survived so intact that it could be easily read and translated. Moreover, the tale the runes told was an "utter absurdity," with suspicious and unconvincing detail, including the anachronistic assertion that party members had been found not just dead but "red with blood" besides. The language, Breda wrote, "appeared to be a peculiar jumble of ancient forms, old-fashioned spellings and entire modern Swedish words intermixed with occasional English forms" (Blegen 1968, 166).

In May, nonetheless, the museum committee produced a favorable report (of which Holand was an uncredited coauthor), which it delivered to the historical society's executive council. Philologists such as George Flom, who had written a 43-page monograph scathingly refuting the KRS, criticized the report as too full of errors to warrant comment. But however flawed, the museum committee conclusions would provide support and ammunition for KRS advocates for decades to come.

After this second round of controversy in the first decade of the twentieth century, the KRS slid into the background, kept alive mostly in articles and books by Hjalmar Holand, who took on all critics, continued to maintain that Ohman, Fogelblad, and Anderson were simple, unlearned rural folk incapable of fashioning a bogus rune stone, and placed the KRS in the context of other allegedly pre-Columbian medieval Norse artifacts in America. Still, the controversial artifact lay largely in obscurity until

February 1948, when no less than the Smithsonian Institution in Washington, D.C., put it on exhibition. During its yearlong stay, it sparked a positive reassessment, including a statement by the director of the institution's Bureau of American Ethnology, M. W. Stirling, who told a reporter from the *Washington Times Herald* that that the KRS was "probably the most important archaeological object yet found in North America" (quoted in Wahlgren 1958, 5). In 1951, the Smithsonian released a translation of a monograph, *Two Runic Stones from Greenland and Minnesota,* by Danish scholar William Thalbitzer. Comparing the two, Thalbitzer concluded that the disputed KRS bore so many likenesses to its Greenland counterpart that the former must be as genuine as the latter. (By the 1970s, however, the institution had reversed its opinion in response to outcries from historians. In 2002, in a book written to accompany a traveling Viking exhibit, the Smithsonian stated bluntly that the KRS "is universally considered a hoax by scholars today.") On March 15, 1949, the KRS was unveiled in St. Paul to honor the centennial of Minnesota's admission into the union as a territory of the United States. In 1956, Holand's pro-KRS *Explorations in America before Columbus* was widely praised, even in ordinarily more cautious publications such as the *New York Times Book Review* and the *Christian Science Monitor.*

But in 1958, Erik Wahlgren's *The Kensington Stone: A Mystery Solved* laid out the first book-length exposition of the skeptical case. A professor of Scandinavian languages at the University of California at Los Angeles, Wahlgren theorized that John Hedberg, the Kensington man who in January 1899 had sent a sketch of the runes to a newspaper publisher and thus announced the KRS to the outside world, played a role in the hoax. There were so many telling differences between the sketch on paper and the runes carved on stone, Wahlgren argued, that it seemed likely Hedberg had produced an early draft. He had said that the markings looked like "old Greek" only to allay suspicion, according to Wahlgren. (Later, KRS chronicler Theodore Blegen disputed the idea, on the grounds that it is not clear which sketch was actually Hedberg's original and which "originals" were in reality flawed copies made by others.) Wahlgren also noted that a phrase from the inscription was a nearly direct quotation from a Norwegian folk ballad known and sung by nineteenth-century Minnesota immigrants. The ballad is believed to have originated in Telemark, from which—perhaps significantly—Ohman's neighbor Nils Flaten (not a suspect in the hoax) had immigrated. Wahlgren implied that Ohman had heard the song from Flaten. The mention of the year "1362," he thought, was also telling. Exactly 500 years later, in 1862, the Sioux of Minnesota engaged in a bloody uprising that killed hundreds of immigrant farmers in the southern and western part of the state. These events were still fresh in the memories of Minnesotans when the KRS was found.

A second skeptical treatment, Blegen's *The Kensington Rune Stone: New Light on an Old Riddle* (1968), expanded on Wahlgren's work, endorsing most of his interpretations, differing from others, and presenting a wealth of new material.

In a major development, the Minnesota Historical Society released the transcript of a conversation tape-recorded on August 13, 1967, but kept confidential at the participants' insistence until 1976, when the relevant portions were printed in *Minnesota History*. It not only provided new evidence of a hoax but brought a heretofore-ignored figure into the scheme: Olof Ohman's neighbor and close friend John P. Gran. The revelation came from Gran's son Walter, in an interview with his nephew, also named Walter Gran. Also participating was the older Walter Gran's sister and the younger Walter Gran's mother, Anna Josephine, who was able to clarify some aspects of the testimony. Anna Josephine had heard the same story from her father.

Walter Gran the elder related that in the late 1920s, when his father appeared to be dying (he subsequently recovered and lived for a few more years), he had come down from Canada, where he was then living, to sit at his father's bedside. In the course of that vigil, the supposedly dying man confided that he had helped Ohman carve the stone. Ohman's motivation, he said, was to fool the learned, whom he despised. John Gran was left-handed, which interestingly confirmed (though not in a way he would have liked) Holand's observation that it would have taken two persons, one right-handed, the other left-handed, to carve the stone. Walter Gran the younger recalled that Ohman, whom he knew well, had never confessed personally to him, but that one of Ohman's sons, also named Olof, had more than once dismissed the KRS as "humbug" in conversation with his friend.

In the wake of these allegations, Elden Johnson, Minnesota's state archaeologist and also a University of Minnesota professor of archaeology, disclosed that on three occasions (in 1953, 1964, and 1975–1976) researchers had quietly conducted excavations at Kensington-area sites where, if Viking artifacts existed and the KRS were authentic, they would almost certainly be found. In each case, nothing of relevance or interest was unearthed.

KRS defenders have reemerged in recent years, notwithstanding nearly universal rejection among mainstream scholars of the stone's authenticity. In 1982 and 1995, Robert A. Hall, Jr., professor emeritus of linguistics at Cornell University, published two editions of a book arguing that the stone is "authentic and important." He insisted that the language on the inscription "is purely medieval, with no modernisms and indubitable archaisms." But deep skepticism remains among most authorities. In a 1997 e-mail exchange with a KRS proponent, University of Stockholm runic scholar Jan Böhme succinctly summarized the doubters' case:

"The whole stone is modern. The decimal notation is modern (never otherwise used with runes), the umlauts are modern, the usage of nationality adjectives, rather [than] provincial adjectives is modern, the whole idea of a stone commemorating a journey is modern, the usage of a calendar year without reference to Christ is modern, the dating with just the calendar year without an account of the day in question is modern, the absence of cases is modern, the absence of difference between masculine and feminine gender is modern, the usage of 'man' as an unchanged plural of 'man' is *very* modern, . . . the mentioning of 'red with blood' before the more important message, that men were dead, is hardly conceivable before Romanticism, etc., etc., etc." (Böhme and Kuchinsky 1997).

The most potentially interesting new research has come not from linguistics but from a geologist, Scott Wolter. Wolter, president of the St. Paul–based American Petrographic Services, was hired by a Wisconsin man, Barry Hanson, to determine how long ago the stone had been carved. At a KRS conference held in the Twin Cities in October 2000, Wolter reported that he had found a strong pattern of mica degradation where the runes were carved, of a kind that would have taken considerable time to show up and also require an environment of moist soil. To Wolter and his sponsors, the Runestone Society (with which Wolter had no association before his research), this meant that the KRS was buried soon after it was carved, and it was carved probably centuries ago. Wolter said that further research could narrow the time frame in which it was created, but his work had already shown that in "absolutely no way was it carved in 1898" (Meier 2000). Asked to comment, retired University of Minnesota geology Professor Paul Weiblen noted that around 1908 another university geology professor who had examined the stone had expressed doubt that it could have been carved as recently as ten years earlier. Weiblen added that he shared that impression from his own study of the stone (ibid.).

At the same conference, a Canadian archaeologist and authority on early Norse artifacts rejected the findings on the familiar grounds that no such inscription could have been written in 1362. Birgitta Wallace, a Canadian of Swedish ancestry, remarked, "If you know Swedish, that is the way my grandfather would write, not my ancestors from the 1300s" (ibid.).

In late 2000, the KRS found a prominent new defender, the celebrated Norwegian anthropologist and explorer Thor Heyerdahl. In a book published in his native country, written with Swedish map authority Per Lilliestrom, Heyerdahl argued that the Norse had a much larger presence in North America than generally believed and that Norse colonists in Greenland had left that island to join countrymen who were already in North America. According to Heyerdahl, a seventeenth-century Roman Catholic document, found in Vatican archives, refers to fourteenth-century annals (now lost) that relate, in the words of Icelandic Bishop Gisli Oddsson, that "the inhabitants of Greenland, of their own free will, abandoned the true faith and the Christian religion, having already forsaken all good ways and true virtues, and joined themselves with the folk of America" (quoted in Gibbs 2000).

As Heyerdahl and Lilliestrom reconstructed this allegedly lost history, the Vinlanders' numbers grew suddenly when as many as 10,000 Norwegian crusaders on their way home from the Middle East got swept up on a westward current that took them to Vinland after their ships turned north, toward what they thought would be northern Europe. Eventually, Vinland grew to encompass "the area from Hudson Strait in the north down through the Gulf of St. Lawrence and all the way down to Long Island" (ibid.). Among the items of evidence the two draw on is the KRS. They believe that the Vinlanders eventually married into the native Indian populations and essentially disappeared. Only occasional reports of light-skinned, blue-eyed Indians by later European explorers attested to their existence. The same Birgitta Wallace who had scoffed at the KRS at the Minnesota gathering in 2000 dismissed this new theory as "not much more than a fantasy" (ibid.).

In August 2001, the KRS was in the headlines, at least in Minnesota, once more. A group calling itself the Kensington Runestone Scientific Testing Team announced in a news conference (held, appropriately, in Kensington) that the previous May it had found another rune stone on an island a quarter-mile from the site where Ohman's stone had been dug up. The new stone, like the old, bore the Latin letters "AVM" (a reference to the Virgin Mary in the Scandinavian language) and carried the runic date 1363. On July 11, the stone, weighing more than 2,000 pounds, had been taken to a secret location for further study. An excavation conducted at the site on July 25 turned up no related artifacts. Minnesota archaeologists expressed their usual skepticism, remarking that since the KRS is an almost certain hoax, any ostensible supporting artifact has to be automatically suspect. Even some KRS supporters responded cautiously to the new development (Meier 2001a, 2001b).

Later developments proved that caution and skepticism were warranted. In late October, two of the five individuals responsible for the "discovery" wrote the Minnesota Historical Society to admit creating and planting the artifact. Kari Ellen Gade, now chairwoman of Indiana University's Department of Germanic Studies, and Jana K. Schulman, associate professor of English at Southeastern Louisiana University, said that the incident had grown out of a field trip to Kensington to observe the KRS. The participants, University of Minnesota graduate students participating in a seminar on runic inscriptions, did it for "fun" and as a way to discredit the KRS. "One of the reasons we came forward" after the discovery, Gade told a reporter, "was we saw that people were being asked to make financial contributions to have the rock tested. We didn't feel it would be right to carry this further" (Meier 2001b).

Today, the KRS itself is on display in the Kensington Runestone Museum in Alexandria, Minnesota. It attracts as many as 12,000 visitors a year. Lecturing there on May 12, 2003, a Texas engineer and KRS student insisted, "Everything on the Runestone can be shown in fourteenth-century documents" (quoted in Albert 2003).

The KRS controversy resurfaced in 2004, when Scandinavian linguists reported that the runes were written in a secret runic alphabet that Swedish tradesmen of the latter nineteenth century used. Longtime KRS critic Michael Michlovic, an archaeologist at Minnesota State University Moorhead, stated, "This new evidence is really devastating," but he predicted that the stone's advocates would keep the debate alive (quoted in Meier 2004).

Sources:

Albert, Joe, 2003. "Linguistics Expert Says He's Solved Runestone Mystery." *Echo Press* [Alexandria, Minnesota] (May 16).

Blegen, Theodore C., 1968. *The Kensington Rune Stone: New Light on an Old Riddle.* St. Paul: Minnesota Historical Society.

Böhme, Jan, and Yuri Kuchinsky, 1997. "Kensington Stone & S. Williams: Debunking Went Wrong?" http://www.lysator.liu.se/nordic/archive/1997/08_06/1849-23.html.

"The Case of the Gran Tapes," 1976. *Minnesota History* 45, 4 (Winter): 152–156.

Fridley, Russell W., 1976. "Debate Continues over Kensington Rune Stone." *Minnesota History* 45, 4 (Winter): 149–151.

Gibbs, Walter, 2000. "Did the Vikings Stay? Vatican Files May Offer Clues." *New York Times* (December 19).

Hall, Robert A., Jr., 1995. *The Kensington Rune-Stone: Authentic and Important.* Lake Bluff, IL: Jupiter Press.

Holand, Hjalmar R., 1908. *De Norske Settlementers Historie.* Ephraim, WI: Forfatterens.

———, 1956. *Explorations in America before Columbus.* New York: Twayne.

Meier, Peg, 2000. "Geologist Thinks Kensington Runestone Not a Hoax." *Minneapolis Star Tribune* (October 29).

———, 2001a. "Kensington Runestone Supporters Find Another Carved Rock." *Minneapolis Star Tribune* (August 11).

———, 2001b. "2nd Runestone a Hoax, Say Two Who Claim to Have Carved It." *Minneapolis Star Tribune* (November 6).

———, 2004. "Kensington Runestone Looking More Like a Fake." *Minneapolis Star Tribune* (April 8).

"More on the Rune Stone," 1977. *Minnesota History* (Spring): 195–199.

Nilsestuen, Rolf M., 1992. "Evidence Shows Kensington Runestone Is No Fake." *Minneapolis Star Tribune* (July 12).

Wahlgren, Erik, 1958. *The Kensington Stone: A Mystery Solved.* Madison: University of Wisconsin Press.

Late 1800s and After: Old Oscar

According to an article in a 1949 issue of the *St. Paul Pioneer Press,* a monster known affectionately as "Old Oscar" haunted the glacial Pine Lake 10 miles northwest of Sandstone. The first white people to note its presence were settlers in the late nineteenth century. Before that, however, the Ojibway who lived in the area feared a "strange, evil spirit" in the water.

Louis Uldbjerg claimed to have seen the creature one day in 1938. He and his wife "watched it move ever so slowly against the wind and current. It rode about two feet out of the water, and what we could see of its body was as long as a rowboat." He ran into his house and emerged with a .22 rifle. Though he fired repeatedly on the monster, it continued undisturbed on its straight course until at last it sank on the other end of the lake. Before it did, however, two farmers saw it. William Sprandel said he "thought it was a giant log at first. Then I noticed it against the current."

In 1941, father and son Vincent and Francis Laska were casting off the lake's northern shore when, as Francis told it, "something grabbed my bait and wouldn't budge when I tried to reel it in. It took the hook and swam around in one-hundred-foot circles. You might as well have tried to pull in a sunken locomotive." He tied the line to a tree, and the two tried repeatedly to spear the creature, but it managed to break away and vanished under the water in the center of the lake. Vincent said that he had had a similar experience in 1931.

Pine Lake is the source of the Pine River, presumably affording the creature entrance and egress. Oscar, if it ever existed, has been forgotten today. And if it was ever real, the one printed source on the subject provides no clues even to what the animal is supposed to have looked like. It is described only as a "black hulk."

Sources:

Dunlap, Roy J., 1949. "Monster of Legend Lives in Pine Lake." *St. Paul Sunday Pioneer Press* (May 15).

Mississippi

1843: The Man in the Moon

MR. JAMES D. KING, a respectable citizen of this county [Penola], and a gentleman of undoubted veracity, called at our office on Wednesday [in May], and gave the following details of a most remarkable appearance of the moon for about an hour between seven and nine o'clock on Tuesday night last.

"He stated that, being in the habit of noticing the appearances of the moon at this season of the year, with a view to the common prognostication of whether it would be 'wet or dry,' he observed, while looking at that object, on Tuesday night, that it appeared at first much larger than common, nearly three times the usual size, and more like a circular sheet of fire than like an ordinary moon. In a few moments a very black spot was plainly discernible about the center of the moon's disc, which immediately commenced playing up and down, backward and forward, on the surface, and as the spot approached the upper edge, it grew less, and a faint light distinctly shone through it. This spot became stationary in the center, when the moon divided into three separate fragments, each giving distinct and separate lights, being of irregular forms, and appearing as though the spot had split them off. Then the moon gradually returned to its original appearance, and from that again looked natural.

"What he had already seen was so remarkable that Mr. King, with his family, continued the observation, and but a few moments had elapsed before the black spot again appeared, and again the moon divided—this time into four distinct irregular parts or fragments; and immediately a light resembling the tail of a comet shot from the lower fragment at the southeast corner, apparently some three or four feet downward, while another, much larger, from the upper portion, or northwest corner, struck off directly upward, to the length of between five and six feet. This last now went off and left the corner apparently four feet or more, and turned into the shape of a man standing erect. The figure was of the most perfect imaginable symmetry, of about the medium size and height, clothed in the purest snow-white, and the back alone presenting itself to view. It was visible for a few moments, when gradually the figure changed to the simple light, the lights returned to the fragments, these again came together, and the moon resumed a natural appearance.

"The family of Mr. King, consisting of his wife and a daughter thirteen years of age, with another young lady, all witnessed what is now above related. Mr. King lives about five miles east of this place [Penola]. He protests that, in calling on us to make public these facts, he has no motive but to tell a plain, unvarnished tale of truth, and leaves others to judge of its import—that he was not in the least alarmed or agitated, but as much in his sober senses as he ever was in his life. And in order that no one should have occasion to doubt the sincerity of his narrative, he has authorized us to give his name and to refer to his family as witnesses of the scene with himself.

"He avows that his statements, from which we have deviated, if at all, in no essential particular, is true, and will at all times be maintained to be true on his honor and character as a man, as he will convince any one more fully who may choose to inquire of him further in relation to it."

Source:

"Wonderful Phenomena," 1843. *New York Express.* Reprinted from the *Penola* [Mississippi] *Register.*

1876: Mystery Platform

AN 1876 ISSUE OF THE *Louisville Journal* told of a fantastic archaeological discovery in southwestern Franklin County. If true, it suggested that an advanced race predating the Indians once lived in Mississippi.

The find was described as a "neatly polished," perfectly level platform or floor that lay 3 feet beneath the surface and stretched 108 feet along a north-south line.

"The masonry is said to be equal, if not superior, to any work of modern times," the newspaper claimed. "The land above it is cultivated with oak and pine trees, measuring from two to three feet in diameter. It is evidently of remote antiquity, as the Indians who reside in the neighborhood had no knowledge of its existence to its recent discovery. Nor is there any tradition among them from which we may form any idea of the object of the work, or of the people who were its builders."

Diggers found a canal and well associated with the platform but did not explore them further. For reasons unspecified, they thought a "subterranean passage" might lie underneath.

Nothing further has ever been heard of this alleged find, leading anomaly compiler William R. Corliss to observe that "the structure is so remarkable that, along with the nature of the source, one is tempted to doubt the story."

Source:

Corliss, William R., ed., 2001. *Ancient Structures: Remarkable Pyramids, Forts, Towers, Stone Chambers, Cities, Complexes.* Glen Arm, MD: Sourcebook Project, 25–26.

1877: Mississippi Monster

A FEW WEEKS AGO we published the particulars of a sea monster, as related by a tow-boat captain. The captain of the tow-boat described the monster as resembling an immense snake with a bull-dog head and a pelican bill about ten feet long. It lashed the water into foam with its tail, and spouted oblique streams of water forty feet high. The monster attacked the barge and the captain found a splinter from its tail embedded in the timber, which he said resembled ivory.

"At the time of publishing the above, we felt a little inclined to doubt the monster story, but now, after ourselves interviewed two gentlemen who have seen it, we really think there is a big sea monster in the Mississippi river.

"The gentlemen whom we interviewed say that on the night of the ninth [of December], while floating down the Mississippi on Capt. Ed. Baker's produce boat, when near Island No. 95, they were startled by a loud splash in the water, and as they had heard of the great monster they were much frightened. They saw a dark object not more than eighty yards from the boat, and for the first time saw the huge monster. It was swimming at a pretty fast rate toward the boat, and it made as much noise as the steamer R. E. Lee. It came on, and as it neared the boat it turned to the right, striking the stern oar and knocking it overboard. John Coughlin and Dud Kelley alone remained on the roof, the balance of the crew taking refuge in the cabin. The monster came near enough to enable two gentlemen to get a full view of him. They judged him to be about 65 feet in length. His body was shaped like a snake, his tail forked like a fish, and he had a bill like that of a pelican. His bill was fully six feet in length. He had a long flowing black mane like a horse. When he swam his head was eight feet above the water. It was a grand sight to see him move down the river. Messrs. Coughlin and Kelley tell us that it was impossible to induce the crew to come out that night. The pilot, Mr. McCune George, was finally led out by his wife, she assuring him that the great monster had departed.

"Captain Baker's boat is now moored at our landing at the foot of Main street. All of his crew, except one man, has abandoned her, and Captain Baker says it is impossible to get a crew, as the men think the monster is still following them."

Source:

    "That River Monster," 1878. *Stevens Point* [Wisconsin] *Daily Journal* (January 12).

1878: Rumors of a Monster

A MISSISSIPPI NEWSPAPER took note in August of "rumors of a monster snake" seen "on Yalobusha" just north of Yazoo City. Estimated to be 18 feet long and 8 inches wide, the snake, when coiled up, was said to be "as large as the largest wagon wheel."

Source:

    Untitled, 1878. *Atlanta Constitution* (August 17). Reprinted from *Yazoo Valley* [Mississippi] *Flag*.

1889: Giant Snake, Gold Coin

THE STORY COMES from Leighton, Miss., that John Davis[,] while hunting the other day, shot and killed a snake of the moccasin variety that measured 16 feet 5 inches in length and 21 inches in circumference. It was such a monster that Mr. Davis concluded to skin and stuff it. While performing the operation he found a Mexican gold coin secreted among its vitals that was used in 1624."

Source:

    "Large Sized Snake Story," 1889. *Reno* [Nevada] *Weekly Gazette and Stockman* (August 29).

1896 and Before: Otherworldly Goat

THREE MILES WEST OF New Albany the Rocky Ford road crosses a creek which was originally named Big creek, but was more appropriately named Hell creek by persons who have been compelled to cross the adjacent bottoms in recent years. Just beyond this is another little run called Mud creek, which stream is grown up with thicket and heavy underbrush, and on cloudy nights the blackness that surrounds the traveler could be sliced into chunks and sold for ink. The bottom or lowlands adjacent to the stream is [*sic*] of unusual width for one so small, and at the best is exceedingly uninviting.

"Some years ago a gentleman passing through the bottom at night was almost thrown by his horse shying to one side, and when he looked ahead was confronted by a monster goat of white color rearing upon his hind feet as if to annihilate the animal and rider. One look was sufficient, and, making a sudden turn, he galloped out of the bottom at the risk of his life, swearing that he would drink no more New Albany blind tiger liquor. Not wishing to put himself up as a target for the jeers of the public, he held his counsel and heard or saw nothing more of the apparition for some time.

"About a year later his goatship was again on the warpath and confronted a gentleman of known sobriety, who, not daunted, urged his animal forward despite the warlike attitude of the ghostly visitor. The goat kept in the middle of the road and when the small bridge was reached disappeared as mysteriously as he had appeared.

"The gentleman related his experience, which became noised abroad and gave courage to the man who first sighted the vapory animal to relate his experience, and the two coincided so well that the people began to give them credit for having seen something to disturb their peace of mind. The story was given enough credence to cause an uneasy feeling to enter the mind of the traveler who crossed the bottom at night and cause a chill to ramble up and down his spinal

The night was intensely dark, and a slight rain was falling. As he drove through the impenetrable gloom, trusting to the instinct of the mules that drew the rank which he was astride to find the road, the misty and uncertain form of the giant goat suddenly appeared in the road ahead of him.

column as he passed the spot where the ghost had been seen.

"Last year Mr. —, who is not a believer in things uncanny at all and has a supreme contempt for a man who has seen spooks, had been beyond the creek harvesting hay and was detained until after nightfall on his return home. The night was intensely dark, and a slight rain was falling. As he drove through the impenetrable gloom, trusting to the instinct of the mules that drew the rank which he was astride to find the road, the misty and uncertain form of the giant goat suddenly appeared in the road ahead of him. The mules reared and plunged[,] very nearly upsetting the rake. Leaping to the ground, he grasped the bits and was gratified to see the phantom recede as the team moved forward. The mules, trembling in every nerve, car-

ried him along, and when the bridge was reached, he disappeared as on former occasions, much to the relief of the gentleman who did not believe in spirits and unnatural apparitions.

"Since that time a number of thoroughly reliable witnesses have been placed in positions to vouch for the truthfulness of the existence of the phantom goat. Persons who travel that road to and from town make their arrangements to pass that spot before nightfall, and very few have the temerity to invade the territory of his goatship after darkness has fallen."

Source:

"Scared by a Phantom Goat," 1896. *Indiana* [Pennsylvania] *County Gazette* (July 1). Reprinted from the *New Albany* [Mississippi] *Gazette*.

Missouri

1873: "Strange Spiritual Manifestations"

SEVERAL MONTHS AGO we copied articles from the Benton County Democrat relating to strange spiritual manifestations attending a young lady, residing near Lincoln, in Benton county. Many of our readers doubtless judged them to be purely sensational, and without foundation in fact. A gentleman who resides near Lincoln, informed us this morning, that the ghostly manifestations still attend the young lady, and the mystery with which it is surrounded is no nearer solved than at first. A change, however, has come over the ghost in making its presence known. Formerly there was loud knocking on the window, and afterward a noise as a dog lapping water. Now there is a shrill whistling, and bonnets, hats, etc., are thrown promiscuously from one side of the room to the other. Lately, the mother attempted to lie down on the bed with the daughter, who is thus strangely affected, but found it an impossibility.

"Our informant gave us one occurrence as stated to him by a gentleman whose veracity is unquestionable: While the young lady was lying on the bed in an unconscious state, with her face uncovered, she slowly raised her arm and pointed directly to each occupant of the room and then to the door. When all had left, the bed clothing was thrown promiscuously about on the bed, and the bedstead raised [sic] up and down and moved to other parts of the room.

"It is said that a love affair is connected in some manner with this very strange case."

Source:

"Benton County Ghost," 1873. *Sedalia* [Missouri] *Daily Democrat* (May 8).

1873: Rocks and Gravel

IN NOVEMBER, AT HER residence 3 miles southwest of Canton, Mrs. Walker Lillard and (in the words of the newspaper report) "a little negro boy" were repairing a brick walkway leading from the gate to the house. At one point, as Mrs. Lillard bent over to look at their work, she was startled to see sand and pebbles falling all around her. She immediately accused the boy of playing at trick on her. He denied it with such evident sincerity that finally she believed him.

As it turned out, this was only the first mysterious event to befall the house. On subsequent nights, other objects fell—apparently—out of the sky. It was reported, "Rocks, from the size of a partridge's egg to that of [a] man's fist, and even larger, nightly fell upon the roof and sides of the house, sometimes striking with such violence that the house was made to tremble." However closely the vicinity was observed, no one saw a human perpetrator. The stones fell even when the little boy was locked inside the building.

"Mr. Walker Lillard is too well known in this community for us to add that he is a gentleman of high standing and perfectly reliable in all his statements," a correspondent attested. "He takes the matter of his strange proceedings around his premises very coolly, and, though he has done all he can to unravel the mystery, and failed, he is determined not to be run away from home . . . by flying rocks, showers of sand, or any other strange manifestations."

Source:

"Ghostly Manifestations," 1873. *Sedalia* [Missouri] *Daily Democrat* (December 12).

1874: Talking Specter on a White Horse

WE DO NOT KNOW WHAT is the matter in this county but one thing is certain, that if all reports be true[,] ghosts are getting thick here. The last one has made its appearance in Nine Mile Prairie Township. A certain young man down there[,] so the story goes, had been complaining that a woman, a perfect stranger to him[,] had been troubling him very greatly for some time past by suddenly appearing at different places and advising him in regard to things which he thought did not concern her. She would suddenly appear, volunteer her advice and as suddenly and mysteriously disappear after telling him when she would again visit him. This sort of thing became so frequent that he grew tired of it. A talking ghost was more than he could stand.

"He spoke of it one day in Williamsburg and his hearers made sport of him. To satisfy them that it was true he told them that the apparition had notified him that he might expect a visit that day, and invited the incredulous to ride a short distance north of town with him. His brother accepted the invitation and they started to meet the strange visitor. A few miles north of the town the spectre suddenly appeared between them, this time riding a white horse. He, she or it, as the case may be, rode with them a mile or two, talking and advising them[,] and suddenly disappeared—vanished—he, she or it and the horse going they knew not where. The young man is daily expecting another visit."

Source:

"Ghosts in Callaway County," 1874. *Sedalia* [Missouri] *Daily Democrat* (October 4). Reprinted from the *Fulton* [Missouri] *Telegraph*.

1877: The Dragon and the Bull

THOUGH THE WORD "DRAGON" was never used in the account, there can be no question that it was a dragon starring in a wild story told as true in a St. Louis newspaper. The alleged event occurred on a Friday afternoon in early October a few miles southeast of Cahokia.

Jabez Smith, "a gentleman well known in his locality as an upright and thorough[ly] reliable citizen," had sent his son to a pasture to drive home a bull. Though not a large animal, the bull made up for its small size in sheer ferociousness, and the boy knew better than to approach it directly. He entered the pasture at a safely distant spot, but then he noticed a strange sound emanating from adjacent woods. The bull, hearing the sound as well, looked in its direction. Then, so the story went:

"From the edge of the woodland there appeared a hideous head upon a swaying neck at least twenty feet in length. The head was that of a wolf or dog, save that there was a prolongation into a huge bill or horny jaws. This bill the monster opened at intervals, displaying a row of immense fangs upon each division, while as he opened it upon each occasion was emitted a hissing noise loud enough to be heard for a great distance. From the back of his head and down the neck depended a mane of coarse redish [*sic*] hair. The monster retained its position for a moment or two, swaying its head gently back and forth, when its eyes fell upon the bull; then it at once showed signs of great excitement. It snorted fiercely, the hissing sound became almost continuous, and it would repeatedly open its immense jaws and snap them together with a sound like the report of a rifle."

Within seconds, its head rose even higher, and in "swift, billowy undulations" it came over the fence, revealing the creature's whole body. Its long neck was connected to a smaller body supported by four short legs with great claws. At the end of the body was a tail as long as the neck, and at the end of the tail was a "huge barb, hard, apparently, as iron, and having the bright red color of a boiled lobster. The neck and entire body were sheltered in scales of a dark blue color, and as large as dinner plates. Most remarkable of all was a pair of huge membranous wings, which were folded along the body on either side. The appearance of the great reptile was indescribably fearful and repulsive."

All the same, the bull stood firm, its head lowered, bellowing occasionally, as the creature circled it. Again, the monster lifted its head high into the air, then plunged toward its victim. Observing from a concealed position, the boy watched as an awful fight ensued between the bull and the dragon. A cloud of dust rose, obscuring his view, and the ground shook beneath his feet as bellowing and roaring filled his ears.

"There could be heard the snap of the monster's jaws and the rattle of the bull's horns upon its mailed sides. Flashing here and there through the dingy nimbus, could be seen the blood-dart upon the reptile's tail as it sought to transfix its active opponent. It was a panorama of desperate battle, a volume of sound, of fierce encounter. The tide of the battle shifted insensibly to the vicinity of a huge oak stump which was near the center of the field. Then the boy saw the tail of the strange monster dart the great barb downward with the speed of a thunderbolt. There was a crack like the report of a cannon. The barb had again missed the bull, and this time encountered a harder substance. The boy, peering through the fence, gave a wild yell of satisfaction. The barb had buried itself in the stump! Then suddenly rearing its head again it unfolded for the first time the great membranous wings folded along its sides and rose in the air like a gigantic bat. With a wild, hoarse cry it darted upward to the height of hundreds of feet and took a southwesterly direction toward the Mississippi. A few moments later, from the direction of the distant river, came the sound of a tremendous splash and swash of waters, as though some heavy body had fallen into the river from a great height. The baffled monster had reached again his native element.

"Battered and bloody, the bull still stood, glaring angrily in the direction of his attacker's escape. The ground was littered with scales and hair over half an acre. The oak stump was split in two from the creature's furious effort to free itself."

Source:

"Fish, Flesh and Fowl," 1877. *Burlington* [Iowa] *Hawk Eye* (October 25). Reprinted from the *St. Louis Republican*, October 7.

1878: Shower of Fish, Lizards, and Leeches

AT 11:30 O'CLOCK ON Monday night [April 22] the storm burst in all its fury at the Tuscan mills on the St. Charles rock road [in St. Louis], and for almost an hour rained fish, crawfish, lizards and leeches, and when day dawned and the good denizens of that part of the city looked out over the surrounding neighborhood[,] it was literally alive with fish, both small and large, ranging from the wee innocent fresh hatched one to the five pound cat with a head as large [as] a saucer, and as cruel as a steel-trap. They were not confined to one species alone, but consisted of catfish, bass, salmon, croppies, sunfish and a fish similar to the Louisiana goggle-eyes, and are only found in the bayous and lagoons of the South. They covered a tract of ground extending from the St. Charles rock road pond on the east to the Tuscan mills on the west, and from the German Lutheran church and school on the south to the St. Charles rock road on the north, about thirty acres in all. During the forenoon the citizens worked incessantly with baskets, buckets, wheelbarrows and carts, and when a reporter visited the neighborhood at a late hour yesterday afternoon there were still tons of them. . . .

"The first man found who knew anything positive about the miraculous shower was James Kavanagh, the conductor of street car No. 1, of the Citizens' railway. He was the last man in, and after turning his car into the stable, he started home. The rain poured down in torrents, but he was already wet to the skin, and pulling his hat a little down he strode on. When opposite the German Lutheran church and school he was struck on the head by some heavy body that partially stunned him, and in another instant he was pelted all over at once by a host of slimy, wiggling creatures. He at first thought that he was afflicted with the jim jams, but catching a large catfish in the neighborhood of the waistband of his indispensables, he grappled it and carried it home through the live shower, and then ascertained that it had been raining.

"Capt. Powers, an old river man, who took the first steamboat up Red river, was not caught in the shower, but saw the ground covered in front of his residence at daybreak yesterday morning. The captain had heard of such occurrences before but had never seen them with his own eyes, and now expresses himself as both ready and willing to be called home. Mr. Weber, a butcher on the St. Charles rock road, gathered up five barrels of the fish, and Mrs. Flemming, who resides just east of the Tuscan Mills, secured a cart load.

"There are always some doubting Thomases around, and as the St. Charles rock road neighborhood is no exception to the rest of the earth, there are a few people who make no bones of saying that owing to the heavy rainfall the pond west of the car stables overflowed and the fish swam out with the current, and as the volume of water spread out and receded, the fish, crawfish, lizards and leeches were left on the sward; but of course that idea is preposterous and will not be entertained for a moment by a sane person."

Source:
"A Strange Phenomenon," 1878. *Sedalia* [Missouri] *Daily Democrat* (April 26).

1883: Exploding Ball of Fire

A FEW MINUTES AFTER 12 o'clock, a few days ago, during the thunder storm, S. Haggard, who lives in what is known as the Irwin House, on the Paxton land, north of town, was sitting in the east front room of the house looking west, a door leading into the west front room being open, and also a window fronting to the west. Suddenly he noticed a small ball of fire, apparently about the size of an ordinary base ball, coming toward the house, and before he had time to stir[,] it had entered the house through the open window, and apparently in the center of the room exploded with a deafening report,

scattering a shower of sparks in every direction, which, strange to relate, went out at once without setting fire to anything or leaving a trace behind."

Source:
"A Ball of Fire in a Missouri House," 1883. *Trenton* [New Jersey] *Times* (June 29). Reprinted from the *Joplin* [Missouri] *News*.

1887: "A White and Airy Object"

AT A PLACE ON THE Turnpike road, between Cape Girardeau and Jackson, is what is familiarly known as Spooks' Hollow. The place is situated four miles from the Cape and is awfully dismal looking where the road curves gracefully around a high bluff.

"Two drummers, representing a single leading wholesale house of St. Louis, were recently making the drive from Jackson to the Cape, when their attention was suddenly attracted at the Spooks' Hollow by a white and airy object which arose in its peculiar form so as to be plainly visible and then maneuvered in every imaginable manner, finally taking a zigzag wayward journey through the low dismal-looking surroundings, disappearing suddenly into the mysterious region from whence it came.

"More than one incident of dreadful experience has been related of this gloomy abode."

Source:
"Drummers See a Specter," 1887. *St. Louis Globe-Democrat* (October 6).

1887: "Ghostly Cable Train"

IN 1887 THERE WAS a story afloat that at 12 o'clock each night a ghostly cable train glided down the incline between Walnut and Main streets [in Kansas City] and disappeared into space. In the grip car, guiding the train, was the ghost of a gripman who had died a short time before, after having been insane for some time, the result of grief over the fact that this train had run down and killed a pedestrian. Crowds congregated at the junction nightly to see the strange sight. For the most part they went away disappointed, although there was [*sic*] plenty who declared they had seen 'it.'"

Source:
"Kansas City Ghosts," 1895. *Bangor* [Maine] *Daily Whig and Courier* (May 27).

1890: "Pink Eyes and Yellow Wings"

JEFFERSON RAWBONE, A well-to-do farmer living near St. Louis, drove into that town with a load of turnips last week, sold his produce, drank thirteen milk punches and started for home. When a few miles from Green Stuff, Mo., Mr. Rawbone witnessed a remarkable phenomenon. The country seemed to be alive with small white snakes with pink eyes and yellow wings. This was Mr. Rawbone's second attack. He thinks it is a warning of some sort."

Source:
"Fresh Snake Stories," 1890. *Freeborn County* [Minnesota] *Standard* (April 17).

1890: Paralyzed by Electrical Terror

A MARVELOUS CASE of electrical paralysis is reported from the Hosmer road, near Lockport, and it takes the edge off most of the fictitious stories of lightning's freaks ever told. The house of Jasper Brown was the scene of the phenomenon. The electricity was so strong that the air was luminous and of a distinct blue color. Though there was no lamp burning in the house every person and every article could be outlined in the cerulean atmosphere as if through a haze; great balls of [St. Elmo's] fire, such as is sometimes seen on the rigging of ships, played and danced about the furniture, now perching upon the back of chairs, now sliding easily along the mantel top, now skating up and down the chandeliers.

"The family were so frightened that they could not speak. The electricity in the atmosphere made their hair stand on end like the quills of a fretful porcupine, and the strangest thing of all was that they could not move. The whole family had lost the power of locomotion. Every member was temporarily paralyzed by the electricity. Few mortals have ever experienced such a night of terror. Their minds were filled with fear and apprehension lest their power of locomotion was gone forever and they would be permanent cripples. But at last, after a duration of about four or five hours, the storm began to subside, the use of their limbs gradually returned to them, and when the storm was over they were able to move about as freely as ever."

Source:
"Held Prisoners by Electricity," 1890. *Marion* [Ohio] *Daily Star* (July 7). Reprinted from the *St. Louis Globe-Democrat*.

1890: Fatal Electrical Anomaly

A SINGULAR ELECTRICAL phenomenon is reported from Americus.... From a small cloud in an otherwise clear sky a single vivid flash of lightning descended, striking two men standing against a barbed wire fence, instantly killing one and paralyzing the other from the waist downward. A number of horses were hitched to the fence, and one was killed and several injured. The shock was felt all over the village, several persons being stunned, while a boy in a house some distance from the point where the bolt struck, was knocked down."

Source:
"Peculiar Electrical Freak," 1890. *Marion* [Ohio] *Daily Star* (July 29).

1891: Eaten by a Giant Snake?

TRAVELING THROUGH THE eastern part of De Kalb County in early October, a salesman for a patent-medicine firm passed along Grindstone Creek. The day was warm, and he was tired, and he had fallen into a drowsy state when suddenly the team of horses guiding his wagon snorted and stopped in their tracks. Staring down the road, he saw nothing out of the ordinary except a log lying across the path some 30 or 40 yards away. He urged the horses to resume their trek, but they refused to move.

At first puzzled by this unusual behavior, he eventually began to wonder if road agents were not hiding behind the log, ready to jump out at his approach and rob him. The salesman stepped down out of the wagon and turned his horses around, then secured them. He took a big revolver from under the seat and walked cautiously toward the log, fully expecting to see a gun pointed at him any moment. In due course, however, he got close enough to peer over the log. No one was waiting there. His relief, unfortunately, proved short-lived because he then noticed something odd about the log. It didn't look right. So he put his hand on it, and it was then that he learned the "log" was a monstrous snake.

Flying back some twenty paces, he aimed the revolver and shot at the creature. The bullet bounced off it and ricocheted through the leaves. Letting out a hissing roar, the snake raced through the underbrush. Its roar could be heard for a considerable distance. In the words of a local newspaper, "In its maddened speed it broke off bushes as thick as a man's wrist, and as it dashed through the creek the water went churning and boiling into the air as if sent up by a cyclone."

When the witness had recovered his nerves, he investigated the site. From appearances, he deduced, the snake was at least 25 feet long, its body the diameter of a flour barrel. Its color was ashen, with some slight greenish hue. Most incredible of all, though, was a discovery he made near where the snake's head had lain. It consisted of a short pencil and a scrap of paper on which some illegible writing could be seen. The salesman deduced, supposedly, that a human victim in the course of being swallowed tried to take down his final thoughts.

Source:
"A Missouri Snake Story," 1891. *Bismarck* [North Dakota] *Daily Tribune* (October 7). Reprinted from the *Maysville* [Missouri] *Herald*.

1897: Airship Landing

ON THE NIGHT OF the 19th . . . near midnight, F. R. Pryor, James Freels and Wm. Barnes saw the much-talked[-]of airship, near said F. R. Pryor's residence on Shoal Creek. The ship came from the direction of Chillicothe, and after alighting near Mr. Pryor's residence[,] where it remained for a few minutes[,] seemingly for the purpose of generating electricity[,] it again took flight, going in a northwest direction. From the best view the parties could get of the ship it was a monster; the sails or wings being spread as those of a South American Condor, and the head light as dazzling to the eyes as the rays of the noon day sun, while the green and red lights shown [*sic*] as the lights at a church festival. The ship seemed to move through the air with as much ease and rapidity as the fleet winged carrier pigeon."

Source:
"The Air Ship Seen," 1897. *Hamilton* [Missouri] *Farmer's Advocate* (April 28). Reprinted from the *Cogwill* [Missouri] *Chief.*

1911: Spotted Snake with Wings

A WHIRRING SOUND ATTRACTED the notice of Mrs. John Bishop and her children in the yard of their St. Charles house one day in September. They looked up expecting to see an airplane (still a novel sight) in the distance, only to see—to their astonishment and terror—a flying snake passing overhead. It was 3 feet long and had spots, its wings beating so fast that they were barely visible.

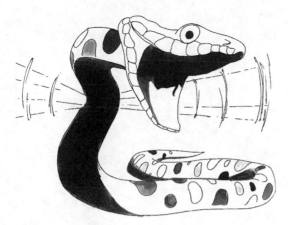

After flying over their yard, the snake turned around and approached them. Mrs. Bishop grabbed the children and fled inside the house, locking all the doors. From this vantage they watched the creature fly and maneuver about the yard for half an hour. Eventually, it flew off to the east, in the direction of Alton, Illinois.

Source:

"Weird and Unusual Happenings Galore in the Animal World during Past Summer," 1911. *Washington Post* (October 29). Reprinted from the *Chicago Inter-Ocean*.

They looked up expecting to see an airplane (still a novel sight) in the distance, only to see—to their astonishment and terror—a flying snake passing overhead. It was 3 feet long and had spots, its wings beating so fast that they were barely visible.

Montana

1864: "A Messenger from the Celestial Regions"

ACCORDING TO A ST. Louis newspaper account a year later, James Lumley, described as an "old Rocky Mountain trapper," claimed that in the middle of September 1864, while trapping near Cadotte Pass, in the mountains 75 or 100 miles above the great falls of the upper Missouri River, he spotted a brilliant object in the early evening sky. He watched it for five seconds as it moved rapidly toward the east, then suddenly burst into particles like a sky rocket. Soon afterward, an immense explosion shook the earth, followed by a rumbling sound such as a tornado roaring through a forest might cause. When that sound stopped, Lumley noticed a sulfurous odor in the air.

He went to sleep. On waking the next morning, he discovered that 2 miles from his camp, something had cut a swath through the trees. According to the press account in the *St. Louis Democrat,* the path was "several rods wide—giant trees uprooted or broken off near the ground—the tops of hills shaved off, and the earth plowed up in many places. Great and widespread havoc was everywhere visible."

On waking the next morning, he discovered that 2 miles from his camp, something had cut a swath through the trees.

Searching for the cause of the devastation, Lumley ascended until he came upon an enormous rock driven into the mountainside. This was remarkable enough, but what he found when he examined the rock took the incident into the realm of the fantastic. The newspaper wrote that the rock "had been divided into compartments. . . . In various places, it was carved with curious hieroglyphics." There were also "fragments of a substance resembling glass, and here and there dark stains, as though caused by a liquid." Lumley said he was certain that the "hieroglyphics" were not natural squiggles but the work of intelligent beings, who were also responsible for the object that had crashed into the mountain. He speculated that it was but a small portion of a much larger structure that "animated beings" had built.

The *Democrat* concluded its account with these words: "Strange as the story appears, Mr. Lumley relates it with so much sincerity that we are forced to accept it as true."

Sources:
Ahrens, Dan, 1999. "UFO Crash in 1865 [*sic*]—Craft Wreckage Still There?" http://www.rense.com/ufo5/crashwreck.htm.
"A Strange Story: A Messenger from the Celestial Regions," 1865. *St. Louis Democrat* (October 19).
Young, Kenny, 1997. "1864 UFO Crash?" http://home.fuse.net/ufo/LUMLEY.htm.

1867: Unknown Animal

DANIEL ROBERTS, AN EXPRESS messenger on the Central Pacific Railroad, states that in Montana wahhoos are not uncommon. . . . A man named Thomas, well known to Mr. Roberts, . . . killed a wahhoo on the Montana road, near the Camas Creek station. It weighed about seventy pounds and fought wickedly after being wounded, until it was finally dispatched. Not long after that, Mr. Roberts, in company with Mr. Bassett, a superintendent in the employ of the Western Union telegraph company, saw two wahhoos together near a place called Summit Station. He states the animal is well-known all over Montana. It is very shy, nocturnal in its habits, and abides in the wilderness away from the habitations of man. It is owing to these reasons that so little is yet known of the wahhoo."

Source:
"The Wild Wahhoo!," 1879. *Reno* [Nevada] *Weekly Gazette* (September 18).

1886: A Fatal Case of Venomous Snake

FOR SIX MONTHS A young man named Ferguson, residing at Great Falls, Mt., had been in failing health despite the care of the physicians, none of whom could fathom the cause of his decline. Among many odd symptoms was that of a peculiar choking sensation, which was not understood until a snake thrust its head out of the young man's mouth. The sick man called his sister, and when next the hissing head appeared she seized it, and with a quick pull landed the venomous reptile at her feet. Her action killed her brother. The tail of the snake had grown into the young man's body, and in tearing it away a blood vessel was broken, and the young man bled to death."

Source:
"The Latest Snake Story," 1886. *Mitchell* [Dakota Territory; now South Dakota] *Daily Republican* (April 21).

1901: "Huge Star Fish" Overhead

A FLYING GHOST IS the latest wonder to agitate the good people of the hill city. Last night while the sky was clear a whirring, whizzing, swishing sound in the heavens attracted the attention of the meek and lowly who 'rubbered' and beheld a mighty dark object sailing or soaring high in the air, its course being northerly. It was estimated that the thing was about 36 feet from tip to tip, but its shape was not that of a bird or bat or butterfly. It resembled in general outline a huge star fish. It was calculated to be about 1,000 feet high and it moved slowly. The strange sound grew less in intensity until only a mild whisper likened unto a zephyr toying with the tender leaves of a quaking asp told of the passing of the mystery. Men stood out of doors until long after midnight watching the heavens hoping in vain to catch another glimpse of the strange object."

Source:

Untitled, 1901. *Anaconda* [Montana] *Standard* (April 11).

1915: "A Huge Black Mass" Overhead

AN AIRSHIP FLEW OVER Bigfork last Tuesday morning [February 9] about five o'clock. Charles Russell, at that hour, heard a buzzing noise over head and looking up saw a huge black mass just over the garage building and not very high in the air. It was headed in a southwesterly direction and went out over the lake. Charlie said it sounded like the buzz of a huge automobile, but the whir through the air sounded more like the flight of a million ducks. While it was moonlight at that hour, yet the monster had not gone a great distance before it faded from sight."

Source:

"Mystery Airship over Bigfork," 1915. *Flathead Courier* [Polson, Montana] (February 11).

Nebraska

1884: "Belonging to Some Other Planet"

NEAR BENKELMAN, IN remote Dundy County, in the south-central part of Nebraska, cowboys in the employ of rancher John Ellis were engaged in a cattle round-up on June 6 when they heard a loud, strange whizzing sound over their heads. A blazing object flashed through the sky and plunged to the ground on the other side of a bank. An area newspaper, *Nebraska Nugget* (published in Holdrege), reported that "fragments of cog-wheels, and other pieces of machinery lying on the ground, scattered in the path made by the aerial visitor, glowing with heat so intense as to scorch the grass for a long distance around each fragment and make it impossible for one to approach it."

The light and heat were so intense that one of the observers, Alf Williamson, fell off his horse, blinded, burned, and driven into a senseless state. The cowboys quickly retreated and watched from a distance. According to the *Nebraska Nugget* account, "The sand was fused to an unknown depth over a space of 200 feet wide by 80 feet long, and the melted stuff was still bubbling and hissing." Soon the men left to seek medical help for their fallen comrade.

Again according to the newspaper, news of the bizarre visitor spread quickly. That night, curiosity-seekers saw that the object was still aglow. The next morning, district brand inspector E. W. Rawlins led a party to the site. By now, the *Nugget* reported, "the smaller portions of the vast machinery had cooled so that they could be approached, but not handled. One piece that looked like the blade of a propeller screw, of a metal in appearance like brass, about 16 inches wide, three inches thick and three

and a half feet long, was picked up on a spade. It would not weigh more than five pounds, but appeared as strong and compact as any metal. A fragment of a wheel with a milled rim, apparently having a diameter of seven or eight feet, was also picked up. It seemed to be of the same material and had the same remarkable lightness. The aerolite, or whatever it is, seems to be about 50 or 60 feet long, cylindrical, and about 10 or 12 feet in diameter."

A leading state newspaper, Lincoln's *Daily State Journal,* reprinted the account in its June 8 issue. An accompanying editorial suggested that unless the story was "greatly magnified or distorted," the object "must be an air vessel belonging to some other planet." Perhaps the vessel had been caught accidentally in earth's gravity and brought to its fiery end. The editorialist looked forward to new developments "with the liveliest interest."

Two days later, the same anonymous Benkelman correspondent reported the dissolution of the mysterious object. He and a dozen other observers, he wrote, had watched the craft melt in a blinding rainstorm, "dissolved by water like a spoonful of salt." The subtext was obvious: The story was to be viewed with a healthy dose of sodium chloride. Those, if any, who still didn't get the joke were informed that Williamson, except for the small detail of permanent blindness, "otherwise . . . does not appear to be seriously injured."

Exactly eighty years later, in 1964, a Holdrege resident discovered the first story, though not its clearly comic sequel, and forwarded a copy to the *Omaha World-Herald.* Intrigued, reporter Russ Toler asked his mother, who belonged to the Dundy County Historical Society, to check

Cowboys in the employ of rancher John Ellis were engaged in a cattle round-up on June 6 when they heard a loud, strange whizzing sound over their heads. A blazing object flashed through the sky and plunged to the ground on the other side of a bank.

out the claim. Ida Toler, who had lived in the county since her birth in 1897, knew many of the oldest residents, but none recalled anything like this event, though John Ellis had indeed owned a ranch near Benkelman. Mrs. Toler thought that the yarn had been concocted because the *Nugget* needed to fill space. A few years later, Nebraska folklorist Roger Welsch conducted his own inquiries. "Nobody had the foggiest notion" about it, Welsch found.

Sources:
"A Celestial Visitor," 1884. *Lincoln* [Nebraska] *Daily State Journal* (June 8).
"A Celestial Visitor," 1884. *Lincoln* [Nebraska] *Daily State Journal* (June 9).
Clark, Jerome, 1986. "Spaceship and Saltshaker." *International UFO Reporter* 11, 6 (November/December): 12, 21.
"The Magical Meteor," 1884. *Lincoln* [Nebraska] *Daily State Journal* (June 10).
Toler, Ida, 1986. Letter to Jerome Clark (January 9).
Welsch, Roger, 1986. Interview with Jerome Clark (January 2).

1889: "Ghost Like Object"

A GHOST LIKE OBJECT, which is said to haunt the evening train on the Elkhorn, between Lincoln and Fremont, Neb., is creating great excitement among the trainmen. The train leaves here at 6:30, and it is said to make its appearance from one of the many thickets along Salt creek, when it keeps with them until Davey is reached, about ten miles out. It was first observed by Fireman Henry Conley, who declares his belief that it is the ghost of a man killed by the cars some time ago. It disturbed him so much that he resigned. The new fireman confirms the story as told by Conley."

Source:
"Some Unusual Occurrences," 1889. *Morning Oregonian* (December 24).

1897: Raining Airships

MONDAY EVENING [APRIL 19] was prolific in airships in Nebraska. One was seen in the east at Table Rock, Pawnee county, another at Juniata, Clay county, in the northwest, and a third at Bradshaw, York county, at about the same hour. Three of these heavenly messengers at once do pretty well and it is possible that there are about ninety counties where one was seen but not reported on account of the bashfulness of the beholders. It fairly rains airships these days."

Source:
"It Rained Airships," 1897. *Nebraska State Journal* (April 21).

1920s: Alkali Lake Monster

ALKALI LAKE IS NOW known as Walgren Lake. It is located in the northwestern part of Nebraska 5 miles southeast of Hay Springs. In the 1920s, Alkali was said to house a monster. Joe Swatek's article on the monster in 1980 described the sightings and summarized several newspaper accounts that appeared at the time.

The first sighting to see print appeared in the *Hay Springs News* for August 5, 1921. Arthur Johnson, who owned land near the lake, was alerted to its presence when his horses suddenly became nervous. Looking 300 feet out into the water, he saw an animal—the account provides few details concerning its physical features—emitting a spout of water 10 or 20 feet straight into the air. It was swimming away from him. Johnson thought it was perhaps 10 feet long and 2 feet wide.

On September 16, the *News* related another sighting, attributed to a small group who had watched the creature from the shore. They described it—again vaguely—much as Johnson had. Two subsequent stories (October 21, 1921, and August 11, 1922) also dealt with the alleged monster. The last account concerned a multiple-witness observation of something "apparently the shape of a huge fish . . . twenty or more feet long" (Swatek 1980).

It is worth noting that in 1889 and 1890 a drought reduced the lake to little more than a mud puddle. Obviously, no monster existed in the Alkali then, and it is hard to imagine how one could have arrived between then and the early 1920s. Though still reported in some modern-day cryptozoological writings, the crea-ture probably is the creation of a notorious journalistic hoaxer, John G. Maher. Maher was employed by the *Hay Springs News* but also contributed to newspapers all over the country, including the sensationalist *New York Herald*. He concocted fantastic stories for the publications to which he contributed—one, for example, involved the alleged discovery of a petrified man in Wyoming—and the Alkali-monster yarns were widely published.

Nevertheless, there must have been enough local belief in the monster to spark a scheme to drag the lake in hopes of capturing it and putting it on display for profit. In 1923, however, the so-called Hay Springs Investigation Association broke up amid much acrimony about how the potential profits were to be distributed. The animal, if it ever existed, was probably a larger than average catfish, though it surely was not 10 or 12 feet long.

Source:

Swatek, Joe, 1980. "Nebraska's Famous Lake Monster." *INFO Journal* 36 (January/February): 2–4.

Nevada

1867: Rolling Stones

DAN DE QUILLE (1829–1898), a Nevada newspaperman and humorist, was a good friend of Mark Twain's. On October 26, 1867, he published a story in the *Territorial Enterprise* reporting a remarkable discovery. In the Pahranagat Valley in the southern part of the territory, he said, there were stones that would roll toward each other. According to De Quille, "These curious pebbles appear to be formed of loadstone [*sic*] or magnetic iron ore. A single stone removed to a distance of a yard, upon being released[,] at once started off with a wonderful and somewhat comical celerity to rejoin its fellows."

Though written with what sounded like scientific authority, the story was a prank. Not everyone got the joke, however, and De Quille was besieged with inquiries. One inquirer, a German scientist, accused him of lying when he insisted that the stones did not exist. A showman offered him $10,000—an enormous sum of money in those days—to tour with the rocks. Realizing that denials were useless, he would respond to correspondents with the suggestion that they contact Twain, who "has still on hand 15 or 20 bushels of various sizes."

Eventually, exhausted by the seemingly unkillable legend for which he was responsible, De Quille published an unequivocal retraction. On November 11, 1879, he published a statement intended to put the matter to rest forever: "We solemnly affirm that we never saw or heard of any such diabolical cobbles as the traveling stones of Pahranagat." The inquiries continued.

Sources:

Boese, Alex, 2002. *The Museum of Pranks, Stunts, Deceptions, and Other Wonderful Stories Contrived for the Public from the Middle Ages to the New Millennium.* New York: Dutton, 76.

Lillard, Richard G., 1944. "Dan De Quille, Comstock Reporter and Humorist." *Pacific Historical Review* 13, 3: 251–259.

1868: "Beautiful Celestial Phenomenon"

THE NIGHT GANGS OF men employed in the silver mines near Virginia City, Nevada, witnessed a beautiful celestial phenomenon at four o'clock on the morning of the 27th November. The sky was perfectly clear in every direction, and the eastern horizon was peculiarly blue and bright; not the slightest sign of mist about the ridges of the distant eastern ranges; yet the morning star rose of a blood-red color, and with a bright white halo, apparently five or six feet in diameter, surrounding it. From the lower part of this halo extended downward a tail apparently eight feet long and two feet in breadth at the upper part. This tail was slightly curved, of a sabre shape, bluntly rounded at the lower end, and both it and the halo appeared to be filled with thousands of small and exceedingly brilliant stars. This strange light lasted some fifteen or twenty minutes, or until the star had risen so high that the tail of the halo appeared to be two or three feet above the crest of the distant range, when it suddenly faded out."

Source:

"Phenomenon," 1868. *Petersburg* [Virginia] *Daily Index* (January 19).

1877: Turbaned Ghost

WE WERE CREDIBLY informed to-day[,] by a gentleman just from Pyramid, of a sensation which has taken the peaceful district by the ears. The Infant Mining Company are working two shifts of men on their 300 foot tunnel. One night last week F. Dennis and Mr. Schoonover, who were at work in the tunnel, fired the fuse to a blast and were leaving the tunnel when they saw passing ahead of them and in the same direction, a ghost about 6 feet high, a turban on its head and huge black whiskers attached to its face. Dennis said its eyes were as big as his fists. When the men wheeled dirt out this ghost went ahead of them, and then followed them in. For five nights these men stood fire, and then Messrs. Jones, Williams and Brown spent part of an evening in the tunnel, waiting the appearance of the aforesaid ghost. Sure enough, along came the monster. The men were well armed and meant to do him much bodily harm, but their revolvers and shot guns proved of little service. The ghost disappeared, but left no part or parcel of himself behind. The next night Dennis and Schoonover saw the same ghost, not a particle mutilated, but as vigorous as ever.

"This aroused the camp, and ten men, armed to the teeth, and carrying about their persons sundry articles, such as a blanket, a pack of cards and several drinks, proceeded to the ghostly spot. They played cards and were so hilarious that the ghosts did not put in an appearance. These proceedings so exorcised the evil one that he has not been seen since that time."

Source:
"Local Affairs," 1877. *Reno* [Nevada] *Evening Gazette* (March 15).

1879: Mystery of the Wahhoo

A. A. ADAMS IS A well-known and highly respected German citizen. His residence is near the corner of Fourth and Chestnut streets. Mr. Adams is a bachelor, and lives in a small cottage [in Reno]. He is not superstitious, and never clouds his mind with ardent spirits. . . . So when he heard some heavy animal walking slowly up and down his verandah last Friday night [August 29], he did not suppose that the noises were in any way supernatural. He simply wondered what kind of a beast could have got into his yard. He listened, and could distinctly hear the tread of some four-footed creature, as it slowly trod to and fro. He could even hear the scratch of the creature's claws on the boards. His curiosity at length aroused, Mr. Adams determined to take a look at the pedestrian on his portico.

"He rose from his bed, and, putting on an additional garment, he stepped forth into the rheumy and unpurged air of midnight. The moon was shining from a cloudless sky, and by its sight seemed a large black bulldog. The animal stopped in its walk, and turned two brilliant, fiery eyes upon Mr. Adams. They glowed with an unnatural brightness; looking more like hot coals than visual organs. He noticed, too, that its 'snoot' was very long, like a pig's; and its tail of surprising length, stuck straight out behind. Mr. Adams clearly saw that, whatever his strange visitor might be, it was no bulldog. After looking at him steadily for over a minute, the beast slowly retreated to the fence, which it climbed by means of its claws, after the manner of a cat. Perched upon the top of the fence, the creature sat, and resumed its survey of the astonished German. Mr. Adams thought he would direct a stream of water from the garden hose upon the animal, and thereby induce it to retire. He did so, but, to his amazement, the creature remained immovable, merely presenting its snout to the stream, its body enveloped in a cloud of spray. Mr. Adams persevered in the hydropathic treatment for about ten minutes, but still the beast kept its position, its eyes glaring at him like the red lights of a railway train.

"Mr. Adams owns to a feeling of dread at the creature's peculiar persistence under the circumstances, and began to think that ghost stories might be true after all. He retreated into the house and locked the door. He looked out of the window, and there still sat the animal on the top of the fence, its baleful eyes throwing a lurid glare into the room. Mr. Adams now felt a great fall in the temperature. He was shivering with cold. He thought he would fire a charge of shot at the brute, and got out his gun for the purpose. He loaded the weapon hastily, but after he had put on the caps, concluded he wouldn't fire after all. He pulled down the blinds, and went back to bed.

"He listened for a long time, but heard no more footsteps. The cold continued, and Mr. Adams shivered nearly all night, and got very little rest in consequence. At daylight next morning he was out, and made a careful search of the yard, closely examined the fence and porch, but discovered no trace of the strange beast. He put his gun in careful order, and made up his mind that should the creature come again he would fire upon it at sight. He was surprised to find, on cleaning the weapon, that in loading it the night before, he had put the shot in first, and the powder afterward. . . .

"Mr. Adams carefully loaded the gun that day, putting in the powder first to avoid the mistakes apt to attend that operation when performed in the dark. The following night he sat up late in company, with a countryman of his, but saw nothing unusual, nor has he since been visited by the mysterious beast. No animal answering to the description is known to exist in this section of the country.

"In addition to the peculiarities already described, Mr. Adams said it had a long, slender neck. Could it have been one of those ferocious hybrids called wahhoos, which are said to prowl at night in the neighborhood of Halleck and Deeth? . . . The mystery may yet be cleared up, but now the question is, among many who hear the story, was it a wahhoo?"

Source:

"Was It a Wahhoo?," 1879. *Reno* [Nevada] *Weekly Gazette* (September 4).

1879: Nature of the Beast

A RECENT NUMBER of the *Gazette* contained some account of an animal found in the neighborhood of Halleck and Deeth, [Reno,] Nevada, known in that section of the country as the 'Wahhoo.' It appears that the creature is not known to naturalists, and finds no place in the catalogues of writers upon zoology. Some readers of the article referred to for this reason supposed the whole story to be a hoax. But it must be remembered that every day the researches of scientists are bringing to light hitherto unknown animals and plants, in every quarter of the globe. Animals well-known locally, in some remote localities of the earth, often prove entirely new and strange to the world of science. . . .

"The account of the wahhoo, which follows, is plain and unvarnished, and it may yet be found that this strange beast will possess a high scientific interest to workers in the wide field of natural history. . . . Richard Smith, the agent of Wells, Fargo & Co.'s express, at Reno . . . while hunting in the neighborhood of Halleck, heard from the residents of that locality many stories of the Wahhoo and its peculiarities. He did not succeed in catching a glimpse of one, but his brother who was with him succeeded in obtaining the dressed hide of a Wahhoo, and took it with him to Los Angeles on his return from the expedition. Mr. Smith's Reno friends, to whom he repeated some of the stories he had heard of the creature, were skeptical of their truth. But the publication of the article in the *Gazette* brought to light some additional testimony concerning the curious beast. Before proceeding with entirely new evidence, it will be well to state what the people about Deeth say of the animal. It is there known both as the Wahoo [*sic*] and the man eater.

"The former appellation is supposed to have been given it in imitation of the peculiar noise it makes. The latter designation originated in the known propensity of the beast to dig up and devour the bodies of the dead. Wahhoos which have been killed near Deeth exhibit a peculiar structure. The legs are short, and the paws very large proportionately, furnished with strong projecting claws of great length. The formation enables the creature to dig with ease and rapidity. The body is long and slender, the tail of medium length and usually curved over the back, the neck short, the head broad, and the jaws provided with formidable teeth. The skin is covered with long, fine hair. Its prevailing color is black, spotted with white. In weight it varies from fifty to seventy-five pounds. The creature is larger than a coyote, and in appearance, when seen at a distance, not unlike a large dog.

"In conversation with Mr. Smith, a young man said that he carefully measured each specimen, and found that the left legs of each were somewhat shorter than the right legs. Although his informant persisted in the assertion, the statement must be regarded with great caution. The probability is that he measured a malformed specimen and jumped to the conclusion that the others showed the same peculiarities. The young man stated, as an explanation for the inequality in the length of the creature's legs, that the Wahhoo was found only upon the hills, along the sides of which it was constantly traveling. The unequal length of its legs would be advantageous to the animal in transversing [sic] the hillsides. It would indeed be strange if nature had provided such a marvelous adaptation of structure to fit the creature for ranging upon the sides of hills. Incredible as such a statement certainly is, it might possibly be true."

Source:
"The Wild Wahhoo!," 1879. *Reno* [Nevada] *Weekly Gazette* (September 18).

1882: Green Storm

ONE DAY IN JULY, while he was hunting and prospecting in the Reese River Valley, a man on horseback riding down the canyon observed what he first took to be a whirlwind suddenly darkening the air. A second glance, however, revealed, he said, its true nature: an immense congregation of grasshoppers. Then he noticed a similar cloud coming from the opposite direction. The roar being emitted was like what would be produced when a hailstorm and a thunderstorm were mixed.

The two clouds collided, and dead grasshoppers fell from the sky like rain, nearly suffocating the witness. While all of this was happening, a real rain occurred. The result was a mess of water and slippery, crushed grasshoppers.

"In less than fifteen minutes," a newspaper stated, "they covered the ground for over a hundred acres to the depth of from six inches to three feet. [The witness's] horse fell twice before he could get beyond the confines of the storm. [Railroad] passengers ... describe ... thousands of crows and buzzards ... in the locality ... and the scent of the putrefying hoppers is terrific."

Source:
"A Hopper Cloudburst," 1882. *Daily Nevada State Journal* (July 16).

1882–1883: Footprints in Stone

THE NEVADA STATE PRISON, built in 1862 on a sandstone hill, used convict labor to quarry rock from a site of 1.75 acres. The quarrying went to a depth ranging from 15 to 32 feet. In the course of this work, the prisoners uncovered evidence of prehistoric life, including the remains of a mastodon as well as tracks of various kinds. Some were clearly from mastodons, deer, wolves, bison, and birds. The most curious of them, however, looked like huge human prints.

These discoveries occurred when C. C. Bateman was in charge of the prison. Bateman paid little attention to them, but his successor, William Garrard, found them intriguing. So did Storey County Sheriff W. J. Hanks. On a visit to California in June 1882, Hanks spoke of them to paleontologist Charles Drayton Gibbes, who subsequently entered into correspondence with Garrard. Garrard wrote back to say, "I know of no place in the world that affords so interesting a field for the scientist as this. ... Well defined human foot-prints, 21 inches long, covered by 34 feet of sandstone, are found nowhere else that I know of" (quoted in Mangiacopra et al. 1996, 211–212).

Soon afterward, members of the California Academy of Science went to the site. They determined that where the prison quarry now was, there once had been a lakeshore. Examining the alleged human tracks, some uncovered by further digging, they noted six series, each involving between eight and seventeen prints, apparently made by individuals wearing sandals.

Since none of the tracks was a perfect representation of what the investigators were looking for, they created a composite sketch.

In an August 7 presentation to the academy, Dr. H. W. Harkness argued that these were not the prints of giants but of normal-sized individuals who had created large sandals to keep mud and sand from soiling their feet.

Between 1882 and 1883, a furious controversy about the prints' true identity raged in the press and in scientific journals such as *Nature* and *American Journal of Science.* In the September 19, 1884, issue of *Science,* W. P. Blake argued that for technical reasons it was virtually certain that sandal-clad humans, giant or otherwise, could not have made the tracks, which "all have the appearance of being made by an animal with short legs" (quoted in Mangiacopra et al. 1996, 219–220). Others noted the utter absence of any independent fossil evidence of human habitation at the location during the period. Today, no paleontologist disputes the conclusion, arrived at in the 1880s, that the prints are from one or more giant ground sloths.

Still, the tale occasionally surfaces in creationist and fringe literature by writers trying to disprove conventional views of human antiquity. One exotic interpretation was proposed by maverick biologist Ivan T. Sanderson, who theorized that these were the prints of Sasquatch-like creatures.

Sources:
Corliss, William R., ed., 1978. *Ancient Man: A Handbook of Puzzling Artifacts.* Glen Arm, MD: Sourcebook Project, 645–647.

Mangiacopra, Gary S., Dwight G. Smith, and David F. Avery, 1996. "Homo Nevadensis of Carson City." In Steve Moore, ed. *Fortean Studies* 3: 211–222. London: John Brown Publishing.

1886: After the Rain

WEDNESDAY MORNING [July 14] the State Prison [in Carson] was visited by a deluge of rain and the heaviest part of the thunder storm. In the morning the officers of the Prison were astonished to find the ground covered with sulphur left after the rain had dried or soaked in the ground. The deposit varied from a sixteenth of an inch to an inch and a half thick and was everywhere. It was swept off the stone walk like snow and was also found in quantities on the roof. It was as fine as flour and of a pure yellow color. There were also millions of young frogs about half an inch long. Quite a number of parties went out to the Prison yesterday to verify the story and all were satisfied that it was sulphur. The explanation of the phenomenon is that a water spout scooped up the frogs and sulphur from some lake and deposited them at the Prison."

Source:
"A Shower of Sulphur and Frogs," 1886. *Daily Nevada State Journal* (July 18). Reprinted from the *Carson Appeal.*

1888: Fish and Frogs from the Sky

A NEVADA MAN named John T. Reid had the good fortune to witness not one but two instances of a mysterious natural phenomenon: the rain of living things from the sky. Reid reported his experiences in correspondence with the well-known American anomalist and author Charles Fort (1874–1932).

The first incident occurred in July 1888 after Reid and his father had driven a freight wagon from Elko to Tuscarora. At 2 p.m., a heavy black cloud appeared in the sky over Tuscarora, and a storm quickly ensued. As soon as the rain fell, so did fish, all of them minnows, each about an inch and a half long, all alive. They covered approximately 2 square miles of the town center, with as many as a dozen per 10 square feet. The rain continued for twenty minutes to half an hour.

Reid doubted that the fish had been scooped by a waterspout out of a nearby river, lake, or stream. A small branch of the Owyhee River was some 6 miles away from the site, but the storm had come from the opposite direction, the southwest. From that direction, Reid wrote Fort, "there is no stream of any kind or lakes for many, many miles." He went on: "My wife

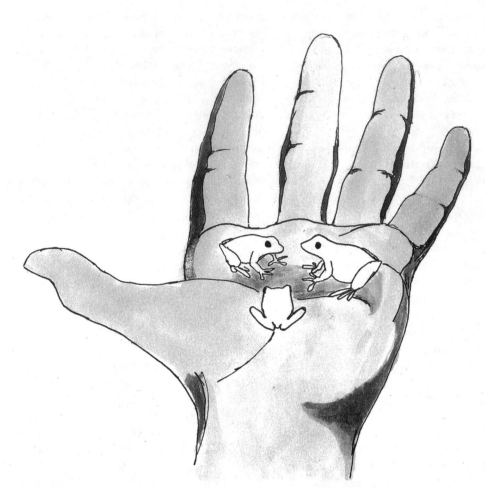

Within minutes a furious rainstorm erupted. With it frogs poured out of the sky, pelting men and horses. The frogs, Reid recalled, were "perhaps a little larger than the first joint of a man's thumb."

remembers that same year, at her fathers [*sic*] Ranch in Elko County, Nevada, which is about ninety miles away from Tuscarora, in a southeasterly direction, that some fish fell in a storm that occurred there, and that ranch is some forty miles away from any river or other body of water except a small lake . . . twenty miles away eastward. . . . However, the storm come [*sic*] from the southwest . . . and it would have been impossible for this lake to have played any part in the matter."

Reid's second experience happened in late August 1891. He, his father, two of his brothers, and another man were conducting freight-hauling business in Hamilton, in White Pine County. On the west side of Pancake Mountain,

in a white-sage desert landscape, at 2 in the afternoon of a very hot day, a heavy wind began to blow from the southwest. Within minutes a furious rainstorm erupted. With it frogs poured out of the sky, pelting men and horses. The animals panicked, and the drivers had to jump off the wagon and restrain them until the storm passed.

The frogs, Reid recalled, were "perhaps a little larger than the first joint of a man's thumb." Like the minnows from the earlier incident, they were all the same size, and they were all alive. There were, if anything, more of them than there had been of the fish. The nearest body of water was about 60 miles away, "the opposite direction from which the storm came."

Source:
"The Charles Fort–John Reid Correspondence."
1994. *INFO Journal* 71 (Autumn): 9–20.

1892: Ghost of the Red Barn

A CARSON DISPATCH SAYS that city is at present excited over a ghost scare at the old building known as the 'Red Barn.' During the last few evenings a hundred or more people have tried to rout the specter from his quarters. It is said to dress in scarlet and to disappear occasionally into the earth. Many people claim to have seen it, but have been unable to corner it."

Source:
Untitled, 1892. *Daily Nevada State Journal* (February 17).

1894: "Mysterious Visitor with a Fanlike Tail"

LAST NIGHT JUST at two o'clock a strange phenomenon appeared in the southwestern horizon, almost, if not quite, in line with the course of the meteor seen in that part of the heavens the other evening. A bright light similar to the tail of a comet could be seen beginning at a point about 20 degrees above the mountains and extending to the zenith, spreading out to a considerable extent as it reached its highest point. It resembled a large comet, but there were no signs of the star or head of such a mysterious visitor, except the fanlike tail. The light seemed stationary for some time, changing its color from a bright white light to a red glow, then gradually fading until it disappeared. During the time it was in view many shooting stars could be seen. It may have been an electrical display purely of a local character [to Reno] and it may be explained by some of our local scientists, but it was very interesting to those who saw it."

Source:
"What Does It Mean?," 1894. *Daily Nevada State Journal* (August 1).

1896: Airship Inventor

IT IS REPORTED AT the N-C-O office that L. T. Stephenson, agent at Chat, saw the much talked of airship light between Chat and the Summit, and had a talk with the inventor. Mr. Stephenson, so it is said, wanted to take a ride on the airship, but the operator would not let him get aboard. The *Gazette* does not get this story from Mr. Stephenson but from the train hands that were on the road."

Source:
"The Airship Again," 1896. *Reno* [Nevada] *Weekly Gazette and Stockman* (December 10).

1907: Sky Ship

IF CLAUDE WHEELER . . . did not have a reputation for veracity, his friends would not believe him when he told yesterday of seeing an airship floating across the sky. Wheeler is positive that it was some sort of a sky craft and not a phantom cloud. He gazed so long at the traveling object that when he started delivering loaves at 7:30 his neck would hardly resume its normal position.

"'I was down on the Southside when the ship first appeared,' said Wheeler. 'My attention was first called to it when I noticed a great crowd gathered, looking into the air. There it was for sure. The cage hung down, but it was so high above the ground that I could not make out smaller objects or tell whether a man was steering. I guess the style of craft would be called dirigible. It was going eastward at the rate of 30 miles an hour. I watched it for 15 minutes, and then it was lost from view.'"

Source:
"Says He Saw an Airship Here," 1907. *Nevada State Journal* (November 14).

New Hampshire

1868: Killing a Giant Water Snake

THE GREAT WATER SNAKE at Willoughby Lake, New Hampshire, the existence of which has been recently denied, was killed, on Wednesday of last week [August 5], by Stephen Edmonds, of Newport, a lad of twelve years. Rushing boldly upon the monster, he severed its body with a sickle. On actual measurement the two pieces were found to be twenty-three feet in length."

Source:
Untitled, 1868. *Blairsville* [Pennsylvania] *Press* (August 14).

1882: Vanishing Star

A SCENE SUGGESTIVE of ye olden time was witnessed one day recently by some of the operators in a sawmill at Bedford, N.H. About four o'clock in the afternoon, when the sun was shining brightly, a star of remarkable brilliancy was seen in the east, at a point about midway between the zenith and the horizon. It remained for about thirty minutes, and then suddenly disappeared "

Source:
"Singular Phenomena," 1882. *Grand Traverse* [Michigan] *Herald* (December 7).

1894: Merbeings

A SHORT ACCOUNT MENTIONED that some fisherman had seen three mermaids and a merman.

Source:
"Brief Bits," 1894. *Frederick* [Maryland] *News* (March 6).

1909: Nocturnal Lights

REFERENCES CONCERNING the Concord 'Double Meteor' have been appearing lately in the press. As to this may I recount the following, seen while in camp on the shores of Lake Winnipesankee last August. Shortly after dark one evening we saw approaching from Meredith way, two bright lights in the sky a fixed distance apart, high in the air and drawing near with lightning speed. Passing our camp, whatever it was, it disappeared over toward the Ossipee hills. Only the great speed of the lights marred our belief that it was an aircraft. All doubt was dispelled the next morning by news received from two vacation people a half-mile distant—Dr. Frank Chapman of Grovetown, N.H., and Dr. Walter Westwood of Beachmont, who saw them returning about an hour later. Thus the meteor theory is disposed of."

Source:
"Air Ships Seen at Night," 1909. *Boston Globe* (December 23).

New Jersey

1866: Jersey Ghost

JERSEY CITY IS HONORED at present with a most mischievous ghost, which has recently made its appearance in a tenement house. At first it was content to wander in unoccupied apartments, but recently went into a room where a man was sleeping quietly and robbed him of his shirt without unbuttoning it or awakening the owner. His ghostship also pulled off his stockings and carried them to the roof of the building, and pinned them to the eaves. The occupants of the house treat their guest with consideration, and are moving out so that he, she or it may have the whole house to roam in."

Source:
"A Ghost," 1866. *Morning Daily Oregonian* (November 14).

1873: No-Show Ghost

A CROWD OF 500 persons, who patiently waited through the midnight hours to see the apparition of an engineer emerge from the engine-house in New Brunswick, N.J., and were after all disappointed, have lost all faith in ghosts."

Source:
Untitled, 1873. *New York Times* (February 17).

1879: Attacked by Ball Lightning

A MOST SINGULAR PHENOMENON happened at Paterson, N.J., at nine o'clock Saturday night [March 22]. Edward Braine, 15, and Robert Durac, 12, were on their way to a grocery store across a vacant lot, when Robert, looking up, saw coming down in an oblique line a small bolt of fire, which struck Edward under the left breast, passed under his coat, and exploded in a mass of flames over the boy's breast and side. In the boy's left hand, which was parallel with his right at the time, he held a silver quarter dollar.

"Both boys were frightened, and the night being dark and stormy, they ran through into Mechanic street into a grocery store kept by a man named Cox, who immediately stripped off young Braine's blazing clothes, thereby saving his life. The boy's coat was burned off on one side, and his underclothes burned down to the skin. His side was badly scorched and blistered, and the end of his thumb, including the nail, was pared off. His left hand was badly burned, so that it is now swollen up to twice its usual size, and the twenty-five cent piece which he held in his hand was partly melted.

"One of the most mysterious things about it is the character or composition of this meteor [*sic*], whether it was a wholly gaseous substance or partly solid. The Braine boy saw nothing. He only heard a hissing noise over his head just as he was struck. The other boy is an intelligent little fellow, who was fifteen feet from him. He saw the ball of fire coming. Search was made for anything having the appearance of the meteorite stone, but none was discovered. The ground was wet and soft at the time, and if any solid substance fell it probably went into the earth out of sight."

Source:

"As We Get It," 1879. *Atlanta Constitution* (March 28).

1885: Red Suspenders, No Head

THIS PART OF BURLINGTON County has in process of production a ghost story with notable modern improvements. A few nights ago a party of four persons driving along a country road near Ellisdale saw the figure of a man standing by the roadside ahead of them. It neither moved nor spoke as they passed it, and each noticed that it was headless. . . . The others looked back but there was nothing to be seen on the spot where the headless man had been standing. All agreed that the figure had on a white shirt and red suspenders, and was without a head. It was distinctly seen in the moonlight."

Source:

"The Cream Ridge Ghost," 1885. *Atchison* [Kansas] *Daily Globe* (October 7).

1885: Whizzing, Fast Giant Ghost

THE USUALLY QUIET and unruffled tenor of the village of Moorestown . . . has lately been aroused to the highest pitch of excitement about a ghostly apparition which has appeared at unequal intervals along the railroad just above the East Moorestown Station. A short time ago as the fast line from Long Branch was dashing down the road the fireman noticed something white just ahead of the train. The fireman said it did not jump off on the left of the track and the engineer said it did not get off on his side. Both

thought the train ran over a man. It was stopped and a thorough search was made for the mangled remains of the victim. The station employees and some citizens walked along the track and carefully examined the ties, but no marks of blood and no body was [*sic*] found. The fireman was certain that he saw a man in his short sleeves in front of the train, and his sudden disappearance from his dangerous position was an unsolved mystery.

"Ignorant of this affair, Mrs. Cassidy, an intelligent lady of Moorestown, and her sister walked over the track at this point a few nights after. She said yesterday: We were walking slowly along and . . . I suddenly cast my eyes to the right and cried out: 'Oh, my! Look there. What is that?' and my blood stopped in my veins and I was paralyzed with fear, for right in front of me stood a tall, thin figure, dressed in black, with a crape veil over its face. It . . . must have been nearly seven feet tall. My sister said: 'Oh, that is some widow,' but it suddenly whirled away with a whizzing noise like pheasants make when they rise and fly away. Then my sister was frightened and we both went down town. I met my brother and told him all about it. He said we were cowards, but came back with us to the place we saw the ghost. It was quite light, as the moon was just breaking through the clouds.

"'When we arrived at the place my brother saw it, too, and started to run after it, but it suddenly vanished. It did not seem to have any feet, and, though I don't believe in ghosts, it was the queerest looking thing I ever saw.'

"The same night Mrs. Cassidy saw the ghost it appeared to others. It has been seen many times since, and probably one-fourth the population of Moorestown have gone up the track well protected to satisfy their curiosity as to the reality of the ghost.

"Aaron Burr, the Town Constable, went out to duel the weird visitor one dark night. He was certain he would prove the mystery a hoax. Armed with a six-shooter he perambulated the track. . . . He met too suddenly the object of his search, but bravely banged away at the mysterious object that confronted him.

"Tom Cassidy, Louis Certain, Tom Morany and William Budd, four incredulous citizens, concluded to ferret out the mystery and bravely face the foe. Friday night, after coming from their work on the railroad, they visited the haunted spot, determined to shoot the rascal. . . . Tom Cassidy described their meeting with the ghost thus:

"'We were walking along the road . . . when a tall, thin form, too large to be a man, stood in front of us. It was dressed in women's clothes and its face was craped. It wore something on its legs that looked like gum boots and its tall form swayed to and fro, like the boughs of trees when the wind blows through them, and made a similar noise. Louis Certain yelled out, "My God, is that the ghost?" Then Bill Budd drew a revolver, and he ran a square following it. He could not get nearer than ten yards to it, and when he thought he was gaining on it[,] it suddenly disappeared in the woods. The strangest part of the affair was that the ghost ran along the sandy road and when we afterward examined the road not a track or trace of one could be found.'

"William Budd said: 'I am a very fast runner, but I couldn't keep up to that thing. I don't believe it is human, for it does not run like a man. I wanted to fire at it when it ran away from me, but the other fellows told me not to shoot, and then when I was going to draw[,] the infernal thing disappeared. I never did believe in ghosts, but this is one.'

"One man said: 'John Dargan's horse saw it on the road, near Gillingham's house, on Thursday night, and frightened at it. People say that they can not get a horse in Gillingham's stable on account of it. Their house is empty and some think it stays in there.' Isaac Lawrence said he saw it come out of Gillingham's yard and it looked like a man seven feet tall. He passed it and it disappeared, but he was certain it was not a human being.

"The ghost has been seen at various times by different people along the track, and all bear testimony to its great size, its somber appearance and the whizzing noise it makes as it sails along. Large crowds have congregated to see the mystery, but it generally appears when only three or four are present.

"Some of the citizens say that just two years ago a similar apparition clothed in white was seen at the same place. It manifested to a great many in the community, and some think it but a reappearance of the old ghost, clothed in black."

This early entity-in-black also has features in common with nineteenth-century legends and reports of the elusive Springheeled (or Springheel) Jack, sometimes theorized to be a before-his-time UFO being, elsewhere held to be no more than a folkloric concoction cobbled together from various ghost traditions.

Source:
"A Jersey Ghost," 1885. *Atchison* [Kansas] *Daily Globe* (October 31).

1886: "Strange-Looking Object, Clad in Black"

THAT MOORESTOWN 'GHOST,' from which nothing has been heard or seen for over two months, has again made its appearance. Last evening [April 3] Mrs. McCubben, who lives on Third-street, near Chester-avenue, started with a friend about dusk to walk to the West Moorestown station to get an express package. While they were walking from the schoolhouse corner, to Church-street, down toward the station, they noticed a strange-looking object, clad in black, carrying in one hand a head and in the other a key, coming toward them at a rapid gait. Both ladies hesitated a moment as if undecided what to do, and, as the figure was still approaching them, they became alarmed and ran as fast as they could for the station. The 'ghost' followed them nearly to the station, and then disappeared behind a lumber pile."

Source:
"The Moorestown Ghost Again," 1886. *New York Times* (April 5).

1886: "The Spirit . . . Was Wide Awake"

SOUTH ORANGE HAS A spook. Several credible witnesses are sure that they have seen it. It hovers, the people say, around Ridgewood road, near Plunkett's lane and South Orange avenue. I. H. Ball said that he saw it on Monday night [July 12], but when he approached it[,] it dissolved into thin air. Michael Preston and James McNulty say that they saw a tall figure dressed in white in the same place as Ball says he saw it. The Irish residents of South Orange say it is a banshee. An expedition consisting of about a dozen people who claim that they are not in the least superstitious started out last night for the purpose of surprising and capturing, if possible, the disembodied spirit. The spirit, however, was wide awake and didn't turn up, even at the witching hour of midnight. The search will be resumed tonight."

Source:

"A Strange Apparition," 1886. *Trenton* [New Jersey] *Times* (July 15).

1886: "Instantaneous Photography by Lightning"

DURING A HEAVY THUNDER storm in Plainfield, N.J., a young lady was removing a polished old gold Japanese tray from the bay window, when a sudden flash of lightning sent a blazing light over the tray. The next morning the young lady's picture was found to have been handsomely printed or burned in the lacquer of the tray, over which she was stooping when the lightning flashed. Electricians are now investigating this new phenomenon of instantaneous photography by lightning."

Source:

Untitled, 1886. *Stevens Point* [Wisconsin] *Journal* (September 11).

1887: Apparition with a Gashed Throat

THE THOUSANDS OF Trentonians who have attended picnics in Morrisville Grove will remember the old stone mansion that stands in . . . the Grove near the Delaware river. The house is in somewhat [*sic*] of a dilapidated state and has not been occupied for years. In this mansion the ghost is said to hold nightly séances.

"One day last week Mr. Peasley and his family took up their residence in the broken down house. Nothing unusual transpired during the first night of his stay. The second night Mr. Peasley was seated alone in the sitting room reading, the family being upstairs. Happening to raise his eyes from the paper Mr. Peasley was surprised to see an aged man standing in the center of the room. He asked the silent intruder what he wanted, but received no answer. Mr. Peasley left the chair he was sitting in and advanced toward the visitor. Without a word or the least bit of a noise the old man retreated from the room. Opening the door leading into the cellar the figure descended the stairs. Mr. Peasley followed, but when he reached the cellar there was no trace of the intruder. The place was searched, but the mysterious person could not be found.

"Mr. Peasley returned to the sitting-room greatly frightened. Summoning his family they waited for an hour for the stranger to return, but no view was obtained again that night. Mr. Peasley concluded that he had seen a ghost.

"The next night, at the same hour, the apparition appeared to Mr. Peasley in the sitting-room. This time it was attired in snow white garments, and Mr. Peasley noticed that there was an ugly-looking cut on its neck. Mr. Peasley again advanced towards the figure, and again it hustled through the door, down into the cellar, where it again mysteriously vanished. For several nights the ghost made its appearance, each time dressed in different clothes, and every evening it vanished into space in the cellar. The family of Mr. Peasley became alarmed and refused to occupy the old mansion. They moved out this week and told the neighbors of the haunted house.

"Every night since[,] a crowd of men and boys have [*sic*] watched for the ghost. It shows itself every evening just before midnight and is dressed in a new costume nearly every night. The face of the ghost is that of an old man. His hair and flowing beard are snowy white. On the right side of the neck is a long, deep, ugly looking gash. Every evening the specter wanders about the house and grounds surrounding the same. Several ineffectual attempts have been made to seize the apparition, but each time it hurries into the cellar and is lost to sight."

Source:
"Morrisville's Ghost: The Old Mansion in the Grove Said to Be Haunted," 1887. *Trenton* [New Jersey] *Times* (February 19).

1889: Ball and Stars

I WILL DESCRIBE WHAT appeared and disappeared on Thursday, [February] 7th, at about 5:20 o'clock P.M. Ascending to the summit of the mount in the valley [near Lake Hopatcong] I saw a light coming in the face of a gale likened unto the headlight of a locomotive. In front it appeared, as it came nearer, like molten iron, and in the rear it was funnel shaped and it gave forth a blueish [*sic*] green blaze and was about twelve feet in length. It came very near at an angle of about forty-five degrees. When near the earth there appeared three small star shaped balls, in a moment a cluster and then another cluster and in a second a final collapse. All the stars followed in the wake of the funnel or large ball, which diminished in size at every discharge. It seemed to consume itself. There was a terrible combustion of gases or some material fanned by the gale of wind. There was a whirring sound and trees and fences cast shadows like the sun, which faded into a dim, misty

twilight. The final collapse took place in a field 150 feet from where I stood, but when I went over the ground there was nothing visible."

Source:
"A Fulton Street Meteor," 1889. *Brooklyn* [New York] *Daily Eagle* (March 17).

1891: Hoop-Snake

A HOOP-SNAKE CHASED architect Rufus B. Van Iderstine down Orange mountain, in New Jersey. It was twelve feet long, and had a big head and a barbed stinger on its tail half a foot long. It started after Rufus, and he took to his heels when he saw the monster move down the rocks. He got half a mile start, but looking back at a bend of the road he saw the snake grasp its tail in its mouth, forming a hoop, that swiftly rolled after him. The hoop-snake came on with electric speed. When the frightful creature was within a hundred yards of him, Van Iderstine darted behind a big oak. The hoop-snake swerved, and, coming on in blind rage, drove his spear-like tail into the tree.

"It was trapped. Rufus hurried to Orange for help to return and destroy the monster. Two hours later half a dozen men, with shotguns, ascended the mountain. The hoop snake was gone. It had pulled its tail out of the hollow, horny stinger, which was left buried in the tree. It was so deeply imbedded that only chips from its edges could be broken off.

"Poison from the stinger was killing the tree, and its foliage was curled and withered when the hoop-snake hunters first saw it. An ugly green liquid dripped from the hollow stinger like sap from a sugar tree-spile and deadened the grass it fell upon."

Source:
"A Snake Story," 1891. *Reno* [Nevada] *Weekly Gazette and Stockman* (September 17).

He got half a mile start, but looking back at a bend of the road he saw the snake grasp its tail in its mouth, forming a hoop, that swiftly rolled after him. The hoop-snake came on with electric speed.

1892: Mysterious Balloon

A MAMMOTH BALLOON, containing four persons, passed over the city about 11:30 p.m. It descended until within about 20 feet of the earth, then rose and drifted off in a southeasterly direction. A brilliant electric searchlight was fastened on the edge of the car."

Source:

"Mysterious Balloon Sighted," 1892. *Trenton [New Jersey] Times* (June 2).

1894: The Apparition on Eagle Rock

ON EAGLE ROCK, AT THE summit of the Orange mountains, where hundreds of Newarkers rush daily—or nightly rather—for a breathing smell, there has been for a week much excitement owing to the appearance of what is said to be a ghost. It appears nightly, so it is said, and as the rock is but a short distance from the spot where Phoebe Paulin, a pretty 16-year-old girl, was murdered 12 years ago, the uncanny apparition is connected with that mysterious and to the present unpunished crime. Not only timid women, but sturdy men say they have seen the apparition[,] and even some of the members of the Eagle Rock company, which has done much to make the place a resort, admit that they have seen it and heard the unusual sounds.

"The first incident that brought the alleged ghost into prominence was the prostration of Miss Mamie E. Keane . . . who has been confined to her home for a week as a result of the shock produced by the sudden appearance of the ghost. . . . Miss Keane, with her sister Lottie and an escort, were [sic] returning from the rock on Sunday evening [July 8], and when near the spot where the Northfield road intersects Eagle Rock avenue they met the apparition. Miss Keane says that she was not at all superstitious, did not believe in ghosts and attributed her illness to a shock produced by mechanical rather than supernatural effects.

"'We were walking leisurely down the avenue toward the electric car,' she said, 'when something white that appeared like the figure of a woman darted across the road. It neither ran nor walked and almost instantly darted back whence it came. The thing appeared only a few yards ahead of our party, and, startled as we were by the sudden appearance[,] our alarm was made more intense a few seconds later when the most unearthly sounds, such as a person moaning from great pain, came from the dense wood at the side of the avenue where the figure had disappeared.'"

Source:
"Spook on Eagle Rock," 1894. *Trenton* [New Jersey] *Times* (July 16).

1896: Shrieking Ghost

Pennsylvania Railroad freight train no. 407 left Jersey City at 11 p.m. on July 30. Between 1and 2 a.m. it stopped and dropped off some freight at Woodbridge. After it stopped, the fireman, a man named Bender, went out to flag any approaching train. A New Jersey newspaper related,

"His hair stood on end when he saw a white form[,] apparently a woman, seated on a stone wall covering a small stream and rapping the stones with a stick.

"He mustered up courage and asked the ghost what it wanted. In response the ghost gave a blood-curdling cry. At this Brakeman Whitehead, who had arrived on the scene, hurriedly took to his heels and sought shelter in the caboose."

The engineer also saw the apparition and quickly moved his train away from the site of the ghost's appearance. "Brakeman Steinmetz, of Bordertown[,] vouches for the story, and says a fireman on another train saw the ghostly figure at the same spot about a week before. Several Woodbridge people also saw the ghost, and the town is excited over the strange apparition."

Source:
"A Ghost Story," 1896. *Trenton* [New Jersey] *Evening Times* (August 4).

1897: "Strange Meteor"

Last night [April 19], while a large crowd was gathered in front of Morey's Hall [in Swedenboro] to listen to the open air services of the Salvation Army, a strange meteor was witnessed in the heavens, towards the north. It then suddenly veered to the westward, and when it reached Main street it was not higher than the tallest tree tops, and was moving slowly through space until it disappeared toward the end of the town. The appearance of this great ball of fire created considerable alarm among the women, causing them to cry out with fright, and not a few of the men manifested some alarm as well. The apparition caused a cessation in the Salvation Army services, and created considerable comment, owing to the fact that something of a similar character was witnessed here about two weeks ago."

Source:
"New Jersey Notes," 1897. *Philadelphia* [Pennsylvania] *Public Ledger* (April 21).

1897: "In the Clouds"

TRENTON HAS NOT only seen the airship, but has had a message therefrom tending to prove that there are four men in the boat, and that they, having lost complete control of their machine, are drifting helplessly about in ether.

"The airship was seen last night by not less than a hundred persons, many of whom had glasses. Some are willing to make affidavit that the mysterious messenger of the heavens is really a bonafide airship. The message from it is in the hands of Peter McAuley. . . . Last night, at about 9 o'clock, when Mr. McAuley was going home from his daily labors, he and others saw immediately above them what appeared as especially brilliant shooting stars. They could not comprehend how they could see the heavens when such a drizzling rain was falling. Naturally their astonishment evolved into amazement. However, no one seemed to be able to volunteer a suggestion that would account for the strange effect. All who saw the 'shooting stars' gave up in bewilderment and retired to their homes.

"This morning, as Mr. McAuley was passing from his residence, he saw his little girl kicking a package wrapped in manila paper, and tied with the ravelings of a blanket, or some other piece of cloth. His curiosity was thoroughly aroused, and he picked up the package to more thoroughly satisfy his curiosity. Inside of the wrapper was a Boston baked bean can, and inside of the can was the following note, which was soaked with water, so much so that the wording, which is in ink, had run together sufficiently to be almost undecipherable.

"'In the clouds; April 26, '97; night—to whom it may interest: A terrific wind and rain storm, which has been raging[,] has just about abated. Our airship revolved like a leaf in a miniature whirlwind. Nearly three weeks ago we descended upon a vast prairie, evidently in the western states (U.S.). Since then we've been sailing, we know not where. We've been out of water frequently, and as often have melted snow for relief. Please notify Maurice Porter, Greenwich, New York. He knows who we are. We hourly expect to perish.' Signed 'We Are Four.'

"Upon reading this Mr. McAuley immediately searched the heavens, but could not detect any signs of the airship. Suddenly he recalled what he had seen last night, and concluded what he took for shooting stars must have been rockets fired as signals of distress from the airship. He is inclined to think the message genuine, and does not think he has been made the victim of a joke. From Mr. McAuley's business standing in Trenton, it is known that he would not himself participate in a hoax of this nature.

"Yesterday, about noon, people in the city hall were attracted by an object about a mile above them that glittered strangely in the sunlight. Some got out their glasses. The object proved to be moving very slowly. It was conical, shaped like an immense cigar, and was evidently made of some light, bright material, presumably aluminum. Many of them were emphatic in their belief that it was an airship. About one o'clock it had worked itself well over toward the west, and had disappeared behind some clouds. It is evident that when the wind storm came up about 6:30 last night that the ship was driven back east again.

"While many scoff at the idea of the object being an airship, many are sanguine that if it is one, that the note Mr. McAuley received came from it. Communication has been had with Maurice Porter, Greenwich, New York. It is hoped that something more definite may be had regarding the passengers."

Source:
"Was It an Airship?," 1897. *Trenton* [New Jersey] *Times* (April 27).

1899: "A Bloody, Mangled Wraith"

THE CROSSING AT INGALLS, N.J., on the northern division of the Ontario and Western railroad is haunted. Nine men are willing to make affidavit to it. The first watchman was placed there last September just after the Chicago express ran into a switch that had been opened by tramps. Three persons were killed. James Coe quit the section gang to take the easy job. After two nights he said there were too many ghosts for him. Another section hand stood it three times. One night was enough for the third man. He said he started to gather wood to make a fire when a bloody, mangled wraith came out of nowhere and called him names. So it went on until nine men had thrown up the job, each declaring that the place was haunted by a multitude of ghosts who shrieked and wailed the night through. Finally the section boss declared that he would prove that it was all foolishness. He would take the job himself. Two nights were all he could stand."

Source:
"A Haunted Crossing," 1899. *Sandusky* [Ohio] *Star* (February 2).

1901: Spirit or Joker

THE SPIRIT OF SOME poor departed soul unable to rest in its lonely grave, or some practical joker with a white sheet about him, is now wandering around the railroad track in the vicinity of Hillcrest, causing great consternation among the railroad 'boys' of the P. & R. They claim to have seen it several times the past week and weird are the tales told concerning its uneasy appearance and gruesome behavior.

"William Barber, baggagemaster on the night train, says that while flagging at Hillcrest as is his custom every evening, he noticed something white coming up the track, and not liking the appearance of the object in question, he started to walk towards Trenton. The faster he walked, the nearer the object approached[,] and when he broke into a run it came at him like a race horse and passed him, making the most fearful noise man ever heard. He says it had eyes that looked like balls of fire and then disappeared as though the earth swallowed it up right in front of him.

"He failed to say anything to fellow workmen as he feared they would give him the laugh. But Monday night [July 22] while the crew was shifting the freight on Prospect Hill it was seen by John Donnelly, the engine wiper, who was taking a ride. He says he was standing on the top of a box car when he noticed it running along the bank. He called to the other members of the crew[,] who noticed it[,] and the bravest

of them feared to go near it. The crew later recovered their bravery and started a search for it but failed to find any signs of a ghost, real or imitation.

"Young Donnelly claims that it is the ghost of a man who was hung in that vicinity sometime ago. Barber says he never did believe in ghosts but he is now convinced that there is such a thing and he does not care to have another experience like he has had the past week."

Source:
"Spook on Rail Road," 1901. *Trenton* [New Jersey] *Times* (July 25).

1905: A Lantern from the Sky

A COLORED MAN, who says his name is Charles Staats, who gives his age as 24 years, and who said he lives at West Philadelphia, is locked up here, having been brought into Bordentown by Martin Ronan, who found him acting strangely. The man is thought to be crazy. He stated he and a ghost had fought in the street here Tuesday night [June 20]. He said the ghost had a lantern he brought from the sky with him. When asked how he made out with the ghost, he said that he got the best of him and took the lantern from him."

Source:
"Colored Man's Strange Tale," 1905. *New York Times* (June 23).

1909: Club-Haunting Ghost

MEMBERS OF THE Riverton County club are excited over the appearance of a 'ghost' that so far has baffled all attempts to capture it. The 'ghost' first appeared at the clubhouse at midnight about ten days ago [late September] and, according to Samuel Conwell, the steward rapped on the door, and as soon as he opened it the figure disappeared.

"Four nights later the 'ghost,' which makes its appearance in the form of a beautiful young woman all dressed in white, again appeared at the house an hour earlier than the first appearance, and after rapping on the door led the steward, his mother, Mrs. Samuel Conwell, and several other persons nearly a square across the links and then disappeared."

Source:
"Ghost Haunts Club," 1909. *Indiana* [Pennsylvania] *Evening Gazette* (October 4).

1911: Canal Serpent

SEVERAL BATHERS IN THE Morris canal [in Richfield] made a hasty retreat from the water when the cry of 'devil fish' was raised by Robert Thompson, a farm hand. Thompson, who had some distance from the other bathers, declares that when he first saw the monster it was following him with its head out of the water. He at first thought it was a small dog, but on looking at it more closely he discovered, he says, that its head was similar to that of a porpoise. The strange creature followed him to the bank, and he says he climbed up the bank just in time to escape it. The monster then turned about and went down stream toward the other bathers.

"Not knowing what the thing was and wishing to warn the others, Thompson called out: 'Look out for the devil fish!' This had the desired effect and the men left the water. Albert Woodrow of Brookdale, who had a good view of the thing as it passed him in the water, says that it was at least 10 feet long, had a round greenish body and a head as big as that of a bulldog, which it bobbed up out of the water at intervals. 'If it was a water snake,' declared Woodrow, 'it was the biggest I ever saw, and I have seen hundreds of them.'

"Several men followed it down the canal for a short distance, but it went so fast that they soon lost sight of it. Word was sent to the lock tender at the Bloomfield plant to keep a lookout for the fish, but so far it has not been seen at that point. The alarm reached Bloomfield and the several bathing resorts on the canal were soon deserted."

Source:
"'Sea Serpent' in a Canal," 1911. *Elyria* [Ohio] *Evening Telegram* (September 19).

New Mexico

1880s and Before: Dinosaur in the Crater

TRAVELING ACROSS THE desert plains on an unspecified date (apparently in the 1880s), a man identified only as Mr. Alexander, "who possesses some mining property in the San Andreas mountains" and was heading toward his property, was surprised when his burro suddenly stopped in its tracks. Its ears pricked up, and then it turned around and ran off in the opposite direction at a very fast clip. Baffled, Alexander was about to set off to retrieve the animal, if he could, when he glanced forward to see what could have occasioned its extraordinary behavior. What he saw froze him to the spot.

It was an enormous reptile a quarter of a mile away, moving toward a crater on the east end of the plain. A press story reported, "He says it appeared to be about 60 feet in length; but what surprised him most was [*sic*] the queer proportions of the creature. The fore parts were of enormous size, its head being fully as large as a barrel. A few feet behind the creature's head two large scales were visible, which glittered in the sun like polished shields; further back were two huge claws on either side, about two feet apart, which were all the monster had in the shape of feet. The rest of the body was comparatively small and tapering to the end of the tail; it traveled at a rapid gait, sometimes rearing its whole body from the ground, and walked on its four claws." He watched it go over a small hill. He chose not to follow it.

The fore parts were of enormous size, its head being fully as large as a barrel. A few feet behind the creature's head two large scales were visible, which glittered in the sun like polished shields; further back were two huge claws on either side, about two feet apart, which were all the monster had in the shape of feet.

Mr. Alexander was not alone in claiming that a dinosaur-like creature lived in the area. According to local belief, it lived in an extinct volcanic crater. It had a ferocious reputation, and Mexican Americans spoke of persons who had entered the crater and never been seen again, presumably devoured by the terrifying beast.

Source:
"A Monster Serpent," 1888. *Atlanta Constitution* (January 2).

1880: On the China–New Mexico–New York Line

IN 1880, GALISTEO Junction (now just Galisteo), the terminus of the Atchison, Topeka, and Santa Fe Railroad, was little more than a railroad depot standing by a few dilapidated shacks in the desert 18 miles south of Santa Fe. If we are to credit an account published in the *Santa Fe Weekly New Mexican,* however, it was the site of a curious event that, from all that is known of aviation history, almost certainly could not have happened.

On the evening of March 26, the train came in from Santa Fe on its usual round, then departed. The depot agent was freed to take a short walk in the cool evening with two or three friends. As they were doing so, they heard voices coming from the sky. At first they assumed that they were coming from the mountains to the west. But the voices, speaking in an unfamiliar language, grew closer. They were punctuated with shouts, laughter, and music. In short order it became apparent that the sounds were coming from a car on the underside of a "monstrous" fish-shaped balloon guided by a large fanlike device. The car carried eight to ten passengers. Along its side were elegant characters.

As the object passed overhead before disappearing in the eastern sky, several items flew from it and plummeted to the ground. One, recovered almost immediately, was a "magnificent flower, with a slip of exceedingly fine silk-like paper, on which were some characters resembling those on Japanese tea chests."

The next morning a search uncovered a cup of "very peculiar workmanship." The depot agent put it and the flower on display. A few hours later, an unnamed curiosity collector, doing archaeological excavations at a church in the community, purchased them. The next evening, a party of tourists visiting the site was shown the items. One of the group was a wealthy young Chinese man who had arrived from San Francisco a few weeks earlier. The newspaper reported that on seeing the flower the visitor expressed great delight and related the background. On April 5, the *Santa Fe Weekly New Mexican* reported:

"For some time past, great interest has been [manifested] in China on the subject of aerial navigation. Hundreds of thousands of dollars have been expended in experiments. . . . Just before he left Pekin strong hopes were expressed that victory had at last crowned these efforts. . . . If such was the case, and he had now no doubt of it, last Friday night's balloon was but the first of a regular line of communication to be established between the Celestial Empire and America. . . . He stated that he was engaged to be married to a young lady of a very wealthy family, who had a sister living in New York, and that knowing he was traveling in this portion of the country she had written the message with her own fair hand and had dropped it from the car . . . evidently with the hope that it might reach her lover's hands[,] and curiously enough her hopes were realized."

The probability that anything like this ever happened is zero. Balloon flights between China and America were not occurring in 1880; they were not even possible. Chinese people, particularly Chinese men, were familiar to residents of Galisteo Junction, however, because many had worked (and extraordinary numbers had died) in the construction of the Western railroads. Racist attitudes—in evidence in unquoted portions of the *New Mexican* account—made the Chinese figures of fun and ridicule. There was also a great deal of popular speculation about the possibilities of human flight. Some of it was expressed in proto-science-fiction tales of balloon flights over long distances. Writing the tale to fill newspaper space, the anonymous *New Mexican* writer did not have to look far for inspiration. Truncated versions of the story—taking it as far as the collector of curiosities but not to the visit of the Chinese tourist—would appear in later decades in UFO literature.

Sources:

"Galisteo's Apparition," 1880. *Santa Fe Weekly New Mexican* (March 29).

"Solved at Last: The Explanation of the Balloon Mystery Which Has Been Perplexing Galisteo," 1880. *Santa Fe Weekly New Mexican* (April 5).

New York

1815: Rising Stones

WE HAVE CONVERSED with several gentlemen, of undaunted veracity, from the county of Ulster, who all agree in the following extraordinary relation:

"That they have conversed with several credible persons from Marbletown, in that county, and they mention the names of persons well known [to] the editor of this paper, & these persons assert and declare themselves ready to make oath, that the stones lying in two fields there, on several successive days[,] rose from the ground, to the height of three and four feet, and moved along slowly and horizontally, from this to thirty-six feet, and that a few of them even mounted over the tops of the trees! That the persons[,] who first beheld these astonishing performances, were disbelieved by the neighborhood; but that all those, who came to see if there was any truth in the account[,] are prepared to swear to them. The last performance was in an open field, without wood or cover near it."

Source:

"Unprecedented Phenomenon," 1815. *Ohio Repository* (November 16). Reprinted from the *New York Courier.*

1822: Mystery Light

AROUND TEN O'CLOCK on the evening of March 4, an observer at Canajoharie noticed a dramatic flash of light that seemed to extend from the sky to the ground. Then the light disappeared, as if passing through a cloud. Soon afterward, it was visible again. It looked like a "blazing meteor . . . appearing to be twenty or thirty feet in diameter."

The light then started expanding in size and no longer seemed to be aflame. It broke in two, and the two detached parts took off toward the northeast and the southwest. According to the *American Journal of Science,* the witness estimated that it "was five or six minutes from the first appearance of the meteor until it finally vanished, and from six to ten minutes from its first appearance before the report of its explosion reached him, which resembled the noise of a distant cannon, and was followed by a strong sulphurous smell, that lasted fifteen or twenty minutes."

If this was indeed a meteor, it was certainly a strange one in nearly all respects, not the least of them its duration. Meteors are visible for seconds, not minutes.

Source:

"The Meteor," 1823. *American Journal of Science* 1, 6: 319. Cited in William R. Corliss, ed., 2001. *Remarkable Luminous Phenomena in Nature: A Catalog of Geophysical Anomalies.* Glen Arm, MD: Sourcebook Project, 279.

1828: Grasshopper on Her Back

THE FOLLOWING RECITAL of a phenomenon which happened about a year since, will be a subject of inquiry among naturalists and physicians. A young woman, the daughter of a respectable farmer in Edinburg, Saratoga county, New-York, while in a field of new mown hay, felt the sting of a large green grasshopper, as she then expressed it. Some time in the following winter, she discovered a tumor on the shoulder between the caracori and acromian process, attended with some pain and uneasiness. After about three weeks continuance, it disappeared from the shoulder, and she felt a pain along the course of the clavicle; and in May, it appeared at the side of the neck, partly under the sterno clenia motoicles muscle. Her physician treated for scrofula with apparent success, for it again disappeared, until July, when it was felt once more at the shoulder—the tumor about the size of a hen's egg, and with evident fluctuation, when it was opened with a small discharge of unhealthy pus, and along with it a living grasshopper, two inches in length, and breadth proportionate. The only conclusion is that the egg must have been deposited the year before, and arrived to maturity by a process of incubation."

Source:

St. John, Eliphalet, 1829. "Remarkable Case of an Insect Supposed to Be Hatched in the Human Body." *Adams* [Pennsylvania] *Sentinel* (August 5).

1840: Between the Earth and the Moon

WE ARE INFORMED, by a gentleman of this city [Troy], that the Moon presented a very singular appearance between two and three o'-clock on Friday morning, the 21st of March. He states that it was obscured for the space of half an hour, by some body between the Earth and the Moon, so as to present the appearance of an annular eclipse, a slight ring of light round the edge of the Moon being alone visible. The same phenomenon was also observed by the watchman on duty at that time. Can any of our astronomical readers explain the cause?"

Source:

"Lunar Phenomenon," 1840. *Adams* [Pennsylvania] *Sentinel* (April 6). Reprinted from the *Troy* [New York] *Whig*.

1842: Extraordinary Aerolite with Bizarre Effects

WE HAVE CONVERSED with Mr. Horace Palmer, who was on his way from Dunkirk to this place [Westfield, on the evening of April 11], when the meteor appeared. He was two or three miles from Dunkirk, when he appeared to be instantly surrounded with a most painful vivid light, proceeding from a mass of florid or jelly-like substance, which fell around and upon him, producing a sulphurous smell, a great difficulty of breathing, and a feeling of faintness, with a strong sensation of heat. As soon as he could recover from his astonishment, he perceived the body of the meteor passing above him, seeming to be about a mile high. It then appeared to be in diameter about the size of a large steamboat pipe, near a mile in length. Its dimensions varied soon; becoming first much broader, and then waning away in diameter and length until the former was reduced to about eight inches, and the latter to a fourth of a mile, when it separated into pieces which fell to the earth, and almost immediately he heard the explosion, which, he says, was tremendous.

"On arriving here this morning, his face had every appearance of having been severely scorched; his eyes were much affected, and he did not recover from the shock it gave his system for two or three days. This is really a marvelous story, but Mr. Palmer is a temperate and industrious man, and a man of integrity, and we believe that anyone conversing with him on the subject, would be satisfied that he intends no deception, but describes the scene, as nearly as possible, as it actually appeared. Probably, however, his agitation at his sudden introduction to such a scene, caused the meteor to be somewhat magnified to him. Witnesses here speak of the sparks which were given off; probably one of these fell and enveloped Mr. Palmer. In addition to its light, Mr. Palmer states that its passage was accompanied by a sound like that of a car moving on a railroad only louder."

Source:

"Brilliant Meteor," 1842. *Norwalk* [Ohio] *Experiment* (May 4). Reprinted from the *Buffalo Commercial Advertiser,* citing the *Westfield* [New York] *Messenger.*

1843: "Nearly as Large as the Moon Itself"

IN A LETTER DATED May 9, 1843, Henry Jones of Patchogue, Long Island, related the following:

"Several credible witnesses residing in the above place have recently informed me at their own dwellings that they and others saw on Saturday evening, April 29, 1843, between nine and ten o'clock, a very brilliant light in the northwest, brighter than the moon, which, increasing and diminishing in size, was sometimes nearly as large as the moon itself, apparently two hours high, and this continued something like three fourths of an hour; and what was most singular, they say that there were streaks of fiery light, resembling red-hot bars of iron, which emanated from it, upward and downward, to a considerable distance, some of which remained for a minute or so, and then vanished. Its disappearance was sudden, and at once, and at a time when it was brightest.

"The same witnesses testify that some six or eight weeks ago they saw very nearly the same image or appearance of the moon in the west, somewhat above the horizon, which continued, as they thought, about an hour. The singularities of it differed from that of April 29, in its having streaks of light more red than itself, of apparently three inches width, which passed directly through its center, forming a perfect cross, and extending beyond the edge of it twice the diameter of the light, with several spots like blood on its surface. Its size, like the other, increased and diminished, with occasional seeming sparks of fire thrown out from it in various directions, and, like the other, its disappearance was sudden, and all at once.

"Barnet Matthias, pastor of the Congregational Church in Patchogue, will witness to the correctness of this matter, if called upon."

Source:

Loughborough, J. N., 1904. *Last-Day Tokens.* Lodi, CA: Pacific Press.

1843: Fantastic Mirage

IN 1843, AS A PARTY of ladies and gentlemen were driving homeward through the swamp [Bear Swamp, near Saratoga Springs] in the evening, (among whom were the late Rev. Francis Wayland and the writer,) they plainly saw, in the middle of the swamp, half a mile in their rear, and, of course, in a direction opposite to the village, a large building with rows of windows brilliantly illuminated. The entire company left the vehicle, and for a long time watched the strange and singular appearance—an appearance seemingly bordering on the supernatural, since they had just come over the road where the spectre-house now stood, and all who were familiar with the country knew of no such house, or any dwelling, indeed, in that direction.

"The next day it was found that passengers coming up in the cars from Ballston at the same hour in the evening had also witnessed this phenomenon, conclusively proving that even if our party all had their imaginations psychologically affected—which is more improbable than the converse—others, miles away, could not have been under the same influence. The explanation undoubtedly was that the apparition of the illuminated building was simply a reflection of a refraction of clouds of one of the hotels at the Springs. . . . Other dwellers in Bear Swamp have related to me experiences of a similar character, though not so marked as the two I have mentioned."

Source:

Stone, William L., 1883. "Letters to the Editor: That Mysterious Light at Amsterdam." *New York Times* (August 24).

1852: Mysterious Sound

QUITE AN ALARM WAS created in the Public and Ward Schools yesterday, by the crashing peal of thunder which startled the City at 2 P.M. The children in the Greenwich-avenue institution were very near a panic, and only the promptness and energy of the Principal Mr. McNally, and his Assistants, prevented an irruption [*sic*] of frightened parents in search of the killed and wounded. As it was, the doors were closed, and the alarm blew over. It was at first thought that the walls had fallen. The phenomenon of the thunder-crash was the more astounding, because the sky was perfectly clear, and the Sun shining brightly. In the Primary Schools of the Eighth Ward, there was for a time much confusion."

Source:

Untitled, 1852. *New York Daily Times* (June 23).

1855: Big Serpent in a Small Pond

IT IS REPORTED, from how reliable a source we are not prepared to say, that a serpent of huge dimensions has recently been discovered in Rias Pond, a small body of water some two and a half miles north of this village. This pond . . . is represented as bottomless, and the impression prevails that it connects by a subterraneous passage with [Silver] lake, and that his snakeship on retiring from that body of water has either accidentally or intentionally made his appearance in this vicinity."

Source:

"A Monster Snake in Rias Pond," 1855. *Hornellsville* [New York] *Tribune* (September 20).

1858: Unmeteorlike "Meteor"

A METEOR WAS OBSERVED at Buffalo on Monday night [July 19]. Its proportions were extremely large, appearing to be at least half a mile in length and exceedingly brilliant. Its course was eastward in a zig-zag manner, and for over five minutes was plainly visible, when it exploded, creating as fine a display of fireworks as could be desired to be witnessed. Those who saw the phenomenon describe it as particularly splendid."

If the duration and trajectory are accurately characterized, this was not a meteor in any conventionally understood sense.

Source:

Untitled, 1858. *New York Times* (July 23).

1860: Snake from on High

DURING THE VERY heavy shower of rain which fell here on the evening of July 3d, I heard a peculiar noise at my feet, and, on looking down, I saw a snake lying as if stunned by a fall from an immense hight [*sic*]. The animal was about a foot long, and of a gray color. I had previously heard of similar occurrences, but now, judging from ocular demonstration, I verily believe that his 'snakeship' had never before seen South Granville, or, indeed, any other part of *terra-firma*."

Source:

Ruggles, William, 1860. "Raining Snakes." *The Scientific American* 3, 7 (August 11): 112.

1866: "Minotaur, Man, Beast, or Demon"

Some vague and ill-defined rumors have reached us of late, concerning a *weird* and mysterious phenomenon, or a succession of supernatural scenes and *outré* developments, occurring some five or six miles from Middletown, in the town of Mount Hope. For reasons which will readily occur to the reader, the parties immediately affected by these strange demonstrations have been very reticent in the matter, and request that names and locality may be withheld from the public until further investigations shall unravel the mystery, or an adequate cause be assigned for what now seems to be unearthly, if not demoniacal, in its origin.

"Some time in the latter part of March last, the residents of a quiet and humble hamlet, standing near the confines of a densely-wooded lot of some four acres, and near the highway leading to the village of Mount Hope, were awakened from sleep by a novel and singular sound, which at first fell faintly, but defiantly upon the ear. It was as if a large choir, or a multitude of voices, were harmoniously humming, or singing with closed lips, a solemn, funeral dirge, the refrain, after regular intervals of silence, resembling, as near as the sound could be rendered, the word *zinzah—zinzah—zinzah—* three times repeated, and, after a moment's pause, succeeded by the same sad, monotonous murmuring as before. This continued for nearly half an hour, when the awed family were summoned to a display of some of the most discordant and unearthly sighs and groans that ever fell on mortal ears. Every phase of suffering, every expression of mental and physical pain which may be conceived of, was uttered forth with a fierce intensity. . . .

"It is not strange that the animals on the premises shared the wonder and amazement which possessed the people. This was manifested by the mingling of their own particular cries and instinctive utterances with the general discord, thus adding an accompaniment, and chorus to the distracting din, and heightening the pandemonic effect of this diabolical overture.

"Amazement, of course, had now almost found its extreme limit, and the people were prepared for any further change in the sardonic program which the invisible performers might still have in reserve. They had not long to wait. Presently the darkness of midnight suddenly changed to the broad glare of noonday. It was as if a thousand waving festoons of electric light had swooped down from the murky heavens to infold [*sic*], with gauzy wreaths of flame and tongues of fire, the trunks and limbs of trees barren of foliage. They sparkled and shone now with a wealth of flashing light, and grotesque, but entrancing forms of beauty. Far off, the northern heavens seemed to catch and reflect from horizon to zenith the auroric scene; and evolving, ever-changing banners, of the tint and shades of the rainbow, flaunted up the broad expanse. . . . It was soon at midnight! But darkness and silence soon settled upon the earth; the rustling of the trees and low moaning of the winds alone broke the oppressive stillness of the hour.

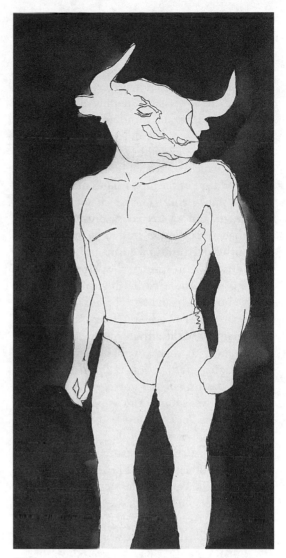

With a deep guttural groan, and still sending forth his fiery breathings, this Minotaur, man, beast, or demon, lifted his disgusting and scaly digits and, slowly and with provoking deliberation, passed them over the face and person of each member of the affrighted family, with a yell that seemed to shake the building.

"But a new and more astounding scene was soon to follow. While the family were exchanging in suppressed and hesitating whispers their surmises, and seeking for some solution of the wonders of which they had been involuntary witnesses, every door in the house was suddenly and violently thrown open by a power unseen, and through the main entrance came stalking, and swaying from side to side, in frightful contortions, an object neither man nor beast, but bearing, in huge and distorted proportions, the shape and form of the upper extremities of the one, and the lower parts, terribly elongated and reeking with mire and filth, and emitting a smell of phosphorous, of the other. The single candle burning upon the mantel went out, and in the darkness which succeeded forth came from the nostril and eye of the monoculous monster a volume of smoke and flame which soon filled the apartment, and so affected the organs of respiration that suffocation seemed to be imminent.

"With a deep guttural groan, and still sending forth his firey breathings, this Minotaur, man, beast, or demon, lifted his disgusting and scaly digits and, slowly and with provoking deliberation, passed them over the face and person of each member of the affrighted family, with a yell that seemed to shake the building; then from foundation to rafters, leaped through the open door at the rear of the house, rushed into the forest, and was soon lost in its pervading gloom and darkness.

"There was no sleep in that dwelling on this eventful night; and the next morning no traces of the phenomenon could be discerned, except that along the track of the strange visitor were found some small particles resembling crushed lava; and the sulphurous and phosphorescent odor which filled the house was proof positive to the inmates that they had not been dreaming."

Source:
"Ghostly Manifestations," 1866. *Titusville* [Pennsylvania] *Morning Herald* (May 15). Reprinted from the *Middletown* [New York] *Mercury.*

1868 and After: Giant

VISITING RELATIVES IN ACKLEY, Iowa, in 1866, George Hull, a cigar manufacturer from Binghamton, New York, found himself in a heated argument. The subject of the dispute, with a Methodist evangelist whom history remembers only as Reverend Turk, was the literal truth of the Bible. Nineteenth-century America saw countless such arguments, as advances in science and technology generated a new spirit of skepticism, but few survive in historical memory. This one does because of what came of it: perhaps the most notorious hoax of its time.

Hull, an atheist, went to bed still irked by the exchange, a portion of which concerned a passage from Genesis 6:4: "There were giants in the earth in those days, and also afterward, when the sons of God had relations with the daughters of men, who bore children to them." As he lay wondering how anyone could believe such a fantastic notion, an idea formed in his head: Why not create his own giant? For one thing, it would show up gullible believers in biblical literalism. For another, it would make him a lot of money. So he would make a petrified man.

In June 1868, on a return trip to Iowa, he purchased a twelve-foot block of gypsum from a Fort Dodge quarryman, remarking that he planned to carve a memorial honoring President Lincoln. Hull and his gypsum went by rail to Chicago, where the hoaxer-to-be hired Edward Burghardt, a stone-cutter, to carve the giant. The two men, with two of Burghardt's employees, worked on the block in a barn on the city's north side. After about two months they had something, but Hull was unsatisfied; it was too clearly a recent creation. Experimenting with ways to make it look older, Hull poured sulfuric acid on it. The acid darkened the statue and gave it the desired effect. He also took a block of wood with needles protruding from it and punctured the statue repeatedly so that there appeared to be pores in the giant's body. It was now ready for burial.

Hull placed the giant in an iron box and took a wandering course around eastern New York, first by train, then by a four-horse team that he and his nephew contracted. In time, apparently, he decided to enlist a relative, Henry "Stub" Newell, in the scheme. At some point—when is not clear—Hull confided the scheme to Newell, who lived on a farm near the tiny village of Cardiff, population 147, 12 miles south of Syracuse. As events would prove, the trip to Cardiff would be fatal to any hopes Hull may have entertained for a hoax he could sustain over the long term. On his way, he and his companion passed through Marathon, well south of Cardiff in the Onondaga Valley, just as the residents were engaged in a raucous community celebration. The wagon and its oversized cargo attracted a great deal of unwanted attention. Many observers asked what the box contained. "Jeff Davis" was Hull's inevitable terse reply, a facetious reference to the president of the late Confederate States of America. Later, many residents of Marathon would recall Hull's passage and his curious secretiveness.

On arriving at Newell's farm on November 9, Hull, his nephew, and Newell dragged the heavy box to the rear of the barn, hiding it under a pile of straw. They waited two weeks to bury it, which they were able to accomplish only with the use of a derrick to lower the 3,000-pound giant into the ground 100 feet east of the house. That done, Hull directed his accomplices to wait a year before launching the hoax into the world.

On October 16, 1869, Newell hired two men to dig a well at the spot where—unknown to them—the giant lay. Two and a half feet below the surface, their shovels struck something. When Gideon Emmons hunched down to discern the nature of the obstruction, he spotted a foot. "I declare," he declared, "some old Indian is buried there." He and Henry Nichols scraped off the earth covering the giant and were duly astonished when they got a full view of the ten-foot figure. In short order, a crowd of gawkers had assembled. One of them, Silas Forbes, remembered the sight: "The form of a man, lying on his back, head and shoulders naturally flat; at hip a trifle over on right side; the right hand spread on the lower part of the abdomen, with fingers apart; the left arm half behind, and its head against the back opposite the other; the left leg and foot thrown over the right, the feet and toes projecting at a natural angle" (Tribble 1998).

Among the witnesses was a traveling evangelist, John Clarke, who after observing the wonder hastened back to Syracuse and spread the word. The city's papers were soon trumpeting the discovery, and before the week was out, the New York City press was doing the same. A pamphlet heralding the "American Goliath" was issued soon afterward. Hundreds of people descended on Newell's farm every day, disrupting his normal work and overwhelming his capacity to deal with them. He built a tent around the giant and started charging twenty-five cents a head admission. He soon doubled the price. Cardiff residents built a makeshift restaurant close to the site.

Only a few were suspicious. Mostly, debate centered on the question of whether the giant was a petrified man or a centuries-old statue. Clergy who took the former view incorporated the Cardiff Giant, as most would call it (though for a while the "LaFayette Wonder" competed with it, LaFayette being a village just east of Newell's farm), into their sermons, citing it as proof of the Bible's inerrancy. The Onondagas, who lived on a nearby reservation, thought the remains were of a prophet who had warned of the coming of the white people. Other onlookers speculated, apparently seriously, that the giant was the twin brother of the Colossus of Rhodes, or one of the oversized men mentioned in the Old Testament. Scientist John Boynton, who examined the statue the day after its discovery, was sure it was a statue. Because it had Caucasian features, it must have been buried three centuries ago by Jesuit missionaries, he reasoned. Lay theorists held the giant to be a statue brought to America by ancient Egyptian explorers. Critics of the statue theory pointed to the absence of a pedestal and to imperfections in the features that no self-respecting sculptor would have let get by. Petrified man or statue, though, most agreed with the pronouncement of no less than the New York secretary of state, E. W. Leavenworth: "It has the mark of ages stamped upon every limb and feature, in a manner and with a distinctness which no art can imitate."

The Cardiff Giant's days in Cardiff were soon over. On October 21, a group of Syracuse businessmen purchased a three-fourths interest in the wonder and hauled it up to the city. The curious flocked to see the giant in numbers large enough to gratify local merchants, and Syracuse's economy boomed as the excitement continued. It was all too good to last, however.

By now, Boynton had begun to question his earlier identification of the giant as a genuine historical artifact. Water, he discovered through experimentation, dissolved the gypsum in reasonably short order; therefore, the giant could not possibly be old. His experiments led him to a precise conclusion: The statue could not have lain in Newell's soil for more than 370 or 371 days. In a letter published in the November 25 issue of the *Buffalo Courier,* Yale University paleontologist Othaniel C. Marsh, who had also examined gypsum's solubility, stated flatly that the giant was "of recent origin, and a decided humbug" (quoted in Tribble 1998).

Boynton's investigation did not stop in the laboratory. He interviewed persons who had seen the box-bearing wagon make its way to Cardiff a year earlier. As the story became national, residents of Fort Dodge, Iowa, recalled Hull's purchase of a large block of gypsum. Bowing to the inevitable, Hull admitted to the hoax in a December 10 interview with Syracuse reporters.

Even after being exposed, the giant managed to draw money-paying crowds. P. T. Barnum, the great carnival promoter, had tried to purchase it before the giant's true nature became known. He was turned down, but he hired a Syracuse sculptor to create a duplicate. It was so successful—notwithstanding Hull and his partners' failed attempt to sue Barnum—that it outdrew the original when both were on display in New York City. Most people were more amused than outraged to learn they had been fooled, and for a time the giant remained popular. When its star faded, it fell into the hands of S. S. Lawrence, who kept it in storage. Around the turn of the century, he sold it to C. H. White, who hoped to profit by displaying it at the Pan-American Exposition in Buffalo in 1901. Unfortunately for him, the giant brought few customers. In subsequent years the giant made infrequent appearances at state fairs in Iowa and New York. For a while, Des Moines publisher Gardiner Cowles owned it, keeping it in his rumpus room.

The Cardiff Giant survives in American memory as a colorful humbug. It was also the inspiration for a quotation usually and mistakenly attributed to Barnum. After a New York court ruled in Barnum's favor in the dispute over the "real" and imitation giants, one of the investors in the former, David Hannum, was inspired to quip, "There's a sucker born every minute."

The New York Historical Association purchased the giant from Cowles in 1947. Since 1948, the Cardiff Giant has been on exhibit at the Farmer's Museum in Cooperstown, New York. Barnum's giant can be seen at Marvin's Marvelous Mechanical Museum in Farmington Hills, Michigan, just north of Detroit.

Sources:
Feder, Kenneth L., 1999. *Frauds, Myths, and Mysteries: Science and Pseudoscience in Archaeology,* 3d ed. Mountain View, CA: Mayfield, 40–54.
"The Great Cardiff Giant Hoax of 1869," n.d., http://www.cardiffgiant.com/hello.html.
Tribble, Scott L., 1998. *Giants in the Head: The Cardiff Giant in American Historical Consciousness, 1869–1998,* http://www.stribble.com/cardiffgiant/.

1874: Blood from Nowhere

WE LEARN OF A RATHER singular occurrence that took place regarding a little girl, which happened last week. The girl is nine years of age and her parents reside about two miles west of this village [Havana] on the road to Townsend Settlement. While quietly sitting on her mother's lap one evening last week, she suddenly started up in great fright, crying with violence, and it was some time before she could be pacified. There was no apparent cause of her terror. But the next night, the household having an unusual amount of company, it was the intention to prepare a place for the little girl to sleep on the carpet; but to this she would not hear, and would only consent to sleep in a bedroom adjoining if the door could be fastened tightly.

"After she had been placed in bed a few hours she started up in wild affright, screaming at the top of her voice. Some of the members of the household hastened with all possible speed to her bedside, and by the light, hastily brought, found her face completely covered with blood. Carefully washing her face and head, no marks which could seem to have drawn a drop of blood could be found about her person. The little creature had persisted that she dare not sleep on the floor, alleging that some one would come and kill her.

"The next night a lady who was visiting at the house slept with the little girl in the bedroom to allay her fears. The girl nestled down towards the middle of the bed, under the clothes, and fell asleep, but after a few hours awoke again with the same terrible affright which characterized the evening before. The lady immediately reached her hand down where the child was and found a substance upon her face and the bed which felt like blood, and striking a light found it to be so, and in such a quantity that she could have dipped it up with a spoon. The child's face was again washed but no marks could be discovered, except a very slight one on the chin and cheek, which had evidently been caused by the scratch of a pin. The house was afterwards searched for some clue, but nothing has yet been discovered to account for the phenomenon. We do not learn whether the occurrence has taken place since or not."

Source:

Untitled, 1874. *Hornellsville* [New York] *Tribune* (June 12). Reprinted from the *Havana Journal.*

1875: Balls of Fire in Church

LAST SUNDAY EVENING [March 28] a Mr. Cozzens delivered part of a lecture on temperance at the Methodist Church in this village [Olean]. The other churches out of courtesy had no evening services, and as a consequence the house was well filled with people of every denomination, who assembled to listen to the remarks of Mr. Cozzens. While the lecture was in progress a heavy thunder storm arose, which at times almost drowned out the words of the speaker. During the climax of the storm a blinding flash of lightning illuminated the building, and after it subsided there were observed for an instant on either side of the room balls of fire which exploded with a hiss, throwing electric flashes all through the house. This phenomenon was immediately followed by a terrific peal of

thunder which shook the building to its very foundation. Then there was a general stampede for the door by the excited and terrified crowd and a wild scene of confusion ensued. . . . The speaker called wildly to the crowd to stay and receive the benediction, but the appeal was of no avail."

Source:

"Excitement in a Church," 1875. *Hornellsville* [New York] *Tribune* (April 2). Reprinted from the *Olean* [New York] *Times.*

1878: Snow Worms

A DRIZZLING RAIN fell monotonously on Lockport on January 10, and as the day turned into night and grew colder, it began to snow. By morning, wet snow covered the ground.

As Union School principal A. B. Evans, who lived on the city's far southeastern side, trudged through it on his way to work, he developed an odd feeling that something about the snow was unusual. Reaching down, he retrieved a handful of it and looked it over carefully—to discover, as the *Buffalo Sunday Courier* reported on the 13th, that "a large portion of what he held in his hand . . . consisted of worms." The worm-infested snow covered an area of a few hundred feet. Each handful, Evans determined, contained anywhere from fifteen to fifty of the creatures. They looked like the sorts of worms one encounters (if one is unlucky enough) inside apples, though he thought they looked a little larger than average.

The paper reported: "Their length ranged from one-fourth to nearly an inch and a half, and in color they were a sort of yellowish white. Their heads were black and their bodies fat, indicating . . . that they must have had plenty to eat, wherever they came from. They appeared to be dead and neither pricking nor heating them [elicited] any signs of life. Under the microscope, however, it could be seen that they had blood which circulated—or what answered with blood" (quoted in Whalen 1985). As the snow started to melt, the worms became shapeless, liquefied, and vanished, except for the few specimens Evans was able to preserve.

Oddly, however, worms fell in a snowstorm that very same morning in Batavia, also in the western part of the state. According to the *Batavia Times* of January 12, "A singular occurrence took place about eight o'clock Friday morning. . . . In the backyard of Mr. S. E. Hovey on School Street, and over several gardens adjoining him, a shower of light brown worms, alive, about half an inch long, fell in that locality and were thickly scattered over new-fallen snow. Mr. H. brought us a basket of snow with the worms in it. . . . It is a most remarkable incident and naturally causes the inquiry as to where these little wigglers came from" (quoted in Whalen 1985).

Source:
Whalen, Dwight, 1985. "Wigglers in the Snow." *Fate* 38, 6 (June): 46.

1878: Spirit Fire

JOSEPH B. SHEPPARD WAS a night watchman who patrolled Third avenue between Thirty-fifth and Fortieth streets [in New York City]. In March, 1878, he aroused a sleeping man at Thirty-ninth street and First avenue. Soon afterward a heavy body was heard to fall with a sickening splash into the East river. Sheppard was missing from that very moment, and investigation led to the belief that he had met with foul play at the hands of the wayfarer whose slumbers he had interrupted. His body was soon recovered and buried near his son's house.

"Shortly after the funeral the east side became a-flutter with excitement. On the water's edge near Thirty-ninth street hundreds watched a mysterious floating flame. This supposed spirit fire danced aimlessly about and glided just over that part of the river's surface where Sheppard's body was seen to sink. When pursued, this spectral flame vanished like breath in the wind.

"Later a strange figure was seen to haunt the Thirty-ninth street wharf. Many identified it as the ghost of Sheppard. The phantom, it was said, would gesticulate toward those with whom that ill fated man had happened to be acquainted in life, but ignored the many who had

been unknown to him. It always appeared to be arrayed in the clothing and slouch hat worn by him on his nightly rounds.

"These visitations are said to have continued throughout the spring of 1878, and so vivid was the resemblance of the weird shade to Sheppard that many who viewed it grew to believe that it was really the old watchman still alive, masquerading in the darkness, and that his family had been mistaken in the identification of the corpse."

Source:
"Weird Tales of Specters," 1901. *Fort Wayne* [Indiana] *News* (December 28).

1880: Sailing on Bat's Wings

IN ITS SEPTEMBER 12 edition, the august *New York Times* reported, to all indications with a straight face, that the previous week a decidedly odd something had made an appearance over Coney Island.

Flying at about 1,000 feet, a "man with bat's wings and improved frog's legs" was seen heading in the direction of the New Jersey coast. Though it is not clear how such a determination could be made at such a distance, witnesses allegedly claimed to discern a "cruel and determined" expression on the frogman's face. "The movements . . . closely resembled those of a frog in the act of swimming with his hind legs and flying with his front legs," the *Times* asserted.

The witnesses were characterized as "reputable persons."

Source:
"An Aerial Mystery," 1880. *New York Times* (September 12).

1882: Visions of Mary

THE *NEW YORK TIMES* reported on March 5: "The lower portion of this city [Troy] is greatly excited over the alleged miraculous appearance of the Virgin Mary in a house in First-street. Eight women and five men assert they saw the apparition. It appeared beneath a picture of the Virgin Mary and the angels, which hangs on a wall. Superstitious persons are visiting the house and carrying away bricks and mortar. The Rev. Father Quigley scouts the miracle theory, and others say the illusion is caused by sunlight dancing on the wall."

The next day, another article appeared in the *Times*. It said: "The Roman Catholics and many people of other denominations . . . are greatly excited about the alleged daily appearance on one of the walls of the residence of Thomas Jones, in the rear of No. 300 First-street, of a picture of the Virgin Mary with an infant in her arms, and a cross with the letters I.H.S. near by. . . . The *Times*'s correspondent, in company with several other newspaper men, visited the Jones's [*sic*] house this afternoon [March 5]. It is a plain, one-story dwelling, with whitewashed walls in the interior, and the rooms are scantily furnished. Mr. Jones, who appears to be a man of considerable intelligence, was not in a talkative mood, but he finally told the following story: 'I am not superstitious, but I am forced to believe what I see—what I see with my own eyes. I did not at first believe the thing was real any more than you do now, and when the women told me of the apparition, several days ago, they appeared greatly alarmed, and I tried to laugh away their fears. Right here,' (pointing to a space on the west wall,), 'about 5 o'clock every afternoon a picture, at first very dim, but gradually increasing to remarkable brilliancy, appears.'

"The reporters approached the spot indicated, one of them holding a lamp while all of them critically examined the wall. There were traces of a cross and the letters I.H.S., but nothing resembling the Virgin Mary could be seen. As the reporters were about to leave[,] the women went on their knees repeating an Ave Maria. Their action attracted attention, and as the visitors gazed at them some one exclaimed, 'There it is again,' and, sure enough, a remarkable sight was witnessed. Standing out in bold relief was a picture of a cross, and near it was a mass of swirling light, from which no well-defined drawing could be discerned. Every one was mystified, and all expressed an intention to

fathom the mystery. A number of superstitious persons have chipped pieces from the building for relics."

Sources:
"An Alleged Miracle in Troy," 1882. *New York Times* (March 5).
"The Alleged Troy Miracle," 1882. *New York Times* (March 6).

1883: Interplanetary Sword

SOMETIME AFTER MIDNIGHT on April 17, physician T. O. Keator, riding along the bank of Rondout Creek in Ulster County, watched a ball of fire—of the apparent size of a cartwheel—descend from the sky with a whirring sound. It plunged into the ground less than a dozen yards from him, leaving a cloud of steam in its wake. Though he wanted to stop, his horse was spooked and beyond Keator's control.

In the morning Keator spoke with the farmer, Daniel D. Bell, who owned the property. Bell searched without success for the object, by now presumably hidden within the ground. But on May 7, his son, Raloy H. Bell, sixteen, and a playmate noticed something in the creek, 2 feet under the water, at the spot. It looked like a bent, rusted scythe, but when, with some trouble, the object was dislodged and brought to land, it proved to be something more interesting: a long, two-handed sword. Young Bell took it to his father, who straightened the sword and polished it. On the May 16, the youth took it to a law office in New York City. A reporter who saw it described it this way:

"The sword is 5 feet 10 inches long, weighs 17 pounds, and the two-edged blade is 2 3/4 inches wide. The handle, which is thick and almost square, and protected by two guards, is made of three pieces riveted together, the center one being a continuation of the blade. Above the hilt is another guard, which projects on either side 10 inches, and from the center of which are two prongs, such as are found on modern carving-forks to protect the hand from an unskillful thrust of the carving-knife. The lower half of the weapon, including the hilt, is covered with strange and puzzling hieroglyphics, which look as much as anything else like an outline drawing of the map of Greece. They are a succession of dots made by punching the metal with a sharp instrument. Some of the figures are labyrinthal in their character, and, like a wagon tire, seem to have no beginning nor ending. Others look like letters of the alphabet, with just enough dissimilarity to baffle their translation. Several have intricate geometric shape, so mixed with meaningless lines as almost to destroy the outline. The adjustment of the weight is so poor that it would be practically useless as a weapon of attack . . . and it looks as if its artificer had a museum in view when he formed it.

"But this could not have been so if the conjectures of Ulster County are correct. They agree that it came down from some far-off planet, warped by the heat from its rapid journey through space and rusted by its short rest on the river bottom. It is believed, then, that some sanguinary giant had it wrested from its grasp and flung over the edge of his planetary residence, whence it whirled away through millions of miles of space on its trip to this little globe. Others think it might have been forged in one of the volcanic craters on the moon and fell through. The fortunate owner has refused $1,000 for it, and one of his neighbors has offered to give his farm in exchange for the curiosity."

The *New York Times* brought antiquities dealer Gaston L. Feuardent to the office to examine the sword. He concluded that it was, to all appearances, a badly executed forgery. "It has not one solitary point to indicate that it was not made in New York State within the past year," he declared, noting that it was "very poorly made, too." He remarked, "It is of iron and so badly balanced in weight that it could never be used, a blunder that the ancient blacksmith never made."

Source:
"Born of a Ball of Fire," 1883. *New York Times* (June 17).

1883: Lights along the Track

FOR SOME TIME PAST it has been reported that a peculiar and mysterious view occurs on dark nights, on the Central railroad, just west of the river bridge. Shortly after the arrival of the fast mail train from the west, a red and a white light are seen to move slowly eastward as if carried by human hands. After coming a short distance to the river bridge, both lights ascend and suddenly disappear. Engineers on westward-bound trains, while taking a water supply, have noticed the lights and waited for them to disappear before moving on. Night watchmen and chronically-sober men assert that this apparition can be frequently seen, but the cause is a mystery, as the effect is produced apparently without human agency."

Source:
"Railroad Ghost," 1883. *Newark* [Ohio] *Daily Advocate* (August 3).

1885: Mountain Light

ON THE NIGHT OF the recent great snow storm," a *New York Sun* correspondent related, a "curious phenomenon" was seen in the high rocks—hundreds of feet in altitude—opposite the village of Mongaup, 5 miles west of Port Jervis, in Sullivan County. According to the story:

"On the Pennsylvania side of the river the Pike County hills rise steep and rocky. . . . While the storm was at its height a bright light suddenly flashed among the high rocks opposite Mongaup, near the summit of the range. There was no blaze, but the snow-covered rock glowed like red-hot iron for a distance of several feet around. The light resembled a huge bed of live coals, and cast a weird reflection on the snow, throwing the trunks and lower branches of the bare trees into strong relief. The light gleamed through the storm for several minutes, and then gradually grew dim and disappeared.

"As it would be difficult and hazardous for any person to reach the spot where the light was seen, even in the daytime and in summer, it is not thought that any one could have possibly clambered to the spot at night, up the steep ledge, covered two feet deep with frozen snow, and through a blinding snow-storm. The mysterious light is, therefore, not charged to any human agency, and the superstitious look upon its strange appearance as some kind of omen. An old inhabitant says that in 1830, when cholera raged through this county, similar lights appeared on the mountains in the winter and spring. He says it is another warning that cholera is on its way."

Source:
"Mysterious Light," 1885. *Atchison* [Kansas] *Globe* (April 15). Reprinted from the *New York Sun*.

1886: A Soap-Bubble Sky

TWO MEN ON HORSEBACK rode down a hillside as they approached Cherry Valley from the east. It was around sunset on a fairly warm day in April 1886, and the snow was melting. Nothing appeared out of the ordinary until both saw something so strange that neither mentioned it at first to the other for fear that he was imagining something that made no sense. Finally, one of them said, "I wonder whether I'm dreaming." The other replied that he had been wondering the same thing.

Beneath them and on the northwestern side of the road was an apple orchard. A mass of spherical objects, of generally uniform size, some six inches in diameter, but most eight or nine, was descending from the sky onto the trees. The objects, which looked like bubbles, would come to rest on the branches or on fence tops, then roll or slide off to disappear from view.

As they continued toward the village, the sky seemed full of bubbles. They were, the journal *Science* reported, "highly colored, iridescent, gave the same sort of reflections as soap bubbles, and apparently vanished individually in much the same way." As darkness fell, they were no more to be seen.

Source:
Swinnerton, Henry U., 1893. "Aerial Bubbles." *Science* 21: 136. Cited in William R. Corliss, ed., 2001. *Remarkable Luminous Phenomena in Nature: A Catalog of Geophysical Anomalies.* Glen Arm, MD: Sourcebook Project, 187.

1886: "A Marine Monster 100 Feet Long"

SOME INTERESTED attention has been directed of late to various excited reports of a monster sea serpent seen in the Hudson a few miles above New York and during this week seen in the Connecticut river. In the early summer the same serpent was seen by some school teachers off the coast of Long Island. Dozens of people are ready to swear that what they saw was really a marine monster 100 feet long, with a huge head which it from time to time reared aloft ten feet in the air. However, one man says it was a huge log having a single long branch at the end which appeared above the water at intervals. Need I remark that the public, though dying to behold a sea serpent, flatly refuses to believe the dozens who say they saw one, and obstinately pins its faith to the single man who says it was a log?"

Source:
"Our New York Letter," 1886. *Statesville* [New York] *Landmark* (September 16).

1886: Ghostly Boatman

AN ALLEGED WHITE-ROBED spirit from the other world has disturbed the quiet of East One Hundred and Twenty-second street [in New York City], and has been seen by a number of prominent residents in that vicinity. At the foot of East One Hundred and Twenty-second street, and hardly five feet from the Harlem river, stands a two story brick building. The house has stood there for many years, and old residents say that it was formerly the dwelling place of the old boatman who ferried the residents of Randall's Island to and from the city, and whose mutilated body was found about fifteen years ago at the door of his little boathouse.

"Of late years the old house has done duty as a stable, and a stranger can hardly pass it without glancing with a shudder at the heavy iron shutters and massive iron door. On the same side of the street and a few feet away the boatman who now ferries the passengers to Randall's Island has a small boathouse.

"It is here that the spirit of the murdered boatman is said to walk, always after midnight, and there are those who are positive that they have seen the phantom spirit in a phantom boat pull slowly away from the boathouse in the direction of Randall's Island, and lose itself in the shadows on the other side of the river.

"Mr. John Conklin, a well known saloon keeper of Harlem, who lives in that neighborhood, had occasion on Saturday night [probably December 3] to pass the old boathouse. He was accompanied by a young man by the name of Mouhlin, and as they reached the foot of the street they heard some one moving about in the old boathouse. As the last boat for the island had left nearly two hours before, the noise seemed to be suspicious, and they determined to watch.

"Concealing themselves in the grounds of the Eastern Boulevard Club on the opposite side of the street, they eyed the door of the boathouse, and almost fainted from horror when they saw the door swing back and a ghostly form step out on the sand. It was with difficulty that they stood the ordeal, as both felt a desire to put distance between them and the spot where the ghost stood shading his face with one long, skinny hand and gazing with eyes that seemed to blaze in the direction of the boathouse on the opposite side of the river.

"The form turned as though to enter the house, when a hoarse blast of some tugboat on the sound broke the stillness, and the spirit disappeared like a flash and the old door swung back. The two watchers were about to leave the spot, thinking that the spirit had disappeared for the night, when once again the old door swung open and the spirit again stepped out of the boathouse.

"Shading his eyes as before, as he gazed out on the river, and, seemingly satisfied by the stillness, [he] grasped the bow of a long, low and narrow boat, and drew it out on the sand. An old fashioned square lantern stood in the bow and shed a bluish light on the scene. Pushing the boat into the water, the spirit took a pair of oars from the bottom of the boat and slowly rowed in the direction of the island and soon disappeared. The two watchers could hardly believe what they saw, and after waiting some time for the spirit to reappear they went to their homes.

"The story is confirmed by Captain Bull, of the clamboat Maria Louise, who swears that he saw the spirit as it appeared to Mr. Conklin and Moulhin. When seen by a Journal reporter yesterday the captain gave the following version:

"'I had been up to Cow Bay all day, and it was about 12:30 when I ran through little Hell Gate and headed up the river for my anchorage on One Hundred Twenty-first street. There was hardly any wind on the river, and I was drifting slowly with the tide, when my attention was attracted by a long, low, white object, drawn up to the stone steps at the boathouse on Randall's Island. I was almost half asleep; but I tell you I nearly fell off the deck when I saw a white, spectral form slowly descend the steps and seat itself in the boat.

"'My dog Fannie, which had been sleeping on the deck, commenced to whine and darted into the darkest corner of the cabin. As I let go my anchor and lowered my sail, the spectral boat moved away from the island, and as the tide was moving up strong it headed down in my direction as though it wished to make One Hundred and Twenty-second street. It passed within twenty feet of my boat, and to save my life I could not summon up enough courage to hail it.

"'The hands of the oarsman were covered with blood, and its long white hair fell almost to the shoulders, and in some places was stained with blood which flowed from a cut in its hands. I watched it as far as One Hundred and Twenty-second street, when it disappeared.'

"Captain Bull vouches for his story, and although not a believer in the supernatural, is sadly at fault over his vision. Neither of the other two men drink[s], and their story is vouched for by themselves."

Source:
"A Spectral Oarsman," 1886. *Atlanta
Constitution* (December 12). Reprinted from
the *New York Journal*.

1891: Fallen with the Snow?

A CURIOUS SIGHT that may be worthy of
mention was seen by several persons one
morning last week [at Lyndon]. A large number
of worms, yellow in color, flat in shape, from a
third of an inch to an inch in length, and forked
at one extremity, were seen mingled with and
crawling over the freshly fallen snow."

Source:
"A Phenomenon," 1891. *Olean* [New York]
Democrat (January 29).

1891: Enigmatic Fog

A CURIOUS PHENOMENON occurred Monday
evening [January 26] in the upper part of
the city. About 10 o'clock the almost perfectly
clear atmosphere gave way to a sublime-like fog,
which settled down very suddenly in a distinct
and seemingly solid stratum. The descent was
rapid, and every portion of the cloud of vapor
reached the earth at the same instant. The phenomenon continued only about five minutes,
when the body lifted and disappeared heavenward as rapidly as it had descended."

Source:
"A Peculiar Natural Phenomenon," 1891. *Daily
Nevada State Journal* (January 30). Reprinted
from the *Albany* [New York] *Argus*.

1891: Unparalleled Storm

A PHENOMENON WHICH probably has no
parallel occurred at Buffalo, N.Y., on the
23d., when a storm of shot from the clouds fell
for over an hour."

Source:
Untitled, 1891. *Stevens Point* [Wisconsin] *Journal*
(May 30).

1891: "A Ball of Liquid Fire"

AN ENGINEER ON THE Delaware, Lack-
awanna and Western says he was coming
down the Chenango valley when the recent
storm burst. A vivid flash of lightning startled
him, but he was not prepared for what followed.
A huge ball of fire was seen on one of the rails
coming rapidly toward the locomotive. He shut
off steam and reversed the engine. The lightning, which looked like a ball of liquid fire
about the size of a twelve inch football, struck
the driving wheels of the locomotive and, after
running several times around them, crossed over
on the axles to the opposite side of the track and
went spinning away in the direction from which
it came and vanished around a distant curve.
The engine was not damaged, with the exception of the glass oilers on the side rods, which
were broken, and the paint on the 'driver' was
blistered."

Source:
"The Locomotive Met a Ball of Fire," 1891.
Daily Nevada State Journal (June 11).
Reprinted from the *Binghamton* [New York]
Herald.

1892: Lightning in a Ball

DURING A SEVERE thunderstorm Monday
[September 5] the phenomenon of ball
lightning was seen in this village [Lyons]. An inspection of the locality shows that the ball was
located between a telephone wire and a conductor pipe about three feet distant, and was doubtless of the nature of an electric brush preceding
the disruptive discharge. It was of a reddish
color and exploded with a report like a musket,
but did no damage, nor was it attended by any
smell perceptible to those who saw it, although
they were distant not more than five feet."

Source:
"Ball Lightning," 1892. *Daily Nevada State
Journal* (August 13). Reprinted from *Science*.

1894: A Living Cloud

AN UPSTATE FARMER TOLD a reporter from the *Utica* [New York] *Observer,* "While driving home from Oneida the other day, I saw a big cloud moving due north over the fields and woods. There wasn't any wind blowing; the air was still and I was unable to account for the presence of a big dark cloud speeding away across the heavens on such a still, bright day.

"At first I thought it was a cloud of smoke from the railroads, but then when I first saw it the cloud was in such a position that it could not possibly have come from the West Shore Railroad, and even if it had been there never was a cloud of smoke that hung so closely together and so long as that did. As I sat in my wagon it appeared to me to be a mile long and perhaps half a mile wide, but of course that part of it was all speculation, for no one can make a very accurate guess of the size of a cloud. The body in the sky was as dark as the smoke from a locomotive and looked to be quite dense. It traveled quicker than any cloud ever scudded before a thunder shower in this section. When it first attracted attention it was high up in the heavens, but it rose and fell several times, like the scaring of a bird. Once it was but a few feet above the top of some woods. Again it took an upward course and continued onward in an unswerving north course. It was about five o'clock that the cloud passed.

"That evening I noticed a number of reddish-winged wood ants about on the grass and in the roads. It occurred to me that the strange cloud in motion might have been a cloud of these flying ants. The more I pondered over the phenomenon the more I became convinced that it was a cloud of ants that passed over the country. Such a story was too big for me to tell, although there was proof enough of the fact for my mind, so I held my peace and simply spoke to my family of the strange cloud. Others had seen it, too, yet none suspected what it was and we finally dismissed it.

"A day or two afterward I was in Constableville and there the farmers told me they had seen the same thing. There was no doubt about it, either, for a number of them watched the cloud and at that place it passed so low that they caught the insects in their hands. They were the same flying ants. We compared notes and found that it required just an hour for the swarm to move from the place where they were first seen in Constableville. The distance in a straight line is thirty-one miles. They were in Oneida county at five o'clock and at just six o'clock they were seen in the north. The ants continued northward and nobody has told me where they stopped."

Source:

"The Cloud Was Alive," 1894. *Freeborn County* [Minnesota] *Standard* (November 28). Reprinted from the *Utica* [New York] *Observer.*

1895: Toads from Clouds

LAST FRIDAY [PRESUMABLY June 7] during a heavy shower at Dunkle, a lumber hamlet near Morrison, on Kinzua creek, twelve miles north of this place [Kane], thousands of little toads seemed to fall from the clouds, until they nearly covered the ground. They were very frisky and remained hopping about for some time after the shower was over, when they disappeared about as mysteriously as they came. Will our naturalists explain this remarkable phenomenon, which is vouched for its truth by one of the reputable citizens of the town where it occurred, and an eye witness?"

Source:

Untitled, 1895. *Olean* [New York] *Democrat* (June 14). Reprinted from the *Kane* [New York] *Democrat.*

1896: "A Wild Unearthly Shriek of Laughter"

'I NEVER BELIEVED IN things supernatural,' said the Four Mile citizen, 'until I chanced to pass that house on Monday night [February 17]. I had been to Olean and started to drive to Knapp's Creek. The night was dark . . . and a fitful wind sighed through the trees along the river road. I passed through North Pole [a small settlement] at the solemn hour of midnight and proceeded on my lonely way until within a short distance of a log hut which for years has stood near the roadway. Suddenly the air was filled by a wild unearthly shriek of laughter that caused my horse to plunge from the road and which tingled the blood in my veins. As soon as I could control my horse I glanced in the direction of the house and saw a frightful sight. The door of the cabin was open and in the aperture stood a weird figure. The specter was very tall and clad in white. The specter was adorned with uncanny whiskers and serpentine coils of hair was [sic] hanging down its back. As I struggled to control the horse, the specter waved its bony arms and emitted another shriek. This exhibition was more than my shattered nerves could stand and I let the horse skate in the direction of Knapp's Creek as fast as it could go.

"'I have since learned that an aged man[,] who lived in the hut many years ago, was mysteriously murdered and his body buried in the lot behind the house. As I remarked before, I don't take much stock in spirits of the uncarthly kind, but I must admit that the specter I saw that night was a sure-enough spook with all the necessary trimmings.'"

Source:
"A Ghost Story," 1896. *Olean* [New York] *Democrat* (February 21). Reprinted from the *Bradford* [New York] *Era*.

1896: Asylum Phantom

THE BIG LUNATIC asylum at Bloomingdale, N.Y., has a ghost. The spook has been seen by several, driving a ghostly white horse attached to an old-fashioned buggy."

Source:
"Ghost at Insane Asylum," 1896. *Fort Wayne* [Indiana] *Gazette* (April 29).

1896: Floating across the Road

MRS. E. C. SAMUELS and Mrs. M. L. Finkelmeier drove to Bloomingburgh yesterday. On their return in the evening, between 8 and 9 o'clock, they received a very severe fright. Neither of the ladies have [sic] heretofore believed in the existence of ghosts, but they are sure if there is such a thing they saw one last night. It disappeared on one side of the road a short distance ahead of them, and as they drew near floated across the road directly in front of the horse and then faded from sight."

Source:
"Saw a Ghost," 1896. *Middletown* [New York] *Daily Argus* (July 20).

1896: Angel Overhead

THE VILLAGES OF South Butler, Butler Center and Slyburg, all in Wayne county, are greatly excited over an apparition which has appeared in that section recently and for which there appears to be no possible hypothesis except a spiritual one. The visitor is nothing less than an angel, or, at least, it bears every appearance of one, and is seen in broad daylight in a clear sky. Its first appearance was on November 2 about 3 P.M., when it was seen by at least a score of people within a radius of five miles of Butler Center. R. D. Mock, a farmer living near Slyburg[;] M. B. Newton, a mechanic whose home is in Savannah, N.Y.; [and] A. W. Miller, farmer, and D. F. Everhart, school teacher, both of Butler Center, have seen the apparition.

"Mr. Everhart says, 'I was returning from Wolcott at 3 o'clock on November 2, when I noticed an object floating in the air about 100 yards away and nearly over my head. Its appearance was that of a girl about 20 years of age, clad in a long white robe with the arms bare. On its shoulders was a pair of long white wings which were nearly motionless. At first the features were clearly visible, but as the object floated higher, they gradually became indistinct. The day was clear, not a cloud in sight. I saw it for fully twenty minutes. To be convinced that I was awake and not dreaming, I even pinched myself. I said nothing for several days, fearing ridicule, till I heard others speaking of similar occurrences.'

"The story told by others named is the same in all essential details and it has created great wonder and some alarm throughout that section."

Source:
"Saw an Angel," 1896. *Olean* [New York] *Democrat* (November 20).

1897: Lawless Airship

THAT AIRSHIP TO WHICH all the Western papers and some of the Eastern ones are now giving the benefit of several doubts, is visible nightly to correspondents in all parts of Illinois, Wisconsin, Michigan, Indiana, and Iowa. The aerial voyager evidently is a Mahatma of high development, for he exists in a dozen places at once and has a different appearance and does different things in each of them. It is lamentable to observe, however, that everywhere he violates the United States navigation laws in the matter of lights. Usually the Captain of this new craft shows only one light. It is most often white, and occasionally red. If this were the worst of his offenses it might be endured, but the ignorance he displays in the matter of side lights is simply criminal and will get him into serious trouble the moment a United States officer succeeds in bringing him to. While sailing over the town of Kenosha, Wis., Sunday night, the sky traveler shocked and angered all the nautical men living there by showing a green light on both bows. At Waukegan, Ill., at the same hour, both of the side lights were red, which is equally illegal and outrageous. If the airship's owner persists in this lawlessness, he will not only get the reputation of being a pirate, but his vessel will run into itself, sooner or later, and suffer injuries so serious that his next exhibitions will have to be made in several pieces instead of several places."

Source:
"Topics of the Times," 1897. *New York Times* (April 15).

1897: "Marvel of the Skies"

AIRSHIP STORIES ARE now coming in from Irondequoit, a town close to Rochester and not far enough from New York to enable one to believe that the metropolis will much longer be spared by the epidemic which has worked such havoc in the West. Cyrus Wheatley, a man whose very name is a guarantee of bucolic respectability, solemnly declares to the residents of his natal village and to reporters from the city which is Monroe County's pride, that while he and his hired man were returning home Wednesday night [April 14], after ministering to the needs of a neighbor's sick Jersey cow, they saw the now famous apparition. It had the usual brilliant headlight, and green and red lamps were displayed on either side of it in the reckless way which has already been noted as a violation of the navigation laws characteristic of airship Captains. Neither Mr. Wheatley nor his servitor could see the aerial vessel itself, but they know it was there, for the lights couldn't have been careering through space by themselves. This is obviously true, and the inevitable conclusion, unless a farmer can be unveracious, is that the marvel of the skies is now in New York State and will soon be over this defenseless city. Some of Mr. Wheatley's envious neighbors say that what he saw was an electric car on a high trestle, but rumors of a big jug of hard cider go with their end of the story, and we do not care to dwell upon so uncharitable a theory."

Source:
Untitled, 1897. *New York Times* (April 19).

1897: Leaping Ghost

AT AUBURN, INDIVIDUALS awake at 1 a.m. claimed to have seen a tall, white-clad, lantern-carrying apparition. Instead of gliding, it bounded like a kangaroo.

Source:
"Novel Ghost," 1897. *Fort Wayne* [Indiana] *News* (August 4).

1898: "A Perfect Cross"

SEVERAL PERSONS LIVING west of the Nyack Hills witnessed a phenomenon in the sky a little before sunset, Wednesday evening [February 9], that filled them with awe. The sun was a short distance above the horizon, with its face nearly hidden by clouds, when part of the clouds seemed to melt away and the sun shone partly through those that remained with a blood red light in the form of a perfect cross. The sight did not last long, but for a few moments it looked as if a cross of crimson had set in the sky just above the western hills. It was so perfect in form that superstitious persons were very nervous over the strange phenomenon."

Source:

"A Blood Red Cross," 1898. *Middletown* [New York] *Daily Argus* (February 10).

1907: Missing Body

THE BRIGHTON BEACH train leaving for the shore from Sheepshead Bay early in the morning on August 7 was just picking up steam when suddenly the motorman put on the airbrakes. The train groaned to a stop, throwing some passengers to the floor. Nearly everyone but the crew assumed that a wreck had just occurred.

The actual cause, however, was the sudden appearance of a man who showed up in front of the train. The motorman was certain that he had seen the engine strike him and the wheels run over him. The crew, along with a few passengers, got off the train and searched in an ever widening circle, but they never found a body. After fifteen minutes they gave up.

"It was said by one of the train crew," a newspaper noted, "that several of the motormen along the line had had just such an apparition and that they were at a loss to account for it."

Source:

"Car Halted by a Ghost," 1907. *Fort Wayne* [Indiana] *Journal Gazette* (August 4).

North Carolina

1793: Giants in the Mountains

IN A JULY 1793 ISSUE, the *Boston Gazette* published a May 17 letter from a correspondent in Charleston, South Carolina, referring to a communication from "a gentleman on the South Fork of the Saluda River" in western North Carolina. The letter, quoted in an article by Mark A. Hall (2002), described giant creatures said to live on the Bald Mountains:

"This animal is between twelve and fifteen feet high, and in shape resembling a human being, except the head, which is in equal proportion to its body and drawn in somewhat like a terrapin [in other words, no neck], its feet are like those of a negroe [*sic*], and about two feet long, and hairy, which is of a dark dun colour; its eyes are exceedingly large, and open and shut up and down its face, the hair of its head is about six inches long, stands straight like a negroe's [*sic*], its nose is what is called Roman. These animals are bold, and have lately attempted to kill several persons—in which attempt some of them have been shot. Their principal resort is on the Bald Mountain, where they lay in wait for travelers—but some have been seen in this part of the country. The inhabitants call it Yahoo; the Indians, however, give it the name of Chickly Cudly."

Source:

Hall, Mark A., 2002. "True Giants." *Wonders* 7, 2 (June): 46–48.

1806, 1811: Ghost Riders

ACCORDING TO MANY residents of a rural Rutherford County community 25 miles southeast of Asheville, near Chimney Rock Mountain, an extraordinary spectacle revealed itself on the evening of July 31, 1806. The story would pass into North Carolina legend, but the witnesses swore that they had seen it with their own eyes.

What they saw, they said, was a "very numerous crowd of beings resembling the human species," yet hard to discern more than that. In other words, the figures, which sailed in a semicircle around a high rock, had the outlines of human forms, but it could not be determined if they were men or women or a combination thereof. Though there were some taller forms, most were the size of children, "all clad with brilliant white raiment." Their number was estimated to be somewhere between 1,000 and 10,000. After an hour, they "vanished out of sight, leaving a solemn and pleasing impression on the mind, accompanied with a diminution of bodily strength."

One witness, Rev. George Newton, stated that the phenomenon either had a scientific explanation or was a "prelude to the descent of the holy city"—the implication, of course, being that the figures were angels.

This was not the end of the legions of phantoms, however. In September 1811, an old farm couple swore that they had seen a conflict raging between "two opposing armies of horse-men, high up in the air all mounted on winged horses." The commander of one army was heard to shout, "Charge!" before the two legions

The figures, which sailed in a semicircle around a high rock, had the outlines of human forms, but it could not be determined if they were men or women or a combination thereof.

attacked one another, their swords reflecting the dimming sunlight. One of the armies was routed, and the shouts of victors and screams of the wounded sounded through the air. The entire episode occurred over a ten-minute period.

Over subsequent evenings, the couple, this time in the company of "three respectable men," saw the same ghostly army, though this time the troops were not engaged in battle. When the two reported their sighting, the nearby town of Rutherfordton went into an uproar. A public meeting was convened, and a delegation led by two prominent citizens went to the couple's home and had them sign affidavits.

The story attracted national attention and much comment. In 1831, a reporter interviewed the old couple's grandson, who had his own explanation: suspended water droplets in the atmosphere had magnified and distorted swarms of gnats, thus creating the optical illusion of phantom soldiers.

Sources:
"An Extraordinary Phenomenon in Rutherford County, N.C.," 1883. *Statesville* [North Carolina] *Landmark* (June 15).

Gibson, Edmond P., 1950. "The Ghost Army of Chimney Rock Pass." *Fate* 3, 1 (January): 86–87.

Park, T. Peter, 2002. "Sky Visions, Ghosts Riders, and Phantom Armies." *The Anomalist* 10: 48–62.

1850: A Shower of Flesh and Blood

THIRTEEN MILES SOUTHEAST of the Sampson County seat, Clinton, on Thomas Clarkson's farm, on February 15, 1850, blood and flesh fell from a red cloud in the sky. Or so reported Fayetteville's *North Carolinian* on March 8.

Clarkson's children were playing on the property some 100 yards from the house when blood and flesh began to rain on them. Shocked and frightened, they ran home shouting, "Mother, there is meat falling!" Mrs. Clarkson hurried to the site, but the alleged shower was over. What remained, however, was unsettling enough. The material stretched some 250 to 300 yards. "The pieces appeared to be flesh, liver, brains and blood," the newspaper reported. "Some of the blood ran on the leaves, apparently very fresh."

A neighbor, Neill Campbell, rushed to the site shortly after the fall, having been alerted by one of his children. The boy told his father that he had smelled something like blood, and when Campbell went to investigate, he found Mrs. Clarkson examining the strange substance.

A Mr. Holland took samples to Fayetteville, where study under "two of the best microscopes in the place" confirmed that the material was indeed flesh and blood. Where it came from—other than from the mysterious red cloud that had hovered overhead—no one knew.

Sources:

"Great Fall of Flesh and Blood," 1850. *North Carolinian* (March 8).

Hairr, John, 2000. "When Flesh and Blood Fell from the Sky." *The Anomalist,* http://www.anomalist.com/reports/fleshblood.html.

1860: Illuminated Mystery Objects

A SINGULAR PHENOMENON was witnessed by some of our [Washington] citizens on Saturday evening last [November 12], just about sunset. An object about the size of a balloon (or a piece of chalk) was seen moving with great rapidity in a southwesterly direction; and notwithstanding the light of day was still strong and clear, the illumination of the object was brilliant and distinct as a balloon at night. We heard, while gazing at this wonder, that two similar ones had passed previously. The one we saw, after moving south-westerly, at an angle with the path of the sun, took a course directly west, and straight from us; fading gradually and very rapidly until lost from sight."

Source:

"Signs in the Heavens," 1860. *Brooklyn* [New York] *Eagle* (November 16). Reprinted from the *Washington* [North Carolina] *Herald.*

1874: Giants in the Earth

THE WORKMEN ENGAGED in opening a way for the projected railroad between Wildon and Garrysburg, N.C., struck, on Monday, about a mile from the former place, in a bank beside the river, a catacomb of skeletons, supposed to be those of Indians of a remote age, a lost and forgotten race. The bodies exhumed were of a strange and remarkable formation. The skulls were nearly an inch in thickness; the teeth were filed sharp, as those of cannibals, enamel perfectly preserved; the bones were of wonderful strength, the femur being as long as the leg of an ordinary man, the stature of the body being probably as great as eight or nine feet. Near their heads were sharp stone arrows, some mortars, in which their corn was brayed, and the bowls of pipes, apparently of soft soapstone. The teeth of the skeletons are said to be as large as those of a horse.

"The bodies were found closely packed together, laid tier on tier, as it seemed. There was no discernible ingress or egress to the mound.

The mystery is, who these were, to what race they belonged, to what era, and how they came to be buried there. To these enquiries no answer has yet been made, and meantime the ruthless spade continues to cleave body and soul asunder, throwing up in mangled masses the bones of this heroic tribe. It is hoped that some effort will be made to preserve authentic and accurate accounts of these discoveries, and to throw some light, if possible, on the lost tribe whose bones are thus rudely disturbed from their sleep in earth's bosom."

Source:
"Monmouth Remains," 1874. *Helena* [Montana] *Daily Independent* (April 4).

1876: Flesh Fall

THE PHENOMENON OF flesh falling from a cloudless sky occurred in Gaston county, N.C., last Saturday afternoon [November 11]. The shower fell in a cotton field belonging to James Hannah, who lives near Gastonia. The flesh—for such it certainly is—fell for several minutes, descending somewhat in the manner of hailstones falling, and sprinkled a space of ground equal to a square rod. Mr. Hannah, who saw the shower as it fell, says it was perfectly clear at the time, and there was nothing unusual visible in the sky at the place of the singular occurrence. Mr. R. Frank Clark, of Chester, returned from Gaston county, last Monday, and brought a specimen of the flesh with him. A microscopic examination indicates that it resembles the flesh of a cow, and a gentleman in town[,] who had the temerity to taste it, says the taste is similar to that of beef."

Source:
"Another Shower of Flesh," 1876. *Sedalia* [Missouri] *Daily Democrat* (November 15).

1880: "The Shape of a Huge Spotted Serpent"

A METEOR OF SURPASSING brightness was seen about midnight of the 1st instant [December 1] almost 8 miles east of Statesville. It made everything very light about the presence of the observer. It had the shape of a huge spotted serpent, 75 yards long, as large as a pine tree, with eyes very distinct and mouth open toward the north pole. About ten feet back from the head it seemed to rest on the sky and the head part to be elevated; then a little further back it was raised in a kind of loop, and the tail reached down toward the tops of the trees. It was seen by the man and his family about a half hour, and then it gradually passed away. The observer thought that it portended some terrible calamity, and was very much frightened."

Source:
"Brilliant Meteoritic Displays," 1880. *Statesville* [North Carolina] *Landmark* (December 3).

1882: "Comet"-Spooked Horses

ABOUT 4 O'CLOCK LAST Tuesday morning [October 10], a man named Henry was driving into Charlotte with a load of beef, when the comet suddenly appeared in the heavens. The mules started at the phenomenon with every exhibition of fright, and presently turning tail they broke into a run, scattering beef in every direction, throwing Mr. Henry out of the wagon and tearing the wagon to pieces."

Source:
"State News," 1882. *Statesville* [North Carolina] *Landmark* (October 13).

1882: Woman or Creature?

FOR MONTHS PAST there have been rumors of a terrible apparition which has terrorized all the women and children in Clear Creek township, North Carolina. The women say that for two weeks they have been subject to terrifying visits, in the absence of their husbands, from a creature like a shiny, black negress, with long hair and gleaming eyes. She asks, in hardly distinguishable gibberish, for a baby to eat, and makes efforts to get hold of children. The men . . . determined to catch the creature, and for the last few weeks crowds of farmers have been daily and nightly chasing her without success.

"The first effort to catch her was made a week ago by John Roberts, a blacksmith. The wild creature had appeared several times at a fire which it was the habit of Carry Moore, . . . of the neighborhood, to light after dark in the yard of his house for the preparation of his meals. Roberts was put there to watch for her, and she appeared even before the flames were well kindled. She presented such a wild appearance in the half-light, and asked for food in such a wild fashion[,] that Roberts was demoralized. He recovered, however, made an outcry and attempted to seize the woman, but she slipped through his hands and disappeared in the dark shadows of the woods.

"Tuesday night she was again enticed from the gloom of the woods by the kindling of a fire and was chased by thirty men without success. Captain Marsh Allen, later in the day, met her in the neighborhood, with her face torn and bleeding, and a long, bloody knife in her hand. The creature was naked, and her hair reached almost to her feet, but was kinky. . . . Yesterday the farmers, some mounted, and others on foot, assembled for the purpose of surrounding a swamp in which the creature was known to be hid. There were 400 men banded together for the chase. At first it was attempted to run her down with bloodhounds, but they refused to chase her, which deepened the superstitious feeling with which the men, as well as the women of the community, began to regard her. The men then made an attempt to catch her, but without success. The excitement . . . is intense."

Source:

"A Cannibal Negress," 1882. *Atchison* [Kansas] *Globe* (November 16). Reprinted from the *New York Commercial Advertiser*.

1884: A Rectangle of Fallen Flesh

AS SHE STOOD IN A freshly plowed field on the Silas Beckwith property, in Chatham County, the wife of an African American tenant farmer found herself in the midst of an extraordinary and frightening shower—even though the sky was clear. The substance falling on her was not water but blood and flesh, splattering on bushes, trees, and ground.

The incident, which allegedly took place on February 25, 1884, passed quickly through the area, bringing hordes of curiosity-seekers to the site. The *Chatham Record* reported that the affected spot was approximately 60 feet in circumference. "We are informed," it stated, "that a reputable physician of the neighborhood visited the spot and said it was blood." According to the testimony of one visitor, S. A. Holleman, the area was roughly rectangular, the red drops "of sizes varying from that of a small pea to that of a man's finger and averaged about one to the square foot. Smaller drops were instantly absorbed, larger ones, with those on the wood, coagulated. Some fell to the bushes and coagulated upon the limbs" (quoted in Venable 1883–1884).

Dr. Sidney Atwater—presumably the physician mentioned in the *Record* account—collected samples and showed them to Dr. Francis Preston Venable, chemistry professor (and future president) at the University of North Carolina in nearby Chapel Hill. Though initially incredulous, Venable was persuaded to visit the Beckwith farm and to interview the primary witness, Mrs. Kit Lasater.

From his investigations, Venable learned, as he would write in a scientific journal not long afterward, that the wind had been extremely slight at the time of the fall. The witness had first heard "something falling between her and the ground, saw it leave a red splash on the sand, heard a pattering like rain around her, looked up, but it was all over and she could see nothing."

Venable conducted a variety of tests in his lab that persuaded him that the material was, as suspected, blood. He was unable to tell, however, from whence it came. He admitted to his colleagues that the episode was very strange. "As to theories accounting for so singular a material falling from a cloudless sky," he wrote, "I have no plausible ones to offer. It may have been some bird of prey passing over, carrying a bleeding animal, but a good deal of blood must have fallen to cover so large a space. If a hoax has been perpetrated on the people of that neighborhood it has certainly been very cleverly done and an object seems lacking. On the possibility that it is not a joke, I have deemed this strange matter worthy of attention. Other similar observations hereafter may corroborate it[,] and combined observations may give rise to the proper explanation."

In fact, comparable incidents had occurred before (at least one in North Carolina; see 1850 entry, page 237), and would occur again, but more than a century later we are none the wiser as to what causes them.

Sources:
Hairr, John, 2000. "When Flesh and Blood Fell from the Sky." *The Anomalist,* http://www.anomalist.com/reports/fleshblood.html.

Venable, F. P., 1883–1884. "Fall of Blood in Chatham County." *Journal of the Elisha Mitchell Scientific Society* 1: 38.

1888: Underground Turbulence

LATE IN JANUARY, water started boiling and bubbling in a violent fashion in the wells around Glenwood in Johnson County. The effect was especially evident on adjacent properties owned by J. B. Hood and J. W. Rose. In the latter location, the sounds could be heard from as far away as 80 yards.

A newspaper said, "The water rises and falls, and appears to be in a state of violent ebullition.

Sometimes it is quiet, but for very brief periods. The soil there is sandy, and the wells are not very deep. The phenomenon cannot be explained. No unusual amount of rain has fallen, and no other disturbances are noticeable, but all the wells over quite a district are affected as described, this strange disturbance having begun just ten days ago."

Source:
"Here's a Good One," 1888. *Atlanta Constitution* (February 8).

1890 and Before: Specter in the Spring

IN THE WESTERN PART of North Carolina is what has long been known as the haunted spring. While the singular phenomenon may possibly at some time be explained, it has up to this time baffled the most skeptical. The spring comes from underneath a huge rock and frequently tempts the thirsty traveler to dismount. Nothing peculiar is noticed until a stooping posture is taken over the spring for the purpose of drinking. Then a most frightful face appears in the bottom, and as the person's face approaches the surface of the water this specter face, with most horrid grimaces, rises to meet it. No one has ever been known to have the courage to drink the water after the appearance of the apparition."

Source:
Untitled, 1890. *Olean* [New York] *Democrat* (February 6).

1897: "Great Sheet of Flame"

ON THE MORNING of the 21st of March, about 5 o'clock, I was standing in my yard when a flash of light suddenly shone around me. Looking to the direction whence it came, I saw as it were a great sheet of flame spreading towards the earth. It was not more than eight rods from me, and below the tree tops in height. In fact it came to the ground about twenty-five rods from the house. The only sound accompanying it was something like the sound emitted by a spark from forged iron, only louder. About the time the meteor, or whatever it was, came to the earth, I heard a sound as of heavy thunder in the north. I thought it was thunder but others in the community think it was an explosion. The direction of the phenomenon which I saw was from the northeast toward the southwest, and towards the earth. A sheet of fire (it looked like red ashes) remained in its wake until sun-up."

Source:
"More About the Meteor," 1897. *Statesville* [North Carolina] *Landmark* (April 13).

1897: "A Brilliant Floating Mass"

HUNDREDS OF PEOPLE were out on the streets and wharves last night, looking at a brilliant floating mass in the heavens to the west of the city. It was moving very rapidly, and many persons saw the aerial wonder. Some of our very best and most reliable citizens saw so much of the heavenly stranger that they had not the slightest doubt but that it was the air ship which has been reported from other cities. The ship moved to the west at a rapid rate. It seemed to have something like a searchlight, facing earthwards, and created a sensation among all classes of people.

"The ship appeared to come from the ocean and passed opposite Market street dock, going in the direction of the Navarro guano works. Some gentlemen who saw the ship through field glasses inform us they could see wires and ropes of rigging about it. To the naked eye many colored lights were visible. Even those who looked at it without glasses admit of no doubt that it was an airship."

Source:
"Was It an Air Ship?," 1897. *Wilmington* [North Carolina] *Messenger* (April 6).

1897: Fall of Unknown Material

A [WINSTON] JOURNAL reporter learns from a gentleman who was at Boonville last week that on last Monday morning [September 20] while the sun was shining brightly the air suddenly became filled with falling flakes like snow, except they were not cold, but were of a greasy nature and many of them were much larger than snowflakes, being two inches long in some instances. The people became greatly interested and some of them very much alarmed. The school children all left school to witness the strange sight and business for the time being was suspended. No one is able to give any explanation of the phenomenon, but the gentleman who told the reporter vouches for the truth of the story."

Source:
"A Strange Shower at Boonville," 1897. *Statesville* [North Carolina] *Landmark* (October 1). Reprinted from the *Winston* [North Carolina] *Journal*, September 27.

1900: Black Cloud, Black Rain

A 'BLACK RAIN' FELL in this county last Thursday morning [March 15]. At least that's what folks are calling it for want of a better name—'black rain.' And nobody has as yet been able to explain the phenomenon, for no one has been found who ever before heard of such a thing. The rain fell early Thursday morning in a belt extending, so far as is known, from Louisburg, in Franklin county, westward by way of Wake Forest to Morrisville, in this county. It came from a cloud of such intense blackness that just before and during the rain at Morrisville lamps had to be lighted. The water that fell looked like rain water mixed with soot. Nobody has been able to offer any explanation for the phenomenon."

Source:
"A 'Black Rain' Fell," 1900. *Statesville* [North Carolina] *Landmark* (March 23). Reprinted from the *Raleigh* [North Carolina] *News and Observer*, March 20.

1900: Howling and Prowling

IN LATE AUGUST, a mysterious but unseen intruder began to unnerve residents of the South Iredell area, particularly at the residences of William Creswell and J. Y. Templeton.

Witnesses characterized the creature's howls as "unearthly." Some compared them to croaks, others to "hollow" cries. Whenever people ran toward the sounds, expecting to see and grab the intruder, the sound would suddenly ring from a different location. Dogs ordered to chase it would stay in place. "This is no ghost story," a local newspaper said, "but a real fact, and will be substantiated by every member of the families in neighborhood of Creswell springs." The creature, however, left no tracks or other signs of its presence.

Source:
"A Mysterious Visitor in South Iredell,"
    1900. *Statesville* [North Carolina]
    Landmark (September 11). Reprinted
    from the *Mooresville* [North Carolina]
    Enterprise.

1904: Sailing Snakes

MRS. JOHN B. LIPPARD and children report a strange sight. They say that while over on their farm near there [Troutman] last Friday afternoon [September 16] they saw 30 or more large snakes sailing through the air. They say the snakes seemed to have fins and would average five feet long and four or five inches wide. They watched the snakes sail around and alight in a piece of thickety pine woods. . . . Most assuredly these people saw something."

Source:
Untitled, 1904. *Statesville* [North Carolina]
    Landmark (September 20).

1908: Ghosts in Jail

THE AUTHORITIES ORDERED a thorough investigation to be made at the county jail [in Asheville] where all kinds of uncanny sounds are heard in the night and unaccountable objects are seen. The prisoners have been badly frightened and many of them found the following morning by Jailer Mitchell in a high state of excitement. It is declared by Jailer Mitchell that it is not a ruse on the part of the prisoners for he knows it to be a fact that mysterious persons are seen at night at the jail and the most weird cries and sounds are heard. Sheriff Hunter has ordered a night ghost watch to be put on, to try to solve the mystery.

"Another visitation of the ghosts took place last night and the prisoners were wild from fright. Several of them fainted and were found this morning by the jailer in a dull stupor. They declared that two ghosts came to the jail last night, let themselves in through the iron grating, carried heavy iron chains and swung to and from the jail ceiling upon ropes. They refused to talk, but made it known by signs that they would return later and make known their business."

Source:
"Spooks in Jail at Asheville," 1908. *Elyria* [Ohio]
    Chronicle (May 14).

1913 and Beyond: Brown Mountain Lights

IT IS NOT JUST A MATTER of conjecture about how long mysterious lights have been seen in Brown Mountain, in northern Burke County 15 miles northwest of Morganton. Local legend says they were known to the American Indians who lived there before white people came to the area late in the eighteenth century. That claim itself may be no more than a local legend, however.

The first known—if ambiguous—recorded reference appears in a journal kept by the first European to explore the region, German engineer Geraud de Brahm. He remarked, "The mountains emit nitrous vapors which are borne by the wind and when laden winds meet each other the niter inflames, sulphurates and deteriorates." Today, refracted automobile headlights are a favorite explanation for the recurring luminous phenomena on the mountain. But de Brahm's description dates from well before the invention of the automobile.

A more modern printed account appeared in a 1913 issue of the *Charlotte Observer*. That same year, D. B. Sterrett of the U.S. Geological Survey (USGS) conducted an investigation and concluded that locomotive lights were solely responsible for the effect. Yet even when rail traffic in the area was suspended for a short time in 1916 because of flooding, the lights continued to be reported. A second USGS inquiry occurred in 1922, this one by G. R. Mansfield. After conducting observations for two weeks from various perspectives (Blowing Rock, Gingercake Mountain, and Cold Springs), he declared that the lights had multiple causes, including car headlights (47 percent), train headlights (33 percent), stationary lights (10 percent), and brush fires (10 percent) (Corliss 2001, 300).

Not everyone was satisfied with these explanations. In 1925, Robert Sparks Walker, a writer for *Literary Digest*, alleged that sightings had taken place as early as 1850. Of the phenomena themselves, he wrote, "These lights have been described as varying in size, but to somewhat resemble 'toy balloons.'" He said they were "pink, orange, or reddish in color" and added, "They are said to rise into view over the mountain and to hover for periods of one to fifteen minutes duration before fading out. Sometimes, it is claimed, as many as three lights can be seen simultaneously at widely separated points. They are alleged to be bright enough to be visible from Blowing Rock, some twenty miles away. Sightings are said to be frequent in fine weather, but also occur when the sky is overcast or the mountain hidden in mist."

Walker reported that on the evening of September 17 he had maintained a vigil on a nearby hill for five and a half hours. Under damp, cloudy skies he witnessed three varieties of lights. The first, yellowish white, appeared at eight- to eleven-second intervals, visible for only half a second at a time. Another kind of light was reddish orange, showing up every twelve to fifteen seconds, again for less than a second. The third was a violet color, and it could be seen for one-eighth of a second, occurring less often than the other two. "Sometimes there would be intervals of approximately thirty-five seconds," he said. "These bursts would be followed by lengthy pauses with no activity until the next burst. The duration of the pauses varied between five minutes and thirty minutes." He thought they looked like lightning, even if they did not exactly *behave* like lightning.

Sightings of the lights, whatever their true nature, have continued unabated. A particularly dramatic example occurred, according to a report by Michael Loveless, one night in October 1969, when he and several passengers were driving near the mountain. They noticed what looked like a spotlight in the sky. When they rounded a curve and were able to get a full view, they were puzzled to see that the beam "had no source on the ground. It was simply suspended in the sky. . . . It was a white light . . . wider at one end than the other, but at the end you would expect to connect to a ground source, it was simply cut off and blunt. . . . Then it began changing shape. . . . The ends began to close in, so that it became something like a fat toothpick. The end pointing toward the mountain began to open out into an arrow or spearhead shape, as though it were being absorbed by the head, until it was gone."

In 1977, scientists and interested lay researchers conducted a sophisticated, formal investigation into the lights. One group set up a 500,000-candlepower arc light in the town of Lenoir, 22 miles east of the mountain. A second group watched from an overlook on Route 181, 3.5 miles to the west of the mountain. When the arc light was switched on, a reddish orange orb was seen a few degrees above the mountain's crest. The experiment proved that refractions of far-away artificial lights are indeed one source of the effect.

Investigators agreed, however, that the lights floating through trees well below the crest, a far rarer phenomenon, are less easily accounted for. These lights may be associated with seismic activity, though the association has yet to be proved.

Sources:

Bessor, John Philip, 1951. "Mystery of Brown Mountain." *Fate* 4, 2 (March): 13–15.

Corliss, William R., ed., 2001. *Remarkable Luminous Phenomena in Nature: A Catalog of Geophysical Anomalies.* Glen Arm, MD: Sourcebook Project, 299–301.

Devereux, Paul, 1989. *Earth Lights Revelation: UFOs and Mysterious Lightform Phenomena: The Earth's Secret Energy Force.* London: Blandford, 121–123.

Walker, Robert Sparks, 1925. "The Queer Lights of Brown Mountain." *Literary Digest* 87 (November 7): 48–49.

North Dakota

1883: Field of Giants

AN ANONYMOUS CORRESPONDENT informed *Scientific American*'s readers in 1883 that 2 miles from Mandan, on river bluffs near where the Hart and Missouri Rivers meet, lay "an old cemetery of fully 100 acres in extent filled with bones of a giant race."

The bodies of both humans and animals apparently had been dumped into deep trenches, then covered with earth. There were also mounds 8 to 10 feet high, some stretching as far as 100 feet in diameter at their base. Inside these, one could find bones and broken pottery. "The pottery is of a dark material," the correspondent wrote, "beautifully decorated, delicate in finish, and as light as wood, showing the work of a people skilled in the arts and possessed of a high state of civilization. This has evidently been a grand battlefield, where thousands of men and horses have fallen."

Five miles north of Mandan was another, comparable field, this one even less fully explored than the first. The correspondent claimed to have asked "an aged Indian" if he knew anything about the ancient cemeteries, and he allegedly replied, "They were here before the red man."

There is no independent evidence that any such graveyards ever existed outside the writer's imagination. This seems to be one of many frontier tall tales invented for the benefit of gullible easterners.

Source:
"A Prehistoric Cemetery," 1883. *Scientific American* 48: 296. Cited in William R. Corliss, ed., 1978. *Ancient Man: A Handbook of Puzzling Artifacts*. Glen Arm, MD: Sourcebook Project, 682.

1885: "Hurled from Some Planetary Wreck"

DURING LAST EVENING'S storm a beautiful as well as awe inspiring scene was presented to the gaze of those so fortunate as to be upon the street at the time. It was just after the flash of lightning and crash of thunder which made every one believe for an instant that he had been struck. No sooner did the thunder peal than a ball of fire, resembling a meteor, shot athwart the zenith, and rolling rapidly to the west, began to descend. Its starting point was in the east, and it began to descend while passing over the city, filling the startled spectator with feelings of terror. It descended rapidly, and as it passed through the air it produced a whizzing, sizzling sound, and threw off numerous sparks in its flight. With terrific force the glaring missile, hurled from some planetary wreck or produced by some strange freak of electricity, struck the Missouri river about a quarter of a mile below the landing, the gurgle and momentary roar of the aggravated waters being plainly heard on west Main street. The strange visitor appeared and disappeared in an instant, and no one will ever know what it was. But as electricity was playing havoc with the elements at the time, it is believed to have been some electric phenomenon."

Source:
"An Electric Ball," 1885. *Bismarck* [Dakota Territory] *Daily Tribune* (July 8).

Three mighty steamships could be seen plowing a calm sea but with rough waves in the wake of the steamers. Heavy ropes reached from one to the other.

1899: Ships in an Atmospheric Ocean

IN THE EARLY EVENING of August 18, in McLean County, twelve-year-old Edwin Granstrom, herding cattle, happened to look up and saw something that terrified him. He ran home and called to his mother. According to the local newspaper:

"When Mrs. Granstrom went to the door, sure enough[,] up in the sky where the boy was pointing—three mighty steamships could be seen plowing a calm sea but with rough waves in the wake of the steamers. Heavy ropes reached from one to the other. The huge cylindrical smoke stacks on each steamer were puffing up black columns of smoke. Men could be seen plainly about the rigging, and small boats were being lowered, and loaded camels were being hoisted as though being boarded. Everything in connection with the scene appeared remarkably distinct and was in full view of all the members of the Granstrom family for over two hours.

"The phenomenon was no doubt a mirage and the presence of camels would show a stupendous mirage-photo of transactions that had appeared—or was [sic] then appearing—at a place like the Suez canal. It was a strange sight truly[,] these giant steamers suspended in midair."

Source:
"A Startling Phenomenon," 1899. *Bismarck* [North Dakota] *Daily Tribune* (August 29). Reprinted from the *Washburn* [North Dakota] *Leader.*

Ohio

1825: Balls, Balls, and More Balls of Light

WALKING ALONG A road near Newton one day in November 1825, two women watched the passage of what they described as a "meteor"—a word then used to characterize a variety of atmospheric phenomena, not just stones from outer space—moving in the same direction they were. Suddenly and confusingly, they found themselves enveloped in light, apparently emanating from a ball of fire that had fallen partly on them and partly on the ground just behind them.

The ball was several feet in diameter, and it gave off no heat and made no sound. As the light hit the ground, it broke into many—"a thousand," the witnesses thought—smaller balls. These proceeded to roll along the ground, separating into yet smaller balls all the while. When they looked up, they saw the meteor that had attracted their attention in the first place as it disappeared in the distant sky. Within moments, the innumerable balls around them had all vanished.

The ball was several feet in diameter, and it gave off no heat and made no sound. As the light hit the ground, it broke into many—"a thousand," the witnesses thought— smaller balls.

Sources:

Olmsted, Denison, 1834. "Observations on the Meteors of November 13th, 1833." *American Journal of Science* 1, 26: 132. Cited in William R. Corliss, ed., 2001. *Remarkable Luminous Phenomena in Nature: A Catalog of Anomalies.* Glen Arm, MD: Sourcebook Project, 132.

1826: Worlds within the World

BURIED IN LUDLOW Park in Hamilton, a small city directly north of Cincinnati in the southwestern part of Ohio, lies the body of John Cleves Symmes, born November 5, 1779, died May 29, 1829. "Forty-nine years and six months," the epitaph notes. It also states that Symmes "was a philosopher, and the originator of 'Symmes' Theory of Concentric Spheres and Polar Voids'. He contended that the Earth was hollow and habitable within." It does not mention that for a decade or so before he died, Symmes was a famous man, derided by many but held by others to be a genius. Today he is remembered, except by the very few who share his views, only as an American eccentric and the man whose ideas inspired Edgar Allan Poe's proto-science-fiction novella *The Narrative of Arthur Gordon Pym* (1838).

Born in Sussex County, New Jersey, young Symmes took an early interest in science, but in 1802 he joined the army. He remained in the service until 1816. In 1814, during the War of 1812, he distinguished himself in battle with the British along the Canadian border. On retiring, he established a trading post in St. Louis. In his spare time he resumed his studies in science. Among the books he is almost certain to have read is Cotton Mather's *Christian Philosopher* (1721), which remarked on the hollow-earth theory of English scientist Edmond Halley (1656–1742), after whom the famous comet is named.

Halley believed that our planet was a shell 500 miles thick. Beneath that shell, he said, lay three concentric spheres, each smaller than the other, each separated by 500 miles of atmosphere. Not satisfied with this fantastic proposition, Halley went on to theorize that intelligent beings lived on each of these worlds. He explained the aurora borealis as luminosity seeping onto our surface world from the interior. It was visible to us because the outer shell was thinnest at the poles—a notion that would soon lead to the belief that the poles were in reality giant holes, entrances to the interior. Though respected for his other achievements in science, not many of his colleagues took his hollow-earth speculations seriously. Two who did, mathematician Leonhard Euler (1707–1783) and physicist Sir John Leslie (1766–1832), independently arrived at the idea that the earth's molten core served as a sun for the interior's inhabitants. Neither bought into Halley's notion of concentric spheres, however.

But it was Mather's paraphrase of Halley's speculation that persuaded Symmes to pursue the question of worlds within the world. From then on, he became a man obsessed. He immersed himself in scientific, geographic, and historical materials looking for confirmation. In April 1818, in a document he had printed and sent to a staggering number of government officials, scientists, and domestic and foreign universities, he declared the earth to be hollow, in-

habited therein, and "containing a number of solid concentric spheres." Moreover, "it is open at the poles" some 4,000 miles at the north and even more at the south. "I pledge my life in support of this truth, and am ready to explore the hollow, if the world will support and aid me in my undertaking," he stated. What he needed, he wrote, were 100 well-equipped men who would follow him to Siberia in the fall. Sleighs driven by reindeer would take them across the frozen sea until they got to a "warm and rich land, stocked with thrifty vegetables and animals, if not men," inside the polar opening. They would return the following spring (Collins 2001; Kafton-Minkel 1989).

In anticipation of the inevitable questions, he appended a document attesting to his sanity. It turned out to be a strategic mistake. The *London Morning Chronicle,* for one, questioned its provenance. The reaction was typical. Nearly every recipient who expressed an opinion on the subject let it be known that even if Symmes himself were not insane, his theory certainly was.

Symmes was undeterred. If anything, he was more determined than ever to prove to a scoffing world the reality of a hollow earth. After publishing seven more papers on the subject, he moved himself and his family to Newport, Kentucky. From there, he circulated yet another treatise, *Light between the Spheres,* which when reprinted in the widely read *National Intelligencer* attracted considerable popular attention. But scientists weren't impressed this time, either. "A heap of learned rubbish," one harrumphed.

Realizing that the printed word was not going to get his case across, Symmes embarked on a lecture tour. Unfortunately a victim of chronic stage fright, he was a terrible public speaker. Even worse, the gaps in his education became embarrassingly apparent to any sophisticated listener, though he addressed relatively few of those. Most of his audiences consisted of unlettered frontier folk and schoolchildren. Many were moved to sign petitions, at his urging, for Congress to finance an expedition to the north pole, with Captain Symmes at the lead. In 1822, Kentucky senator Richard M. Johnson spoke to his colleagues about Symmes's "desire to embark on a voyage of discovery, to one or other of the polar regions," and his need for two large sailing vessels and other aid. The following January, Representative J. T. Johnson, also of Kentucky, argued to the House that such an expedition would lead to "new discoveries in geography, natural history, geology, and astronomy." It would also open "new sources of trade and commerce." His petition, like its predecessor in the Senate offered by another Johnson, got nowhere.

By this time, however, Symmes had enlisted the support of James McBride, a prominent literary and intellectual figure in southern Ohio. He encouraged Symmes to take up residence in rural Hamilton, where McBride's uncle had left him a farm. He also put together a collection of Symmes's papers and had it published in 1826, with the mouthful title *Symmes' Theory of Concentric Spheres, Demonstrating That the Earth Is Hollow, Habitable Within and Widely Open about the Poles—by a Citizen of the United States.* Most likely out of a fear of ridicule, McBride left his by-line off the cover and title page. If so, it was a wise idea; the book was savaged, no more so than in a long, unsigned piece in *American Quarterly Review* for March 1827. The reviewer judged the book and the theory "completely at variance both with reason and with facts" and proceeded to explain why, point by point.

The ever defiant Symmes shrugged off all criticism and continued to seek supporters. He attracted sympathetic attention from at least one government, Russia's, which responded favorably to his request to join an expedition intended to explore Siberia. A shortage of funds kept him from participating. Still, all was not lost. He had gained an important new advocate, Jeremiah Reynolds, an articulate young lawyer and newspaper editor from Wilmington, Ohio.

Reynolds urged Symmes to take his case to the East, where he believed he could meet influential people who could help him. When Symmes hesitated, Reynolds told him he would accompany him. Beginning in September 1826, they worked their way throughout Pennsylvania, ending up early the next year in Harrisburg, where Symmes delivered a successful address to the state legislature. Fifty legislators signed a letter to the president attesting to their conviction that an expedition would be a wise investment. Symmes's theory, they wrote, made more sense than Columbus's.

In Philadelphia, however, Symmes's health, always precarious, gave out. Reynolds took up the bulk of the lecturing, soon putting more emphasis on polar explanation of a consensus-reality kind than on his ostensible mentor's curious ideas. Soon enough, the latter were going unmentioned. The two men parted company. Symmes shakily, and largely ineffectually, continued to speak out on his own at venues in New York, New England, and eastern Canada, until he could no longer continue. Too weakened to return home, he was forced to take up residence with a New Jersey friend. He would not make it back to his family in Hamilton for two years. Even then, he had to be carried on a mattress. He managed to dictate further monographs on his theories until his death in May 1829.

Symmes's most famous advocate was none other than the great American man of letters Edgar Allan Poe. His Gothic science-fiction novella *The Narrative of Arthur Gordon Pym,* first published as a serial in 1837 issues of *Southern Literary Messenger,* concerns a polar expedition that encounters hollow-earthers from "Symzonia." Poe borrowed the concept of Symzonia from an 1820 novel of the same name, the first literary treatment based on Symmes's ideas. The true identity of the author, the pseudonymous "Captain Adam Seaborn," has never been identified, though he has been misidentified as Symmes himself.

A further literary footnote: Joshua Reynolds eventually got the ear of influential figures in the John Quincy Adams administration and secured limited funding, enough to support three small ships. They sailed for the south pole but got only to the coast of Chile. By then the poorly provisioned crew were starving and in sufficiently hostile humor to mount a mutiny. Reynolds survived, staying in Chile, where he roamed, explored, and conducted scientific inquiries. Returning to America after two years, he wrote an account of his adventures. In them he recounted a story he heard from an elderly sailor, who spoke of "Mocha Dick," a white whale said to exist in Pacific waters. Among those who read Reynolds's memoirs was Herman Melville, who transformed Mocha Dick into the object of mad Captain Ahab's quest in *Moby-Dick* (1851).

Sources:

Clark, P., 1873. "The Symmes Theory of the Earth." *Atlantic Monthly* 31, 186 (April): 471–480.

Collins, Paul, 2001. *Banvard's Folly: Thirteen Tales of Renowned Obscurity, Famous Anonymity, and Rotten Luck.* New York: Picador USA, 54–70.

Kafton-Minkel, Walter, 1989. *Subterranean Worlds: 100,000 Years of Dragons, Dwarfs, the Dead, Lost Races & UFOs from Inside the Earth.* Port Townsend, WA: Loompanics Unlimited, 56–73.

"Symmes and His Theory," 1882. *Harper's New Monthly* 65, 389 (October): 741–744.

1832: Fall of Slimy Material

ON SATURDAY NIGHT [August 11] we [in Chautaqua] had a heavy fall of rain, which continued until late on Sunday morning. Immediately after the rain had subsided, there was observed on the fences, door sills, and door steps of houses facing to the southwest, and also on boards and boxes, &c. lying in yards, and on logs and trees in the woods near this village, a mucilaginous matter of a light color, not ropy, but of liver jelly, apparently having fallen in large drops, and spread from the size of a five cent piece to that of a fifty cent piece, and from the sixteenth of an inch to an eighth of an inch thick. As this slime or matter began to dry or congeal, there was observed to be suspended in it specks of a darkish color, resembling in substance curdles in sap or wine, and some persons who examined it imagined that they could discover live insects in these specks."

Source:

Letter, 1832. *Ohio Star* (September 13). Reprinted from the *Chautaqua Republican.*

1837: Graveyard of Little People

IN 1837, *GENTLEMEN'S Magazine* reported that "on one of those elevated, gravelly alluvions, so common on the rivers of the West," an ancient graveyard of small coffins had been discovered. The bodies were of tiny people, ranging from three to four and a half feet. There were numerous skeletons, suggesting that a huge city had once existed on the site.

"A large number of graves have been opened, the inmates of which are all of this pigmy race," it was said. "No metallic articles or utensils have been found to throw light on the period or the nation to which they belonged."

This story is apparently a hoax. No independent evidence of so fantastic a discovery is known to exist.

Source:

"Pigmies," 1837. *Gentlemen's Magazine* 3, 8:182. Cited in William R. Corliss, ed., 1978. *Ancient Man: A Handbook of Puzzling Artifacts.* Glen Arm, MD: Sourcebook Project, 681.

1859: Spirits

AT VAN WERT . . . THERE is a doggery [dive] (in which a man lost his life, not very long since, through the agency of that article, called whisky which makes one citizen kill another,) which is reported to be 'haunted' by a 'spook' or 'hobgoblin.' It is said that strange noises are heard at night, and empty boxes and barrels, etc., are tumbled around by some unseen power—all of which has the good effect of frightening the keeper so that he will not stay in his den at night, but leaves it at dark[,] allowing the 'ghost' to revel among the *spirits* alone."

Source:

"A Haunted Tavern," 1859. *Ohio Repository* (February 23).

1859: Image of a Rattlesnake

MR. JOS. WRIGHT, AN old citizen of Cincinnati[,] was bitten by a rattlesnake in Vanceburgh, Lewis county, Kentucky, on the 4th inst. [July 4], and died in less than fifteen hours. The reptile struck its fangs in the back of his hand, and by the time he reached home he was entirely blind, and his body and head were covered with spots of the same color as those of the rattlesnake.

"In preparing the body to be laid out, a singular phenomenon presented itself. In addition to the spots referred to, there was a picture of the snake itself—perfect in shape and color, and as distinct as if daguerreotyped there—extending from the point on his hand where the fangs had struck, up the arm and to the shoulder, and then down the side to the groin. To the truth of this, our informant assures us not only himself, but some four or five other citizens, who saw it, can positively testify."

Source:
"Death from the Bite of a Rattlesnake," 1859. *Coshocton* [Ohio] *Progressive Age* (July 27).

1873: Headless Ghost, Phantom Fires

XENIA, O., AS TOLD in a telegram from that city, has developed the ghost of a man whose anatomical peculiarity is that, like that fabled race of men of strange lands who carried their heads under their arms, he is headless. Seen at night upon the streets in the mysterious hours of darkness, he disappears like a shot when approached. His mission seems to be to build phantom fires in the highways of the town, but when the morning comes there are no traces of fire visible there [even though] during the night, a bright light was to be seen. 'Solid men' of Xenia have seen him, and though the night be cold and rainy he wanders to and fro, coatless and wringing forlorn hands. The sensation caused by this visitation is a genuine one in the lucky town, and a committee of the Common Council of Xenia will be appointed to investigate the mystery."

Source:
Untitled, 1873. *Defiance* [Ohio] *Democrat* (February 15).

1877: Rural Poltergeist

QUITE A SENSATION has been created near Caldwell, Noble county, on the farm of William Staats[,] by singular phenoma [*sic*] which commenced two weeks ago [early October], and continued till the 18th, during night and day, with occasional intervals. The manifestations commenced by the overturning of milk-pans and buckets without any visible power. Mr. Staats then communicated the strange affair to his neighbors, who availed themselves of the opportunity of witnessing these unusual and so far unaccountable demonstrations. A looking-glass was dashed to the floor and broken, a bootjack was thrown across the room from side to side several times before stopping, a tea-kettle full of boiling water was lifted from a stove and fell to the floor, pictures were hurled from their places on the wall and dashed to pieces, a box setting in the middle of the floor was seen to move off across the room, a glass was broken from the windows and fell outside of the house[;] one of the pictures was replaced, the cord securely wrapped about the nail, and an additional cord fastened about this and tied. The picture was at once thrown to the floor again.

"Wednesday morning [presumably October 17], while the hired girl was washing the dishes, the knives commenced moving across the table and fell to the floor. The bottom fell out of a glass pitcher full of water leaving the handle and upper part of the pitcher in her hands without a sign of breakage. Many persons, not believers in the supernatural, witnessed these things and were unable to offer any explanation. Mrs. Staats has been obliged to leave the house on account of the noisy and exciting character of the manifestations."

Source:
"A Tumultuous Ghost," 1877. *Grand Traverse* [Michigan] *Herald* (November 15). Reprinted from the *Cincinnati Enquirer*.

1880: Supernatural Flames

A CLEVELAND FAMILY identified only as the Bushes complained of weird blue flames of unknown origin that would suddenly break out and consume clothing right off the bodies of the people wearing it. Persons inside the clothing would get scorched, but no one suffered fatal injury. The flames appeared, burned fiercely, then went out as fast as they had ignited. A press account said, "Partly-burned garments are the best evidence the Bushes can produce in support of their word, but their motive for deception is not apparent."

Source:

Untitled, 1880. *Waukesha* [Wisconsin] *Freeman* (June 3).

1881: "One of the Most Disagreeable Spooks on Record"

AT MINERAL POINT, Tuscarawas county, there is a ghost that is making things disagreeable for the peaceful folks of that region. The ghost inhabits a brick factory and is supposed to have inhabited the body of a fireman at one time. It is one of the most disagreeable spooks on record, having a habit of knocking things about, heating up the boilers until they threaten to burst, and tearing the bricks out of the flue. Besides, the ghost smells bad, which is disagreeable in a ghost, and a habit that it ought to reform, if it expects to retain the character of a respectable apparation [*sic*]. The people are greatly excited."

Source:

"Ohio News," 1881. *Defiance* [Ohio] *Democrat* (February 3).

1881: Fear of a Serpent

THE PEOPLE OF Williams Creek, near Rockville, Ohio, are in terror over an enormous serpent that has infested the place for several years. It is at least twenty feet long and a foot and a half in circumference, of a dark brown color, and is supposed to have escaped from a circus that visited this town."

Source:

Untitled, 1881. *Hagerstown* [Maryland] *Herald and Torch Light* (June 16).

1883: Misogynist Phantom

EARLY IN AUGUST, in Hanover, a tiny town east of Newark, O. Z. Hillery's farm adjoining the village became infested with a dangerous nuisance: an apparently invisible entity that hurled stones at women and children. The attacks were unnervingly accurate, often inflicting injury to the victims. Weirdly, men were never targets.

As the attacks grew in intensity, villagers decided that something had to be done. On the night of August 24, many armed themselves with rifles, shotguns, pistols, knives, or whatever else was at hand and went to the farm. Concerned that nothing would happen if only males descended on the site, the men brought along seven women who in effect served as bait. Shortly after their arrival, the stones—apparently cast from a variety of directions—sailed in the women's direction. Two victims, a Miss Fletcher and a Mrs. Fleming, were hit particularly hard and were seriously injured. Every time a stone flew, the men would rush in the apparent direction of its flight, but they never found any evidence of the thrower. None of the men was struck.

Source:

"The Hanover Ghost," 1883. *Newark* [Ohio] *Daily Advocate* (August 27).

1884: Assaulting a Ghost

A THOUSAND PEOPLE surround the grave yard in Miamisburg . . . every night to witness the antics of what appears to be a genuine ghost. There is no doubt about the existence of the apparition, as Mayor Marshall, the revenue collector and hundreds of prominent citizens all testify to having seen it. Last night [March 24] several hundred people, armed with clubs and guns, assaulted the specter, which appeared to be a woman in white. Clubs, bullets and shot tore the air in which the mystic figure floated without disconcerting it in the least. A portion of the town turned out en masse to-day and began exhuming all the bodies in the cemetery.

"The remains of the Buss family, composed of three people, have already been exhumed. The town is visited daily by hundreds of strangers and none are disappointed, as the apparition is always on duty promptly at 9 o'clock. The strange figure was at once recognized by the inhabitants of the town as a young lady supposed to have been murdered several years ago. Her attitude while drifting among the graves is one of deep thought, with the head inclined forward and hands clasped behind."

Source:
"A Genuine Ghost," 1884. *Philadelphia Press* (March 25).

1884: Lights, Moans, Figures

ONE NIGHT IN NOVEMBER, ten streetcar conductors descended on an empty house, said to be haunted by shadowy forms, on Cleveland's Woodland Avenue. As six of the men stood watch outside, four entered and conducted an extensive search, checking every room and banging walls and floors on the chance that hidden closets or trapdoors lay behind or beneath. Finding nothing, they started to leave the premises when, with no apparent cause, the lanterns they were carrying went out. An eerie, extended moan sounded, encouraging the searchers to take to their heels. The moaning, unfortunately, followed them "for a long distance."

On the morning of Saturday, November 20, Andrew Belden and a hired man visited the house. They conducted excavations with shovels, apparently on the theory that human bones were buried beneath the structure and that a dead person's remains could be tied to the haunting. "They found a number of half-burned bones and shreds of clothing," a press account stated. "A silver ring with the letter 'D' marked upon it was also found." Belden subsequently turned the bones over to an unnamed chemist.

The next night a man named James Wilson, riding by the house, was startled when a bright red light abruptly illuminated all of the windows. Fast-moving white-clad forms were visible through the glass. Wilson's horse panicked and bolted, throwing him from the saddle. On Monday evening a group of boys on their way home spotted a white-garbed figure apparently digging in the backyard. Two of their fathers went to the site but found no evidence of disturbance in the grass or soil.

Source:
"Put to Flight by Spooks," 1884. *Frederick [Maryland] Weekly News* (December 4).

1884: Square White Light

SHORTLY AFTER DARK ON December 27, a large bright light showed up suddenly in the eastern sky not far above the horizon. Square in shape, it moved over the woods outside Newcomerstown. The tail that followed it was so brilliant that it lighted up the trees, branches, and shrubs and made them visible to observers as if in daylight. The initial impression of many witnesses was that it was an unusual meteor, but its rate of speed was so low, and its period of visibility so extended, that the theory seemed a poor explanation for the phenomenon. "The superstitious are troubled," a newspaper account said.

Source:
"A Mysterious Visitant," 1884. *Elyria [Ohio] Republican* (December 28).

1884: "Not a Pleasant Looking Customer"

A STRANGE WILD animal has been creating consternation in the vicinity of Glenford. The beast has been called a panther, a lynx, a hyena, a bear, and dear knows what other names. Several efforts have been made to bring it down with a gun, but without effect. What it is or where it came from are both alike unknown. The beast is not a pleasant looking customer. A combined effort will probably be made to dispatch the roving animal."

Source:

"Ohio State News," 1884. *Fort Wayne* [Indiana] *Daily Gazette* (December 31).

1884: Light amid the Snow

AT DEFIANCE THE other evening, was witnessed a very strange phenomenon, which, as yet, can not be explained. At the time snow was falling when in a flash the earth was brilliantly lighted up with a pale blue light tinted with red. This was instantly followed by a second illumination, the two lasting but a few seconds. In appearance the light resembled that made by the explosion of a Roman candle ball or a sudden flicker of electric light. The general opinion prevails that it was a meteor, but owing to the clouded condition of the sky it could not be seen. Superstitious people were badly frightened."

Source:

Untitled, 1884. *Fort Wayne* [Indiana] *Daily Gazette* (December 31).

1885: Haunted by Fireball

TIFFIN IS REVELING IN a ghost sensation. A ball of fire is hurled in[to] the house of Mr. Sheets every night. Even the police cannot solve the mystery."

Source:

"Local Gossip," 1885. *Marion* [Ohio] *Daily Star* (September 23).

1887: Woman in White on the Track

THE VILLAGE OF REPUBLIC has a ghost, which stops trains, and there is great excitement there. A few nights ago, when express No. 5 (the same train that was wrecked January 4), was approaching the scene of that horrible disaster, the engineer saw a red light, the danger signal ahead, and applying the brakes and reversing the engine, the train came to a standstill on almost the exact spot of the great wreck. Strange to say, when the train came to a stop, the light had disappeared, and could nowhere be seen. Before stopping, both the engineer and fireman noticed that the light appeared to be carried by a woman in white. Puzzled by the disappearance of the signal, the conductor walked over the track for some distance ahead, but could discover nothing.

"The train then backed to Republic station, and the operator was questioned, but he assured them that no signal had been sent out. The train then proceeded on its way cautiously, the engineer keeping a sharp lookout, but nothing more was seen of the mysterious woman. This strange apparition has appeared on three occasions, and has greatly excited trainmen. A posse has watched the place for several nights, but the ghost has not since appeared."

Source:

"Ghost Stopping Trains," 1887. *Elyria* [Ohio] *Daily Telephone* (March 4).

1887: "A Tall, White Object"

IN AN AREA OF eastern Newark called the "commons," apparently an informal park located in a small area on which there were no buildings, some local people swore they had encountered something they believed to be supernatural. They described it as tall and white, with a human appearance. It would vanish if spoken to.

A young railroad worker named Al Warner gave an account of his own encounter. He thought he might have seen it on two other occasions, but on the third, in April, the encounter was sufficiently spectacular to remove any doubt from his mind that some kind of supernatural entity lived in the park area. "On the night of the 13th . . . I stayed till rather late in the evening," he recounted to a reporter. "I finally started home with a Mr. Harrington, who works in the shops. I left him at the corner of East Main and Mill streets, and though I was some afraid, I decided to go 'cross-lots,' through the 'commons[,]' home. It was after 11 o'clock, I think."

When Warner was about halfway across the commons, "a tall, white object arose apparently from the ground" in front of him. Warner said, "Hello, how are you?" to the ghost, which was about 10 feet away. Warner reported: "It turned aside and did not answer. No, I didn't follow it. I went the other way and pretty fast, too."

Local lore held that the alleged entity was the ghost of a man murdered half a century earlier and buried under a tree in the commons area.

Source:
"Spooks. Uncanny and Inexplicable Occurrences in the East End," 1887. *Newark* [Ohio] *Daily Advocate* (April 15).

1887: "Monster Reptile"

FROM SPARTA, MORROW county, comes the monster story. It seems that a few days ago an engineer on the Baltimore and Ohio road, while running from Peerless to Marengo, observed a large black object lying near the track, somewhere near Culver's woods. On nearing the object it moved to one side, raising its head as high as the tender. The engineer saw that it was a large snake, but could not tell its length. Arriving at Marengo he spread the news, and soon a great many were on the chase. The snake was pursued for several days. Its track could be plainly seen in many places. Through the meadows it broke down the grass, leaving a trail; in crossing a road it would spread the dust, making a track eight inches wide. The monster was finally seen about three miles west of Sparta, and was killed by Dave Hunt, of Bloomfield, by being shot in the head. The snake measured seventeen feet and two inches in length. This monster reptile is producing much talk here, and it certainly is a curiosity."

Source:
"A Big Snake Story," 1887. *Marion* [Ohio] *Daily Star* (July 16).

1887: Hazards of Drinking from a Spring

SNAKE 21 INCHES LONG was ejected from the stomach of Michael Wheatley, a Dayton fireman. It had troubled him for five years. It is supposed to have entered his stomach when he was drinking from a spring."

Source:
Untitled, 1887. *Coshocton* [Ohio] *Semi Weekly Age* (July 16).

1888: Warning by a Headless Ghost

ABOUT THREE YEARS AGO a young man, James Donahoe, of Pennsylvania[,] was killed by an engine on the O.C. [Ohio-Chesapeake] Railroad, a short distance from the station [at Centerburg]. His headless body was found next day and buried in the village cemetery.

"A short time ago a young man of our village was appointed night watchman at the above mentioned station. One night as he was sitting at his post of duty, he was startled by the sound of an incoming train, though none was due at that hour. He seized his lantern and rushed out upon the platform; all was dark, the hour was midnight, and the grinding of the wheels could be distinctly heard, although no whistle was blown. Nearer[,] nearer up the track came a shadowy, unearthly shape, like the outlines of an engine, and right in front of it upon the track stood the headless figure of a man, wildly signaling for down brakes. The engine seemed to approach and crush the figure beneath it, when the horrified watchman became so paralyzed with terror that he let his lantern fall and the light was extinguished, leaving him in total darkness and alone, for the apparition had vanished.

"The next day a train was ditched a short distance from the place where Donahoe was killed.

Eight cars were overturned but no one was injured. And now whenever there is danger of a wreck on the O.C. road it is said that the spirit of poor Donahoe gives warning. The watchman resigned his place soon after the ghostly visit, and a man of stronger nerves, from Granville, holds the position."

Source:

"Centerburg," 1888. *Coshocton* [Ohio] *Semi Weekly Age* (February 17).

1888: Rain of Stones and Broken Pottery

HENRY KUSSMAUL, OF the Granville Times, reports a remarkable meteorological phenomenon near the Times office in the shape of a shower of stones and broken pottery from a perfectly cloudless sky."

Source:

Untitled, 1888. *Newark* [Ohio] *Daily Advocate* (April 17).

1888: Circle of Fire

SOME QUEER STORIES have been floating around this town [Upper Sandusky] for some time to the effect that a strange apparition is to be seen of nights on the banks of the Tymochtee, in the vicinity of the place where Colonel Crawford was burned to death by the Indians almost a century ago. The form of the human being is seen, or said to be seen, moving about, inclosed [*sic*] in a circle of fire, and of course the conclusion is arrived at that it is the ghost of Colonel Crawford. Just what has raised it does not appear to be definitely settled."

Source:

"Local News," 1888. *Marion* [Ohio] *Daily Star* (August 4).

1888: "Men of Gigantic Stature"

IN *HISTORICAL COLLECTION of Ohio in Two Volumes* (1888), Henry Howe wrote:

"There were mounds situated in the eastern part of the village of Conneaut and an extensive burying-ground near the Presbyterian church, which appear to have had no connection with the burying-places of the Indians. Among the human bones found in the mounds were some belonging to men of gigantic stature. Some of the skulls were of sufficient capacity to admit the head of an ordinary man, and jaw bones that might have been fitted on over the face with equal facility; the other bones were proportionately large. The burying-ground referred to contained about four acres, and with the exception of a slight angle in conformity with the natural contour of the ground was in the form of an oblong sphere. It appeared to have been accurately surveyed into lots running from north to south, and exhibited all the order and propriety of arrangement deemed necessary to constitute Christian burial."

Source:

Howe, Henry, 1888. *Historical Collection of Ohio in Two Volumes.* Cincinnati: C. J. Krehbiel. Cited in Ross Hamilton, 2001. "Holocaust of Giants: The Great Smithsonian Cover-up," http://greatserpentmound.org/articles/giants3.html.

1889: Unexplained Illumination

WIDOW NANCY REEDY lived on a farm southwest of Plain City. The farm had originally belonged to her deceased first husband, and she lived on it with her son from that marriage, William Atkinson; his wife; one child; and a hired man named Tuck Raynor.

Mrs. Reedy was staying with another son in the area on the evening of March 24, a Sunday. Atkinson and his family, together with Raynor, had taken the carriage to town, 4 miles away, to attend a revival meeting at the Universalist Church. Before they left, Atkinson had closed the curtains on each window, tied the dog to a post in the kitchen, and locked all the doors. On their return, around 11:30, as the house came into view half a mile away, the little boy shouted, "Oh, papa, our house is on fire!" Every window glowed with brilliant light. Deeply alarmed, the party raced to the front yard.

There the dog, almost beside itself with fear, greeted them, somehow having escaped the house. As they stood there, the illumination began to diminish in the north-room window—the one closest to the observers. After it had gone out, the light in the middle room did the same, and then the south room (the kitchen). Now the house was completely black.

Atkinson unlocked the north door and made his way gingerly through the darkness until he got to the kitchen, where the matches were. He lit a lamp, and he and Raynor conducted a thorough search. They found no evidence of intruders. The local newspaper noted, "The curtains were all drawn down, save one in the kitchen window. A pane of glass in the upper sash of this window had apparently been pushed out from the inside, and was found lying, unbroken, on a pile of stones under the window, outside of the house. It is supposed that the dog made his exit from the room through this opening; but why he should have attempted escape at so high a point of egress, nobody attempts to explain."

Interviewed by a reporter the next day, Mrs. Reedy recalled a similar incident a few years earlier. She also said that a while later, as her sister slept in the north room, she had awakened around midnight to observe, in the dim lamplight, a man's hand and arm near her face on the pillow. She screamed, and the hand and arm vanished. A search by other family members immediately followed, but as the reporter wryly noted, "the balance of the man could not be found." Subsequently, a hired man broke off an engagement with Mrs. Reedy on the grounds that the house was haunted by undescribed specters who would not let him get a good night's rest.

Source:
"Boo! Ghostly Midnight Illumination of a Farm House," 1889. *Newark* [Ohio] *Daily Advocate* (March 26).

1889: Snails Falling

A SPECIAL FROM Tiffin, Ohio, records a strange phenomenon that took place there last night [April 21]. It was no more nor less than a heavy shower of snails, from a pin head in size to some as large as a half dollar. The ground on Highland addition, a suburb of the town, was covered with them, and the noise made in their descent was like the falling of hail. In the eastern part of the city snails literally covered the sidewalks last night, although it only sprinkled slightly."

Source:
"Heavy Shower of Snails," 1889. *Newark* [Ohio] *Daily Advocate* (April 22).

1889: Giant Snake Ate Giant Man

A STRANGE DISCOVERY was made Wednesday [probably April 17], by a citizen in the northwestern district of this county. Having occasion to sink a well, one Semms selected a spot in a valley near a ravine of great length and which, during the heavy rains, is transformed into a raging torrent depositing in the valley gravel, mud and other debris.

"After reaching a depth of four feet, and while in a formation of limestone gravel that had continued almost uninterrupted from the surface down, Somms [sic] came upon the bones of an animal. The ribs were about the size of a small pig's and rapidly tapered. Carefully unearthing the bones towards the tapering end, Semms soon came to rattles which[,] when counted, numbered 17, the largest measuring six inches across.

"Attracted by the strange find, neighbors gathered and the work of unearthing the monster was prosecuted with vigor. After laying bare 19 feet of the remains of the monster of other times they found the entire skeleton of a man of tremendous stature in the stomach of the snake. The remains of man and serpent so far as the serpent has been exhumed are as perfect as when first denuded of flesh, and were doubtless covered with lime and gravel soon after death. Near the bones of the man's right hand is a rude stone hatchet, which local geologists of some repute state to be similar to the handiwork of Paleolithic man."

Source:
"This Beats Them All," 1889. *Elyria* [Ohio] *Democrat* (April 25).

1889–1890: The Conductor Returns

TRAINMEN ON THE Lake Erie and Western railroad, between Findlay and Fostoria, are greatly disturbed over what they claim is the ghost of a dead freight conductor. The conductor was killed one night last November, about eight miles east of Findlay, by his train breaking into sections in such a manner that he was thrown to the track from the car on which he was standing, and beheaded by the wheels before the train could be controlled.

"This accident occurred near the little town of Arcadia, at a point where dense woods nearly arch the track above the rails, and here it is, the trainmen assert, the ghost of the mutilated conductor—who was known in life as Jimmie Welsh—makes its appearance nearly every night as the midnight train going west from Sandusky reaches the spot where Welsh met his fate. The engineer and other officials of the train say that when these woods are reached an object looking like a headless man comes walking slowly out toward the track with a lantern in its hand which it waves backward and forward, as if searching for something. The trainmen are positive that it is the ghost of Welsh hunting for the head which was severed from the body by the car wheels. They insist that this object is plainly visible until the engine passes by, when the phantom slowly turns and fades away among the trees.

"Two crews have already abandoned this run, and have been transferred to other divisions of the road on account of this alleged ghost, and the engineer who brought this train through last night was so terrified over encountering this headless conductor that when he reached Findlay it was with difficulty that he was persuaded to stay on his engine until relieved at Lima. He said that he would not pass through another such experience for any sum of money. No other trains are annoyed by this ghostly conductor, but this is explained by the fact that no other crew passes the spot where poor Welsh lost his life at the time of night when he was decapitated by the wheels of his train. The story has thoroughly alarmed all the employees of the road, and unless the spirit of Jimmie Welsh is appeased in some way this midnight train will have to be abandoned."

Source:
"A Headless Conductor," 1890. *Fort Wayne* [Indiana] *Sentinel* (January 30).

Late 1800s: Flying Entity, Black Dog

ELIJAH L. DULING, OF Bacon Run, tells me he saw a ghost one night at the crossing of Bacon Run, as you go towards Lafayette, at the farm of H. W. Duling. There was a foot log across that run there, in the days when Mr. Duling and I used to go out and see our 'gals.' Well, he says one night as he returned home after a good spark with his 'gal,' he got across the foot log at the run, and here stood a man! Young Duling stopped and viewed him from head to foot. He was minus part of his head and his left shoulder and arm. Duling approached him bravely, without dread or fear . . . and was about to demand who he was and for what purpose he was standing there! But, as he approached him, the spectre arose perpendicularly towards the clouds, and slid off in air in a northeast course, and fell again to earth ere he got out of sight of Duling. Mr. Duling says he heard him fall to earth with a great thud as of a saw log! Duling told his uncle Dan Dean about it.

"Dean said he saw the same thing there years before. Also Mr. Dean said that after this spectre left him, a big black dog trotted along side of him, till he got back to the turn of the road at

Rice's corner, when the dog vanished out of sight. It got out that a big, black dog was seen there at night. Joseph D. Workman and Robert Brownfield are said to have seen the dog there one night. The canine followed their wagon a piece, then vanished out of sight. Had those two men declared that they saw the devil there that night they would have been believed!"

Source:

Magness, James, 1903. "Something about Ghost Stories." *Coshocton* [Ohio] *Democrat and Standard* (May 12).

1890 and Before: Big Snake of Upper Sandusky

TWO YOUNG MEN, who, sorry to say, were fishing Sunday evening [June 1 or 8], came rushing into town, about 9 o'clock, with the story that they had seen a monster snake, near the horse shoe bend. They said, while watching their poles for perch, something raised [*sic*] up on the other side of the river, about ten or fifteen feet from the water's edge, and was startling in proportions. They immediately recognized it as a snake, with a head larger than their own, and a body six or eight inches in thickness. They could see its eyes flash fire and a display of fangs that was horrifying. They didn't stop to investigate his snakeship but left for town instanter [*sic*], leaving their fishing tackle toying with the twilight.

"Now, whether these young men imagined all this or actually saw a huge snake it is hard to define. A muley cow may have come to the water's edge to drink, and boys fishing on Sunday have no doubt guilty spirits, and shadows of evening dancing around a muley cow might arouse imagination to such a pitch as to send forth a ton or two of pure reptile in good shape; especially, as in early times, the bottom along the Sandusky in that locality was noted for a monster snake which made its appearance annually about this time of year. This hugh [*sic*] snake was last seen during the [Civil] war, lying across the Lower Ford, in size and form, like the trunk of a good sized tree, an account of which we gave some years ago.

"Forty years ago there were many here who believed that this snake existed and were very careful not to go near the Sandusky bottoms after nightfall. The Indians believed in this snake, and in the months of June and July were always on the lookout. They had a horror and dread of this Big Snake, and by some it was estimated a hundred feet long. It would suddenly make its appearance and as suddenly disappear. There used to be a large hole in the bank north of the Paper Mill, which was generally supposed to be the retreat of this leviathan. All manner of traps were perpetuated [*sic*] at the mouth of this cave by the Indians, to catch or cripple this monster, but of no avail. . . .

"Has this monster appeared again, or were the boys . . . startled at their own shadows and imagined the rest? At all events they were pretty badly frightened."

Source:

"A Wyandot Monster," 1890. *Marion* [Ohio] *Daily Star* (June 10).

1890: Swallowed a Frog

THE STRANGE PHENOMENON of a live frog in a human stomach has just developed here. Mrs. Anna Nickel, who lives with her husband in this city [Columbus], has complained of a peculiar sensation in the stomach, as if something having life was moving about. This continued for six months. A number of prominent physicians in Columbus and elsewhere have been consulted, but none gave the woman relief. Saturday evening [August 9] she complained of a tickling sensation in her throat, and called Dr. Voght, who formed the opinion that the sensation was caused by the presence of an insect. After swallowing a powerful emetic, Mrs. Nickel was relieved by the expulsion of a live frog from her stomach. It was about two inches long, almost white, and the hind legs were missing. The physician gave it as his opinion that the woman, while drinking water, had swallowed the egg, which was hatched by the warmth of the stomach. The frog had been placed in alcohol and forwarded to Prof. Youzer, of the American Medical college at St. Louis, with a view of securing a scientific opinion as to the unusual occurrence."

Source:
"Live Frog in a Woman's Stomach," 1890. *Indiana* [Pennsylvania] *Democrat* (August 14).

1891: Figures in the Glass

IN A LATE EVENING in early January, three men, including a real-estate agent, were sitting and chatting in his office in the small town of Nevada, Ohio, when they noticed the apparitions of three life-sized female figures in the plate glass window at the front of the office. Unable to determine where the reflections were coming from, they got to their feet and looked in all the obvious places. They pointed out the figures to passersby, who were similarly puzzled. All in all, there were nearly thirty witnesses to the phenomenon, which eventually faded away.

Source:
"Ghosts of Plate Glass," 1891. *Mitchell* [South Dakota] *Daily Republican* (January 8).

1891: Headless Horse Horror

A GHOST IN THE FORM of a horse without a head has been seen at a mine at Coalton, Ohio."

Source:
Untitled, 1891. *New Oxford* [Pennsylvania] *Item* (November 6).

1892: Mystery Mist

A PHENOMENON IS causing much excitement in scientific circles at Martinsburg, Ohio. In the rear of a residence in the village there is a continual shower of mist over a space of a dozen feet square. The mist is said to fall whether the weather is rainy or clear, but is most pronounced between the hours of 1 and 2 p.m., when a mirror held up instantly becomes covered with a moisture. Professors of the Ohio University are investigating, but as yet can give no explanation of the phenomenon."

Source:
Untitled, 1892. *New Oxford* [Pennsylvania] *Item* (October 28).

1893: "A Frightful Object in White"

MR. H. L. GRACE HAD quite an experience on the evening of Feb. 12. He was returning home at a late hour when near the residence of Nick Parks there suddenly appeared before him a frightful object in white of huge dimensions. Lee says that he did not speak to it but he believes it was a ghost beyond a doubt."

Source:
Stroup, E. L., 1893. "Tyrone Topics." *Coshocton* [Ohio] *Democratic Standard* (February 24).

1894: Pale Rider

THE GHOST OF OLD man Woodbeck has been seen in Delaware county. The old gent lived there about forty years ago, and when the cholera came around he swore like blue blazes and then died. One night recently two farmers met an apparition on the highway consisting of a man, horse and an old-fashioned vehicle. The man drove the horse through a tall rail fence without touching a rail, and disappeared at Woodbeck's grave. What seems funny about it[,] the men who saw the thing were returning from church."

Source:
"Here and There," 1894. *Marion* [Ohio] *Daily Star* (February 8).

1894: Late-Night Visitant

A SO-CALLED GHOST has been attracting the worthy east enders of Middletown. The ghost, a woman's form draped in black, was followed by a courageous citizen the other night and it entered silently a certain house in the still watches of the night. No explanation can be made of the strange apparition except that the good lady living in the said house is away from home, and her husband is left there all alone."

Source:
"Middletown's Ghost," 1894. *Hamilton* [Ohio] *Daily Republican* (August 11).

1894: "Spirits of the Air"

JACKSONBORO HAS OF late been the scene of wonderful spiritual manifestations. It has its ghouls and ghosts and banshees and the inhabitants are filled with terror. The Jacksonboro ghost rides on the wings of every storm and chases belated farmers. The spectre haunts a deserted graveyard. A character named Lonely Sam recently had his buggy lifted up bodily and appropriated by the spirits of the air."

Source:
"Jacksonboro Spooks," 1894. *Hamilton* [Ohio] *Daily Republican* (November 19).

1894: Lights in the Snow

AT DEFIANCE, THE other evening, was witnessed a very strange phenomenon, which, as yet, can not be explained. At the time snow was falling, when in a flash the earth was brilliantly lighted up with a pale blue light tinted with red. This was instantly followed by a second illumination, the two lasting but a few seconds. In appearance the light resembled that made by the explosion of a Roman candle ball or a sudden flicker of electric light. The general opinion prevails that it was a meteor, but owing to the clouded condition of the sky it

could not be seen. Superstitious people were badly frightened."

Source:
Untitled, 1894. *Fort Wayne* [Indiana] *Sentinel* (December 31).

1895: Sign of the Cross

A CLEARLY DEFINED cross is said to have been seen recently in the heavens from East Liverpool, Ohio. Revival meetings are now being held in that town, and the church people consider that the cross was a good omen. Questions naturally arise. Should the revival meetings now in progress in some thousand other towns where no such omen has appeared be immediately shut off, or should they not? Also, if a faro or poker game is nightly played in East Liverpool may the players consider the heavenly phenomenon a good omen for them? If not, why have the good people a 'cinch' on the omen? The deeper a man goes in these soul-perplexing problems—well, the deeper he goes in them, that's all."

Source:
Untitled, 1895. *Fresno* [California] *Weekly Republican* (February 22).

1895: Finger-Pointing Cloud

OVER NEWARK, AT AROUND 9 p.m. on June 29, a phenomenon "which is not easily explained" appeared for twenty minutes. "It was in the shape of a finger and a thumb," milky white in color, pointing toward the east. Some residents who saw it considered it an omen of "turbulent times in the far east."

Source:
Untitled, 1895. *Newark* [Ohio] *Sunday Advocate* (June 30).

1895: "Rather Peculiar Ghost Actions"

FOR THE PAST FEW nights some of the residents of Tomb street [in Tiffin] have been experimenting [*sic*] rather peculiar ghost actions. The apparition is attired in male attire and with a white cape thrown over its head. Women and children are frightened nearly out of their wits and will not leave their homes after nightfall. A number of people have seen it, and it was first discovered by a lady, who met the being face to face. She was so frightened that she about fainted. The next night glass blowers watched for it and saw it in a cornfield[,] and several shots were fired with good aim, it is declared, but without effect. They followed the strange being through barns, trees and fences and state that it went right along without taking any notice of them. Some state that a man in that neighborhood was refused the hand of a maiden and suicided and that this is his ghost returning to take communion with the maiden."

Source:
"A Ghost Story," 1895. *Columbus* [Ohio] *Dispatch* (August 12).

1895: Lights and Heavy Bodies Falling

TWO FAMILIES LIVING in the Reed House on Thomas Fork, a mile from Pomeroy, moved out in succession after experiencing harassment from unearthly forces. The phenomena consisted of "lights [that] resemble those of an incandescent lamp, and which appear and disappear with frightful suddenness and irregularity," as well as noises that sounded like falling bodies. Over a period of weeks, the sounds grew louder and ever more frightening.

Source:
"Of Course It's a Ghost," 1895. *Columbus* [Ohio] *Dispatch* (November 26).

1897: Dangerous Visitor

AT 9:30 THIS MORNING [March 9] people at New Martinsburg . . . were greatly terrified. They saw a cylindrical shaped light, resembling a huge ball, about forty feet in diameter, from which three columns of smoke were issuing[,] rapidly passing through space, producing a half rumbling, half hissing noise, heard for miles. Suddenly the ball burst, producing a terrific report, heard for twenty miles distant and breaking window lights of some of the houses. When the meteor passed over the house of David Leifure the latter was knocked down and for some time was unconscious. Shortly afterward, upon going to his stable, he found one of his horses dead in its stall with the side of its head blown off. Another horse in an adjoining stall was made deaf by the concussion. Search is being made for the meteoritic stones."

Whatever this object may have been, it is unlikely to have been a meteor.

Source:
"Cylindrical-Shaped Light," 1897. *Manti* [Utah] *Messenger* (March 13).

1897: Crazy about the Airship

POSSIBLY, EVEN PROBABLY, there is no airship sailing about the Western skies, but none the less people believe in it, and so intense was the excitement which the belief created in the mind of Mrs. Eleanor Woodruff of Findlay, Ohio, that she has gone violently insane. Her mania took the form of anger because she could not secure the capital necessary for building an airship on plans of her own devising, and so strong was her sense of regret that other inventors were getting all the glory to be derived from aerial navigation that she violently assaulted the relatives and friends whose imaginary selfishness prevented her from obtaining her share of fame. Now she is in the State asylum at Toledo, and the doctors there say her condition is almost hopeless."

Source:
Untitled, 1897. *New York Times* (April 22).

1897: Log-Throwing Snake

THE LATEST SNAKE story comes from the sawmill gang south of town [Cochranton]. George Brady, one of the gang, was walking through Mrs. Gray's woods when he came upon a large snake. He affirms that it was a monster reptile and he undertook to throw logs on the snake in order to capture it, but the reptile would throw the logs off its back as if they had been mere twigs. Therefore he was unable to capture it."

Source:

"Late Snake Story," 1897. *Marion* [Ohio] *Daily Star* (December 17).

1898: Over Lake Erie

THREE TRUSTWORTHY FARMERS at Akron declare they saw an airship pass over the lake."

Source:

Untitled, 1898. *Delphos* [Ohio] *Daily Herald* (February 11).

1898: Bullet Strikes Ghost

IT WAS FIRST SEEN IN the vicinity of the C. C. & S. station [in Navarre] by small boys. It was tall and stately and wore a robe of pure white. Last night C. B. Blank, telegrapher at the W. & L. E. station, found himself face to face with the ghost while returning home from 8 o'clock. He drew a pistol and fired. The bullet seemed to strike the object, but he heard no cries of pain. The spook then disappeared as mysteriously as it had come. The village is aroused, and plans are now under discussion for organizing some sort of an investigating party."

Source:

"The Genuine Article," 1898. *Massillon* [Ohio] *Independent* (September 15).

1899: Skeleton of "Dazzling Brightness"

THE RESIDENTS OF and near West Huron, O., are all stirred up over the appearance of a ghost in that locality. This is not an ordinary ghost . . . a visitor from ethereal regions, robed in flowing white. It is different from any ever before seen. In the first place, this ghost is supposed to be the visible spirit of a departed bachelor who had large sums of money while he was an occupant of a flesh and blood structure on this mundane sphere. The ghost is not to be seen every night, and his visits are so regular that those who have a desire to view him may go to West Huron at stated times and see him walk through board fences, disappear in haystacks and sink mysteriously into the ground or soar heavenward on fiery wings.

"The ghost is always seen on the Woolverton farm, West Huron, between 11:30 and 12:30 on the thirteenth day of each and every month unless that day happens to fall on Sunday. Farmer Dildine, who lives near West Huron, describes the ghost in glowing words. He says:

"'I first saw him more than a year ago and exactly 13 years to the day after the death of the old rich bachelor. As I was passing down the hill at Statecut I looked across the valley on to Woolverton's farm, and there I saw a sight which baffles description and puts skepticism at a dead non plus. I saw a light kindle up suddenly, as of someone swinging a lantern. I paused to see what it meant. A weird glow, like an incandescent electric lamp, issued from the air or earth. Then there came to my view the form of a human being, but instead of being outlined in white it was invested with a bluish tinge—just as if you would look through a blue glass from the dark into a light room. This seemed to be the outside of the thing. Then the skeleton of the apparition appeared in dazzling brightness. The visitor seemed to be solid, and yet he was not. The face of the thing, for I cannot better describe it, was intelligent looking, and the thing's feelings and emotions were plainly discernible. You could tell by its countenance what was transpiring within its phosphorescent self, as much as an ordinary man or woman's face shows sadness or pleasure. It beckoned for me to follow it, but I was afraid to do so.'"

Source:
"Timely Ghost," 1899. *Fort Wayne* [Indiana] *News* (February 27).

1902: Prowling Ghost

SEVERAL KALIDA PEOPLE claim the distinction of having seen a ghost prowling around, north of the railroad bridge near that town. The ghost appears to them and then in the twinkle of an eye it disappears."

Source:
Untitled, 1902. *Delphos* [Ohio] *Daily Herald* (March 19).

1902: "Dressed Entirely in Black"

THE CURIOSITY OF the people on west Fifth street has been aroused to a painful state by the queer actions of a woman. Those who have seen the mystery say that she is dressed entirely in black and that she wears a heavy black veil. She is said to appear every night at the corner of Clay and Fifth street, where she will stand to a late hour, as though looking for some one. Many of the ladies in that neighborhood are afraid to leave the house after dark. There are theories out concerning her. Some think it a crazy person, others that it is some one trying to raise an excitement while there are those who think it some wandering spirit."

Source:
"The Fifth Street Ghost," 1902. *Delphos* [Ohio] *Daily Herald* (June 26).

1905: Haunting Lights

IN AUGUST, RESIDENTS of a rural area 2 miles north of Marion reported seeing mysterious lights moving through the night. "One light, a very large one, of peculiar phosphorescent color, has been more than the others of smaller size," a newspaper said, "and has boldly approached the workmen employed at the Central Ohio Lime and Stone company at the Norris & Christian quarries. The light is start[ing] from the Hinamon woods."

One employee, Richard Gallamore, saw the lights on a number of occasions. Late in August one approached him so closely that the witness, a Bible reader, spoke from Scripture. The light went out.

A local belief held that the lights were from persons killed in the neighborhood. The largest one, speculation held, was the ghost of a brakeman who had died in a railroading accident.

Source:
"Ghosts Appear, Some Think," 1905. *Marion* [Ohio] *Daily Star* (September 4).

1905: "Blue Bells"

THE HOME OF SIDNEY Veon in Ravenna, Ohio, is said to be haunted and so terrified have he and his family become that they declare they will move. A ghost in the form of an aged man is said to appear each night between 8 o'clock and midnight. Doors and windows open and close, and fiendish laughter is heard. Recently the music of 'Blue Bells' was heard when no one was near the piano. When the window opened the other night and the ghost appeared[,] Veon made a grab for him, but he vanished with a laugh. The family says a bed was turned completely around the other night, and Veon's hat, hanging on a peg, whirled about like a buzz-saw."

Source:
"Family in Terror at Ghost," 1905. *Iowa Recorder* (October 25).

1906: Bolt out of the Blue

AT FRANK WEBSTER'S farm a mile west of Quaker City, the sky was blue and cloudless at 5:30 p.m. on September 18. Nonetheless, a bolt of lightning struck a valuable horse, killing it instantly, and started a fire in a hay mow. The resulting fire spread rapidly and was hard to put out. It consumed seventy-five bushels of wheat, farm implements, and two stacks of straw.

Source:
"Bolt Falling from Clear Sky," 1906. *Cambridge* [Ohio] *Jeffersonian* (September 20).

1908: Nerve-Wracking Ghost

AT HAMILTON IN LATE summer, the *Coshocton Daily Times* reported, sixteen-year-old May Irwin was "suffering from nervous prostration over the supposed presence of a ghost in her room. Unnatural noises, apparently coming from the top of her bed, awakened Miss Irwin, and the strange noises continued until the gas was lit. Members of the family sat in the room with the daughter, and when the light was turned out they also heard the spooky sounds. Miss Irwin is a nervous wreck and is under the care of a physician."

Source:
"Spooks Wreck Girl's Nerves," 1908. *Coshocton* [Ohio] *Daily Times* (September 2).

1908: "Awful Thing" in a Long Black Cape

THE GOOD NEIGHBORS on the street [Spruce] say . . . a legless man . . . dashes out of the yards and jumps across the road only to disappear in an orchard across the street. Mrs. John Gehrke . . . was the first of many residents to discover the 'awful thing' last Thursday night [September 3] about ten thirty o'clock, and she flushed the ghost in the yard and was terrified when the thing[,] which looks like a man dressed in a long black cape, skimmed over the ground across the road and disappeared in an old orchard, which has the reputation of being spookey [*sic*] on account of a suicide which occurred there one winter's night.

"Mrs. Eva Collier . . . was the second party to witness the aeroplane antics of the specter, and although the story told by Mrs. Gehrke was laughed at, the smiles from the lips of the doubting Thomases changed to looks of grave concern when Mrs. Collier described the unnatural flight which she witnessed.

"Mrs. Ella Danforth . . . was the third resident of the street to run across the night flyer and when her story was told among the neighbors, the cold chills began to creep along the spines of her listeners, and to add to her description of the ghostly thing, a young lad burst into the room and said he had seen the gruesome object the night before and had pursued it until it was lost in an old grape arbor near Woodford avenue."

Source:
"Legless Ghost Is 'Walking,'" 1908. *Elyria* [Ohio] *Republican* (September 10).

1908: A Sign from Above

WHILE EVANGELIST WILSON was holding a religious meeting in a large tent at Blackfork, a meteor came through the tent, causing a panic among several hundred people. No one was seriously injured, but great excitement prevails among the people, who believe that the phenomenon was a spiritual manifestation."

Source:
"Meteor Causes Panic," 1908. *Coshocton* [Ohio] *Daily Times* (September 25).

1909: Luminous Enigma

A LIGHT, BELIEVED by many to have been attached to an airship, attracted a great deal of attention in the northeastern skies about 11 o'clock Thursday night [July 22]. One minute it would be large and brilliant and the next small and dim. It would remain apparently stationary for a while and then speed along in an easterly direction at a rapid rate."

Source:
  "Lights in Skies," 1909. *Sandusky* [Ohio] *Daily Register* (July 23).

1909: "The Coils of a Huge Snake"

HAMBDEN IS EXCITED since that locality has become the abiding place of a monster black snake, which stretches the width of the roadway, being nearly fifteen feet long. Hiram Toland, a farmer, saw a dark mass in his pasture and walked within twenty feet of it before he discovered the coils of a huge snake. He said the reptile lay over a slight depression of the ground in three coils and that the diameter of the outside coil must have been three feet. The reptile raised its head, showing its fangs."

Source:
  "This Snake Yarn Is Real Thing," 1909. *Elyria* [Ohio] *Republican* (September 9).

1909: Church Phantom

THAT A GHOST INFESTS St. Bernard's Catholic church [in Springfield] is the belief of John Herzog, an electrician who has been wiring in the church for some weeks.

"Alone in the auditorium, he was startled one night to see a white figure of a woman silently walk down the center aisle. Herzog says his hair stood on end, as the ghost advanced toward the chancel where he was working, passed him, marched solemnly back through a side aisle, ascended the winding stair to the choir loft and was seen no more. In connection with the story as to the church being haunted, members recall the mysterious burning of the great pipe organ three or four years ago."

Source:
  "Ghost's Home Is in Church," 1909. *Coshocton* [Ohio] *Daily Tribune* (October 14).

1909: Flying Bucket

RAY MCCULLOUGH, son of James McCullough, who resides about three miles west of Roscoe, sighted what seemed to be an airship floating in the distance over their farm Thursday evening [October 14] about 7:30. He called his father and they watched the strange object vanish in the southern sky. It looked about the size of a bucket. They are at a loss to know what it could be unless it was an airship."

Source:
  "Sighted Airship Thursday Night," 1909. *Coshocton* [Ohio] *Daily Age* (October 16).

1910: Light on the Lake

MARINE MEN AND residents along the lake shore at Lorain were deeply puzzled on Saturday night [March 26] by the appearance of a strange light in the lake off the port. The light flashed once and then again twice several times, covering many minutes. At first it was supposed to be the signal of a boat in distress. Spectators watched the strange sight for a long time but could find no solution to the problem.

"Manager Campbell of the Lorain office of the Great Lakes Towing company boarded the tug Excelsior and went out into the lake to ascertain the trouble. It was believed that this would be the better course to pursue as in case some craft were in trouble aid could be rendered. The Excelsior went out about seven miles. The farther out the tug went the dimmer the light grew and finally it disappeared.

"No one has yet been able to learn what the light was or why it was flashed in such a peculiar manner. From appearances it is believed it was an electric light but what sort of a craft it was on is still a matter of conjecture."

Source:
  "Strange Light Has Lorain Marine Men Searching on Lake," 1910. *Sandusky* [Ohio] *Daily Register* (March 30).

1911: Unexplained Red Light

A MYSTERIOUS RED light in the northern sky attracted the attention of a small number of Sanduskians about 7 o'clock Friday night [October 6] according to inquiries received at the telephone exchange and The Register office. An investigation failed to reveal the light to reporters who hunted for it or persons aside from the inquirers who had seen it. The opinion seemed to be that the light dangled from the basket of a balloon sent up at Kansas City the other day. According to reports, however, the balloons were carried by winds in directions anything but Sanduskyward. Lights are always seen when a balloon race is on somewhere but as yet, except, perhaps, in one instance, those sailing over Sandusky were not attached to balloons."

Source:

"View Strange Light," 1911. *Sandusky* [Ohio] *Daily Register* (October 7).

1911: Out of the Sky

ON NOVEMBER 24, 1973, a seventy-seven-year-old woman named Elsie Shirley wrote a letter to J. Allen Hynek. For twenty years, Dr. Hynek, chairman of Northwestern University's astronomy department, had been the U.S. Air Force's chief scientific consultant on UFO matters. He and his employer had parted ways in 1969 after he became an ever more vocal proponent of the reality of UFOs and a critic of Project Blue Book's contrary view. He went on to write *The UFO Experience* (which coined the phrase "close encounters of the third kind") in 1972 and to found the Center for UFO Studies the next year. As the most visible figure in the UFO controversy, Hynek often got letters from private citizens reporting their own sightings.

It is by no means certain whether Mrs. Shirley's story was about a UFO sighting, but it certainly was a strange one. It went back to 1911, Union County, in rural west-central Ohio. At the time, her thirty-six-year-old sister, Idella, lived in a house with her two preteen sons

just outside Piqua. The railroad rented the house, and Idella had been hired to cook and provide rooms for out-of-town contractors who were supervising a bridge-building project on the railroad to the south.

One day in the middle of May, Idella took the train to Broadway, approximately 40 miles to the east. She had traveled to visit her family, including her grandmother, who lived just across the street from the depot. Idella entered the house in a state of physical and emotional distress so marked that the grandmother phoned her daughter (Idella's mother), who lived 14 miles to the east in Delaware. The mother and middle brother immediately boarded a buggy and drove furiously toward Broadway. Concerned about the welfare of Idella's two boys, the youngest brother hopped a freight train to Piqua, where he would learn from them and hired hands that Idella had left in a cheerful state of mind. In other words, whatever happened, it apparently had happened on her trip to the train station or during her train ride to Broadway.

Idella, however, was unable to explain what had taken place. She could speak normally on any other subject, but when asked about the source of her physical injuries, she became frightened and grew incoherent. The physical injuries were most peculiar. Doctors who examined her (one was her brother D. C. Fox) found that something had scared her face, and her tongue was badly swollen. "The most puzzling symptom," Mrs. Shirley recalled, "was that just below the largest part of the calf of the legs were creases deep enough to encase three fingers held closely together—fore, big, and ring—in the indentation." The doctors rejected the theory that these were rope marks; something smoother than rope was responsible, they were sure.

Idella's emotional collapse led to her moving into her mother's house along with her sons. Elsie, her brothers, and her mother had to watch her all the time because she kept talking of suicide. On one occasion she was caught just before she was about to jump from a high beam in the barn onto heavy farm equipment below. Another time she jumped into the Scioto River but was rescued by her mother. Finally, in February

1912, she was committed to a state mental hospital in Columbus. The following November, while still hospitalized, she threw herself out of a window in the middle of the night and died.

Mrs. Shirley told Hynek of an odd exchange that had taken place before Idella's hospitalization. "One night in our kitchen," she wrote, "as Idella stood with one hand on the old cistern pump and the other on my mother's shoulder, she spoke with great earnestness of how 'they' had come out of the sky; they were going to destroy all of us and take her boys and my mother's boys. . . . Night after night, she would fill pages detailing her experiences, but my mother, perhaps feeling a little ashamed, burned them each morning."

At the time, no one knew what to make of what Idella was saying, except to see it as further evidence of her growing madness. After World War I, Elsie and a brother spoke of her words and wondered if she was talking about airplanes, which were little known in the rural Midwest in 1911 and 1912. Many years after that, when UFOs were being sighted and discussed openly, she thought of another possible explanation behind the weird family tragedy.

Source:
Shirley, Elsie, 1997–1998. "Out of the Sky? A Curious Story of 1911." *International UFO Reporter* 22, 4 (Winter): 13, 27.

Oklahoma

1912: Raining Toads

WILLIAM W. BATHLOT drove a horse-driven mail wagon out of Floris. One day in late October 1912, when he was about a mile from town on a return trip from his rounds, a bolt of lightning ripped through the sky. A crash of thunder followed. When Bathlot looked out of an open window in the wagon, he saw a black cloud covering the western sky, with its darkest section above him.

The rain fell hard for a few minutes. Then thudding sounds echoed on the wagon's roof. Bathlot assumed that the rain had turned into hail—until he noticed small, living forms bouncing off it and landing on the ground or on the backs of the horses. "They bounced up from the sandy soil like little rubber balls, lay stunned up on the ground for a few seconds and then flopped over on their stomachs as lively as you please," he would recall. All landed on their backs, but none were injured. "I could peer outward for perhaps one-hundred feet through the falling rain," he reported, "and as far as I could see the top of the earth was alive with the little creatures." He stuck his hand out and caught four as they fell. They were small, fat baby toads.

The fall continued for another three minutes or so, at which point it ceased as mysteriously as it had begun. The rain kept falling, however.

Bathlot learned that the toad fall had extended to Floris. He and other witnesses estimated that the fall had covered an area a mile long and a quarter-mile wide. The animals disappeared almost immediately, apparently going on their way in the wet, sandy landscape. The number of toads involved seemed beyond calculating. Bathlot thought that it must have involved "millions." He had no idea where they could have come from.

Source:

Bathlot, William W., 1953. "Does It Rain Toads?" *Fate* (June): 90–92.

Oregon

1869: Screaming Head

AT NIGHT, AT A haunted house in Albany, persons claimed to have seen "apparently a human form, but headless, standing with arms extended, as though imploring help." Other witnesses described a horribly mutilated head stuck on a pitchfork, blood pouring from its bleeding mouth as it groaned and screamed. "Its eyes seem revolving balls of fire, emitting occasional gleams of light sufficient to allow the beholders to see things distinctly for several yards around."

Source:

"Oregon," 1869. *Morning Oregonian* (November 29).

1870: In the Mines

IT IS BELIEVED BY many that the disembodied spirit of John C. Holgate—who was killed during the Chariot Elmore war of 1868—has its abode in the old deserted chambers of the mine, which he died in defending. Not long since[,] a Chariot miner saw what he termed Holgate's ghost standing in a dark passage near the boundary line between the two mines. The ghastly lips moved as if speaking, and the thin bony hand beckoned the observer to approach. He felt his hair standing on end, and turned and fled, reaching his companions in the [illegible] more like a ghost than a live being. Others have seen the same apparition, and declare it to be either Holgate or his ghost."

Source:

"The Territories," 1870. *Morning Oregonian* (November 28). Reprinted from "an Owyhee paper."

1896: Airship Madness

FROM BROODING OVER the sensational stories of the appearance of an alleged California airship and his failure to perfect an invention of what he claimed was an aerial vessel, Henry W. Herne, an eccentric individual living west of this city [Portland], has been committed to an insane asylum. Herne a few days ago became violently excited over the airship and declared that one of his ideas had been stolen. He refused to eat and secluded himself from the neighbors till the authorities took him in charge."

Source:

"Insane over Airships," 1896. *Daily Nevada State Journal* (December 4).

1896: "Immense Ball of Fire Flying through Space"

ONE NIGHT LAST WEEK when train No. 59 was crossing the hill near Blalock, the trainmen were treated to a beautiful sight. A large and very bright meteor was seen falling from the heavens, and when apparently about a mile from the earth seemed to stand suspended for a moment, and then passed on its northerly flight. The meteor was visible for about five minutes and had the appearance of an immense ball of fire flying through space."

Source:
"From Oregon," 1896. *Montesano* [Oregon] *Vidette* (December 18).

Pennsylvania

1834: Eggs of War

WE HAVE BEEN INFORMED by the most credible authority, that a number of hen's eggs have been found in this county, one by a son of Mr. Isaac Jeffries, of Newlin, with the word 'WAR,' in distinct and handsome characters, written legibly upon it. The letters are raised above the level of the shell, like the letters used for instructing the blind to read. Those who have seen the eggs, state they must be natural productions, and that there cannot be any deception in them. We state this singular phenomenon for the curious—what it denotes, no one can tell—every egg containing the letters, is also marked, with equal plainness, '1836.' The eggs have been seen by numbers, and we understand, may be examined in our market-house, on Saturday next."

Source:

"Curious Phenomena," 1834. *Adams* [Pennsylvania] *Sentinel* (September 1). Reprinted from the *Village Record.*

1849: Caterpillars in the Snow

A YOUNG GENTLEMAN showed us, on Friday morning last [February 2], some small black looking caterpillars, about 1/2 an inch in length, which he had picked up on the top of the snow, near the woolen factory a short distance above town [Danville]. It was snowing at the time he found them, and there were large quantities of the same sort promiscuously scattered over a large space, all alive."

Source:

"Something Curious," 1849. *Ohio Repository* (February 21). *Danville* [Pennsylvania] *Democrat,* February 9.

1857: "A Sulphurous Mass"

IN A LETTER TO THE *New York Tribune,* one P. Johnson of Carbondale reported a bizarre and unlikely meteorological or meteoritic phenomenon he claimed to have witnessed the previous evening. The alleged incident occurred apparently in mid-June.

Johnson said he had seen a large cloud float in from the northwest on the strength of a great wind. As it approached the city, "a dark looking cloud, or substance, was seen to leave the cloud and make diagonally for the earth." On hitting the ground, it turned—to the witness's surprise—brilliantly luminous. Then the object passed through the center of a large barn standing in the field on which it had landed. The barn erupted into flame, and the object continued onward, its rate of speed increasing all the while. It streaked on a straight course for some distant woods, "melting stone of considerable size, and burning up brush and under wood, making a complete road, of a rod or more in width, for the distance of three miles, and finally fetching up against a perpendicular breast of solid anthracite coal of 60 feet in thickness, proving rather too much for its cometship, leaving nothing but a sulphurous mass behind."

Source:

"Meteoritic Phenomenon," 1857. *Ohio Repository* [Canton] (July 15). Reprinted from the *New York Tribune*.

1864: "A Beautiful White Rose"

AN UNNATURAL PHENOMENON took place in Rapho township, three miles east of Mount Joy, on Thursday last [February 25], at the residence of Martin Inly. His daughter, aged eighteen years, died a quarter before eight o'clock. After the usual ceremonies were attended to, towards morning the attention of the mother was drawn to the corpse, when she discerned something unusual on the lips of the deceased. Attempting to remove it, apparently a voice seemed to say, 'Let it remain.' Astonishment caused an examination, and the fact was, there appeared in the middle of the lips and teeth, a complete rosebud ready to open and on the left side of the bud a full bloomed rose—what florists call a double levy rose. On close examination it was ascertained that the bud and rose were a hard substance to the touch, and in appearance like ivory. It was perceptible that from the first and last nights of the phenomenon it enlarged. Thoughts were entertained that the removal of the body to the burying-ground (which was in a hearse and over a mile of rough road) might cause a separation, but the bud and rose still remained as when first seen, only increased perceptibly in size. The rose and bud were as white as ivory, and supposed equally as hard.

"Many persons were eye witnesses of the fact. The rose was very natural, and those not knowing the fact, felt satisfied that it was a real rose and bud placed there as an ornament; and reports caused many to witness the fact. The disease of the deceased was measles, a relapse took place attended with a cold, which caused her death."

Source:

"A Strange Phenomenon," 1864. *Adams* [Pennsylvania] *Sentinel* (March 1). Reprinted from the *Elizabethtown* [Pennsylvania] *Trumpet*.

1867: Poltergeist Speaks

IN THE WEST WARD of the city [Williamsport], reside a quiet family, exemplary in all respects, the heads of which are, and have been for a long time, members of the Pine street M. E. [Methodist Episcopal] Church. A short time ago they were surprised at certain, or, perhaps, we should say, very uncertain sounds, as of rapping with the fingers or knuckles, and sometimes a scraping or scratching sound on the floor or wall. These sounds seemed to follow a little girl, about sixteen years of age, a niece of the gentleman of the house. For a time they paid no attention to them; but they increased in such a manner as to attract attention, and were apparently determined to be heard. What was more strange than aught else about it, was the fact that the spirit, or whatever it is, spoke in an audible voice! It called the name of the girl and of other persons. The pious head of the family betook himself to prayer, in order to lay the spirit, but it would not down. On Saturday evening [February 16] the pastor was sent for, and he, after convincing himself that there was no fraud on the part of the girl, called in another aged and well known clergyman. They both prayed, and the unseen visitant spoke audibly during the prayers of each. On Sunday morning the girl attended church with the family. There the rapping was heard by many, and the girl's name was called. Fearing to attract attention, she left the church. At class meeting the same day, the same phenomenon occurred.

"This much, and a great deal more, had occurred up to Saturday morning last. It seems to follow the girl, yet some demonstrations have occurred when she was out of the house. They have talked and rapped at her, or for her or with her, while on the board walks in the street, and the gate, the door-steps and other places, and, as we understand, the talking has continued while the girl was absent from the house. The members of the family, with whom we conversed, say that they are all satisfied that there is no trick or collusion in the matter on the part of the girl. The two clergymen pronounce the whole thing entirely inexplicable, and we are told that they also are satisfied that no member of the family has any voluntary agency in producing the sounds or the talking. The parties are too respectable to admit of the theory of collusion. Indeed, they are all pained at the occurrence, and would gladly be rid of it. What will come of it, of course we are unable to say."

Source:

"Wonderful Phenomenon in Williamsport, Pennsylvania," 1867. *Titusville* [Pennsylvania] *Morning Herald* (February 23). Reprinted from the *West Branch* [Pennsylvania] *Bulletin*.

1869: "It Seemed to Inspire Terror"

A MOST SINGULAR PHENOMENON occurred at midday on Saturday last [August 6], near the village of Adamstown, Lancaster County. About 200 yards north of the village is an open lot and at 12 o'clock, when the villagers were taking dinner, a luminous body was seen to settle near the center of the lot. It is represented by four or five different parties, who witnessed it from several points, to have assumed a square shape and shooting up into a column about three or four feet in height and about two feet in thickness.

"The sun was shining brightly at the time and under its rays, the object glittered like a column of burnished silver. The presence, after reaching its full effulgence, gradually faded away and in 10 minutes time, it had entirely disappeared. Those who saw it were unable to tell what it was. It seemed to inspire terror rather than admiration. After it had disappeared a number of persons visited the spot but not a trace of anything unusual could be found.

"Similar objects have been seen in the neighborhood on several occasions during the nighttime but none before in the daytime or so bright as this. The land in the immediate vicinity is dry, there being no swamp about, otherwise the phenomenon might be accounted for."

Source:

"Singular Phenomenon," 1869. *Reading* [Pennsylvania] *Daily Eagle* (August 14).

1872: Phantom in the Coal

A VERITABLE GHOST has taken up his residence in a shaft of a Pennsylvania coal mine. That he is rather a formidable inhabitant appears from the annexed description. The ghost is represented as being about six feet high, of white, thin, vapory substance, and moves about through the chambers, appearing first to one gang of men and then to another. Pistols have been fired at him and, although many shots have been passed through him, yet he seems invulnerable to pistol-balls. He has frightened the mules, driven men from their chambers, threatened individual persons, and produced a consternation among the workmen. All attempts to catch him have proved abortive, and he is getting to be a decided nuisance."

Source:

"A Ghost in a Coal Mine," 1872. *Petersburg* [Virginia] *Index* (May 11).

1873 and After: Haunted Railway

IN 1873 A TRAIN killed a hobo on the Reading Railroad tracks, just below Port Kenney Station. The death was an accident, but most who knew the details thought it could have been avoided if not for the engineer's negligence. Soon afterward, railroad men began reporting seeing apparitions at or near the site. The sightings continued for seven days, then ceased. It was thought that the ghosts were finished with their business, to the relief of those whose work responsibilities put them in the way of the unknown intruders.

Then, on Christmas night 1882, an "apparition of unusual size," according to a newspaper story, "attacked the nine o'clock freight train, which is managed by Engineer Charles Welch." Every night after that, shadowy entities engaged in an onslaught on train traffic. Trainmen took to carrying missiles and other weapons they could use to fire at the ghosts. On New Year's night, brakeman George Nelson stood on the front platform of the first car and was enjoying the fresh air when the headlights fell on what looked very much like the figure of a man some 50 yards ahead. Nelson yanked the bell rope and screamed to the engineer to stop the train.

"Although he laid hold of the cord at once," it was reported, "he says that it was not until the apparition was passed that the gong struck. Gradually the train neared the person, who seemed to be standing with one of his hands shading his face and the other pointing to the throbbing engine, straining to mow him down. There was a sudden blankness, a cold blast of air which carried off his hat, and Nelson did not know what happened till the conductor opened the door and told him he would catch cold. He was certain that what he had seen was not flesh and blood."

The next evening, Nelson secured a big piece of iron, expecting to use it on the ghost, but it did not show. The following evening, however, he saw it clearly and got close enough to hurl the iron right at it. A passenger armed with a revolver fired two shots directly into the apparition's face, with no effect.

Engineer Welch, who had already had an experience on Christmas, had another on February 24. "He spied it, as usual, ahead, but it looked so different from what it did on the previous occasion that he thought it was a real individual and not an artificial one," a reporter wrote. He hit the brakes and got the steam whistle blowing and the bell ringing—too late. The train rolled over the figure. Horrified, he said to the conductor, "We've killed someone, Jim, and we had better go back and pick up the pieces." When they uncovered not a trace of the victim, they concluded it was the ghost. They got back into the train and went on down the track.

Source:
"A Ghost on a Railroad," 1883. *Stevens Point* [Wisconsin] *Journal* (March 3). Reprinted from the *Philadelphia Times.*

1879: Brimstone Snow

ONE OF THOSE PHENOMENA, occasionally happening and puzzling philosophers to account for them, occurred on Monday [March 17] in eastern Pennsylvania—a fall of sulphur.

At Allentown, Reading, Easton and other points, there was a light fall of snow, with a sprinkling of a yellow substance having all the appearance of sulphur. Being scraped up and set on fire it burned as readily as the common article and emitted the same fumes. Some of our citizens observed the same phenomenon in Gettysburg, but the fall was less marked than east of us."

Source:
"Sulpher [*sic*] Shower," 1879. *Gettysburg* [Pennsylvania] *Star and Sentinel* (March 20).

1880s: Giant in a Vault

ACCORDING TO AN 1885 issue of *American Antiquarian,* scientists from the Smithsonian Institution who had opened an Indian mound found within it the skeletal remains of a "giant measuring seven feet two inches." The figure lay in a vault buried deep beneath the surface. Its hair, dark, coarse, and waist-length, had survived, as had the crown on its head. Nearby were the skeletons of children of various ages. They were covered with beads made of bone.

Once these had been removed, "the bodies were seen to be enclosed in a net-work of straw or reeds, and beneath this was a covering of the skin of some animal. On the stones which covered the vault were carved inscriptions."

The artifacts were shipped to the Smithsonian.

Source:
"Giant Skeleton in Pennsylvania Mound," 1885. *American Antiquarian* 7: 52. Cited in William R. Corliss, ed., 1978. *Ancient Man: A Handbook of Puzzling Artifacts.* Glen Arm, MD: Sourcebook Project, 683.

1881: Vision or Hallucination?

AN EIGHTEEN-YEAR-OLD woman, blind and suffering spinal meningitis, claimed to have seen an apparition of the Blessed Virgin Mary on several occasions over a period of weeks. Though the physician dismissed the visions as hallucinations, some of her neighbors said they believed her because they themselves had witnessed the miraculous visitation.

Source:
   Untitled, 1881. *Gettysburg* [Pennsylvania] *Star and Sentinel* (August 3).

1883: A Ghost, Madness, and Death

LAST NIGHT [JUNE 18] Mrs. Moore, a widow, who lives with her young son at Brush Valley, near here [Snyderstown], was startled by hearing a shriek, followed by agonizing yells, in the room next to her own, which was occupied by her son, a youth about 16 years old. In a fright she ran into the room and found the boy almost dead with fear and trembling in every limb. She quieted his fears and questioned him, and, after considerable delay, he told her that about midnight, while listening to the furiously raging storm, he was startled by seeing through the dim light of a lamp which burned in the room a man raising the window.

"Almost paralyzed with fear, he sat up in bed unable to move until by the aid of a vivid flash of lightning he perceived the features of the man to be similar to those of his father, who was killed in the mines five years ago. With the shriek that had so startled his mother he sank on the bed and the intruder fled hastily. When the mother had heard his story she turned her gaze toward the window and beheld the identical face pressed against the pane. With a loud yell she sprang to the window, and raising the sash, jumped through, and striking the ground 25 feet below with terrible force, injured herself fatally. The shrieks and moans of the young Moore brought a few neighbors to the spot, and they carried the limp body of the woman into the house, and after a few hours' labor succeeded in bringing her to sensibility. As soon as she fixed her eyes on her son she burst into a violent fit of laughter, in which the son joined, and which lasted until both fell to the floor exhausted. On the part of young Moore the fit of laughing was then followed by violent spasmodic attacks. He foamed at the mouth, barked like a dog, and made vicious snaps at those who attempted to quiet him.

"A young farmer named Herrick went up to him, and while attempting to quiet him caught hold of his hand. No sooner had he done so than he too was seized with fearful spasms, and writhed on the floor in intense agony, exhibiting the peculiar symptoms manifested by the others. The few other neighbors who had come to the scene were so badly frightened as to be of little assistance, and they fled precipitately, leaving the three maniacs alone in the room. Mrs. Moore was stark raving mad, and soon the two young men were busy at work, demolishing the furniture and striking one another.

"One of the women who had at first rushed to the house ran home and returned with her father, an old army Sergeant named Billheimer, who ran into the room, and, grasping Herrick, threw him to the floor, and, putting his foot down on his breast, bound him with the bed ropes. He then scoured young Moore in a like manner. Mrs. Moore was bleeding from the wounds received from falling out of the window and lay on the floor insensible. Lifting her in his arms, Billheimer carried the woman to the open air, the storm having ceased. One of the neighbors had mounted a horse about one hour before and ridden at full speed across the rough country road in search of a doctor. After a long search he found one and brought him to the stricken family. The young man Herrick was taken home by his father in the morning, and another physician attended him.

"No hope is entertained of Mrs. Moore's recovery. Her son was unusually violent this morning, and could scarcely be held by four men. The case has occasioned much excitement among the farmers throughout the neighborhood. Mrs. Moore and her son are practical, steady people. Young Herrick is not so violent, but arrangements are being made for his removal to an asylum. Mrs. Moore was in a sinking condition this morning, and she will hardly live until morning."

Source:
   "A Singular Ghost Story," 1883. *New York Times* (June 20).

1884: Headless Man with Lantern

IN LANCASTER COUNTY . . . the Rev. Daniel Whitmer when going home from church, Wednesday night [December 22], was joined in the fields by a headless man carrying a lantern. The frightful apparition accompanied the preacher to his front door, and then disappeared. Other persons returning from church saw the strange phantom, and their description of it agree[s] with Mr. Whitmer's statements. When such a story is backed up by reliable witnesses, it is hard to dismiss it as a deception or an illusion."

Source:
Untitled, 1884. *Atlanta Constitution* (December 27).

1885: "The Head in One of Its Hands"

A GHOST HAS MADE its appearance near a school-house in Manor township, Lancaster county, Pa. It was first seen one night by Rev. Daniel Witmer and M. H. Kauffman, the latter at the time being in the company of two ladies. They were returning from attendance at a religious meeting which had been held in the school-house[,] where Mr. Witmer had preached.

"A short distance away from the fence dividing the fields the reverend gentleman separated from Mr. Kauffman and the ladies and started homeward on a near cut across the fields. He had only gone about one hundred yards, carrying a lantern, when Mr. Kauffman and his lady friends saw approach what they supposed to be another man, who also carried a lantern, walked a short distance, and finally disappeared from view. The following day Mr. Kauffman met the minister, who said, 'After leaving you I started across the field and had gone but a short distance when I saw the approach of what I supposed to be a man carrying a lantern. He came directly towards me and when near enough I was almost paralyzed with horror to perceive that no head crowned the body. The object was a moving body without a head, but when it came quite close to me my terror was increased still further by seeing the head in one of its hands. The fearful object took its place by my side and accompanied me to my very door. How I managed to retain consciousness and reach home I don't know. It was the most fearful experience I ever had, and I hope I may never have the like again. This same object, whatever it may be, has been seen several times.'

"This is the story as related by Mr. Kauffman and corroborated by others of the vicinity. . . . Several other members of Mr. Witmer's congregation also saw the object, but ran for their lives. Other reputable citizens have also been accompanied home by the apparition. It is popularly supposed to be the specter of a rich miser who was found dead recently and known to have large sums of money hidden."

Source:
"A Lancaster County Ghost," 1885. *Gettysburg [Pennsylvania] Compiler* (January 20).

1885: "A Woman, Spectral in Appearance"

JOHN MAGINN AND Thomas Stinson were walking out Kerlin street night before last, and when near the crossing of the Philadelphia, Wilmington and Baltimore Railroad, they were startled by the sight of an apparition near the watchbox. It was the figure of a woman, spectral in appearance, and slowly advanced toward them. Both men saw it, and stopped and watched its movements, while their hair assumed a perpendicular. After walking a short distance the spectre vanished and did not reappear, and as the men were not anxious to see it again, they left for home. It is supposed to be the ghost of the colored woman killed at the railroad crossing a short time ago."

Source:
"A Ghost on Kerlin Street," 1885. *Chester [Pennsylvania] Times* (February 14).

1885: Phantom Freight

A STRANGE STORY IS in circulation along the line of the Branch Railroad. It is said that lately, on several nights, a phantom train, consisting of an engine with head light, and a number of freight cars, passed over the road towards East Berlin. The conductor resembled Mr. Williams who lost his life a short time ago at Red Hill. Some people living near the track declare that they have seen this ghostly train speeding along."

Source:
Untitled, 1885. *Gettysburg* [Pennsylvania] *Compiler* (November 24).

1885: Ghost on the Heights

A GHOST IS CLIFTON'S latest sensation, and the inhabitants of the town speak of it in cautious whispers as something which might be stolen by their envious neighbors, were it known outside the town limits. Clifton is a pretty town on the Medea branch of the Philadelphia, Wilmington and Baltimore railroad, eight miles from the Broad street station. Last Monday [December 14] Mr. Grady was on his way to the village from his mill. Just as he crossed Darby creek to ascend Clifton heights, a ghostly figure appeared before him. It was tall, and its height seemed greater every moment. It was dressed in a plain white sheet, which covered it from crown to toe. A small tomahawk or hatchet, Mr. Grady does not know which, was carried in the right hand, [and his neighbors thought] as soon as they heard the story, that it was a wandering spirit of an Indian warrior. . . .

"The heights is a deep-wooded hill, through which runs a patch, and within the last five days the ghost has been seen by persons passing along there at least half a dozen times, according to their statements."

Source:
"The Ghost of Clifton Heights," 1885. *Trenton* [New Jersey] *Times* (December 21).

1886: Frog in Coal

ONE COLD DAY in 1886, in McLean County, fourteen-year-old Eddie Marsh, son of a local bookkeeper, sat in front of a stove waiting for the coals to fire up enough to warm him. Impatient at their slow progress, he took a poker and began punching at the coals. When he struck a particularly large specimen, it burst open, and pieces of it flew out of the stove. One piece landed in the young man's hand. He was about to toss it back in when he noticed how strangely lit it seemed. On looking at it, he was shocked to see that it was a perfectly formed, mummified frog.

The strange story attracted considerable attention in the area and was reported in the local newspaper. The frog, the paper reported, "had been embedded in the center of the large lump of coal, and its bed was plainly discernible when the lump was laid open. The lump of coal came from the third vein of coal in the McLean county coalshaft, which is 541 feet under ground."

Among those who viewed it was James Stevenson of the U.S. Geological Survey. He managed to persuade the Marshes to give him the specimen, which he subsequently showed to biologist and *Science* correspondent R. W. Shufeldt. Stevenson, who knew the family, expressed confidence in the honesty of all concerned. He asked Shufeldt to study the remains.

"I at once recognized it as a species of Hyla," Shufeldt wrote, "though I am unable to say which one. It apparently agrees in all its external characteristics with a specimen I have of Hyla versicolor . . . though it is rather smaller. . . . It is completely mummified, and in a wonderful state of preservation, being of a dark, snuff-brown color, somewhat shrunken. . . . I am aware that these tree-frogs very often climb into some of the most unheard-of places; but it struck me that it would be interesting to have some tell us if they ever heard of a Hyla finding its way to the vault of a coal-mine 541 feet under ground, and climbing into the solid coal-bed after getting there."

Source:
  Shulfeldt, R. W., 1886. "A Mummified Frog."
  Science 8: 279–280. Cited in William R.
  Corliss, ed., 1980. *Unknown Earth: A
  Handbook of Geological Enigmas.* Glen Arm,
  MD: Sourcebook Project, 706–707.

1886: Slimed

A CURIOUS PHENOMENON happened recently at Shickshinny, Pa. For about fifteen minutes a lively shower took place, and after its cessation the streets, sidewalks, housetops and vegetation were found to be dotted with spots of slime about the color of the white of an egg. The substance succumbed to the rays of the sun, and by noon all had disappeared. Several bottles were filled with the slime, which will be sent to a scientist for examination."

Source:
  Untitled, 1886. *Stevens Point* [Wisconsin] *Journal*
  (August 28).

1887: "Monster Air Snake"

WHILE A NUMBER OF men were standing outside the works [at the Spang, Chalfant & Company's pipe mill, in mid-September] one of them looked up and saw what he at first thought to be a snake about five feet long, evidently more than 2,000 feet away, in the air. He did not say anything to his fellow-workmen until he saw the object coming closer, and it was then that he called the attention of others to it. The men watched it, and it was not long until it was over them, looking to be about 500 feet high, or about as high as the hill at that place. The object was more than five times the length it seemed to be when first noticed.

"As it came nearer it looked to be a monster snake. It was jet black and in thickness looked like an ordinary keg. The ponderous jaws of the reptile were frequently seen to open, from which emerged a large tongue. It sailed in a regular course, but when the jaws opened it then took a downward course and seemed as though it would fall to the ground below. On the descent the mouth remained open, and after a fall of about 100 feet the jaws would close and the snake would raise its head and slowly wend its way up to its former height.

"The course of this monster air snake was in a northwesterly direction. During its stay of about an hour it seemed to long for a visit to every part of Etna. From the mill it moved like a snake on land westward about a mile to a point on the Allegheny river, from where it took a back course to the place where it was first seen by the naked eye. From there it took an upward direction and it was watched until it disappeared behind the mill, sailing somewhat toward the northeast.

"A welder named William Stewart was the first to see the snake."

Source:
  "A Flying Snake," 1887. *Atlanta Constitution*
  (September 23). Reprinted from the
  Pittsburgh Chronicle.

1889: Spectral Driver, Horses Breathing Fire

IN THE NORTHWESTERN corner of lower Marton township, Montgomery county, Pa., a terrible hubbub has been raised by the appearance of a phantom farmer, who is nightly seen plowing a field. The apparition was first discovered about three weeks ago by a farm hand who was returning late from courting a pretty maid. Emerging from a wooded pathway, that skirted an old forest for miles, this rustic was startled to hear a sepulchral voice commanding a team to halt. He looked in vain about the place for a moment or two, and was about moving on again when the same sound fell on his ear. A shiver crept down his spinal column as he heard the creaking of an unseen harness, and this terror was far from being allayed by the whinny of a horse almost directly before him.

"At that moment the new moon stole over the neighboring tree tops, and in its misty light the rural swain plainly saw the phantom farmer. It was clearly outlined against the dark back ground and its two hands held in steady grasp

At that moment the new moon stole over the neighboring tree tops, and in its misty light the rural swain plainly saw the phantom farmer. It was clearly outlined against the dark background and its two hands held in steady grasp the projecting handles of a plow.

the projecting handles of a plow. Before it marched a pair of spirited horses, dimly outlined in the misty light, their heads erect[,] their eyes flashing fire as they moved hastily along. The young man waited another moment to reassure himself, and was about to take to his heels when plowman, horse and plow suddenly vanished. Then he, too, fled in wild alarm.

"At Silas Brown's corner grocery on the night succeeding this the young man, Albert Cooper by name, told the startling story. Brown, like many of the lodgers in the store, smiled incredulously, and advised Cooper to 'reform.' A discussion arose, in which hot words were exchanged and several bets were made. . . . [The party] adjourned to the alleged scene of the ghost's operations to verify or disprove Cooper's tale.

"They had not long to wait. Without the noises that had warned Cooper the night before, the phantom farmer appeared before the seven men who sat upon the fence, or to be more appropriate, almost ran from it in terror. His long white hair and beard streamed in the passing wind. No hat was on his head, nor could any

portion of his face be seen except the glistening eyes. These shot out from a height of more than seven feet from the ground, indicating that the spectral granger was taller than the average human kind. About his body which could not well be traced there was a phosphorescent glow which dazzled the eyes of the terrified spectators and shown [*sic*] far ahead of the steadily moving horses. The plow he leaned on seemed of skeletal frame, but it tossed off the soft, moist earth as easily as a steamer turns the river waves. On he came, the horses seeming to exhale fire, their heads erect and arching, and footfalls as firm and clear as any the watchers ever heard. At the corner of the field they turned obediently at a word from their spectral driver, and again passed before the affrighted spectators who thereupon fled in haste.

"On the following morning a crowd of rustics determined to go to the field and see whether any trace of the farmer could be found. As they came in sight of the enclosure one of the number exclaimed in astonishment, 'I'll be durned if the thing doesn't plow sure enough!' He was right. One half the field had been gone over, evidently by no novice. The furrows were not quite so broad as those made by an ordinary plowman, but they were less ragged and more deep and were as straight as the most experienced eye could make them. A day or two after[,] the same group went to the field again, and this time they found that the phantom had finished his work. The owner of the field was one of the number and he took a solemn oath that he had not turned a sod in the enclosure."

Source:

"A Phantom Plowman," 1889. *Atlanta Constitution* (February 24). Reprinted from the *St. Louis Globe-Democrat*.

1889: Apparitional Black Dog

FOR SOME TIME PAST those trudging home over the mountain about four miles south of this city late at night have seen strange sights and heard peculiar noises. It appears that a large black dog makes his appearance at a certain point along the turnpike and walks with the traveler until a well-known mark, still further up the mountain, is reached, when he suddenly disappears as he came. The dog utters no sound and betrays not the least show of either friendship or violence. There are those who believe that the mysterious animal is the materialized ghost of a murdered man, as he invariably makes his appearance at the exact spot of a tragedy.

"There are a dozen men who claim to have encountered the dog at that exact spot, some of them more than once. They confess that they never pass the spot without a shudder, and they hurry by the 'spook' spot as fast as they can. Women never pass the place at night, and many even make a wide detour to avoid it in the daytime."

Source:

"A Dog Ghost," 1889. *Ogden* [Utah] *Standard* (July 21). Reprinted from the *Philadelphia Inquirer*.

Late 1800s: Otherworldly Reptile

AN EXTRAORDINARILY STRANGE story, told as true, asserts that a fantastic manifestation appeared at a county schoolhouse set at a rural crossroads—traditionally the place where one is most likely to encounter the devil or other supernatural entities and creatures. A folklorist collected this tale from William Johnson, who claimed to be speaking from personal experience. He had lived in the county in his teens, he said, and remembered the event well. Beyond this single anecdote, no other record of the alleged episode has been unearthed.

Toward the end of the nineteenth century, country people would attend social functions at the school. At some point they became aware of the presence of . . . well, something. They never saw all of it, only a portion of the body of what was assumed to be an immense serpent. Neither head nor tail was ever seen, and it was believed that these were hidden under the building. To enter the schoolhouse, people had to step over the body, a foot in diameter, with sharp scales that apparently were electrically charged, or so it was inferred from the effect it caused of hurling to the ground anybody who made the mistake of stepping on them.

What made things worse was that the thing would show up only on nights when there was no moon. Even then, not everybody could see it. No wonder the land's original owner lost his nerve, sold the property, and put as much distance between himself and Somerset County as he could manage. The new owner raised a large family, apparently untroubled by the presence of an otherworldly semi-invisible critter. Others, however, were sufficiently unhappy that, usually under the influence of liquid courage, they would attack it with axes, sharpened stakes, and whatever else might do some damage, though it never did.

The creature was still there when William Johnson moved away at the age of thirty. If it is still there, no one is telling.

"A multi-dimensional reptile, perhaps?" asks writer Michael Winkle (1996), allowing himself the luxury of taking the story seriously for a moment. "Such a creature might project part of its body into our universe while other parts—its head and tail, say—would remain in a higher spatial dimension."

Sources:

Whitney, Annie Weston, and Caroline Canfield Bullock, 1925. "Folk-Lore from Maryland." *Memoirs of the American Folk-Lore Society* 18: 193.

Winkle, Michael D., 1996. "Fabulous Beasts: The Worm Ouroboros." *INFO Journal* (Autumn): 30–31.

1890: Ball of Fire

A BALL OF FIRE WAS seen to pass over Tioga county, Pa., Tuesday evening last [January 21]. It is the fourth similar phenomenon of late in that region."

Source:

"Happenings Hereabout," 1890. *Olean* [New York] *Democrat* (January 30).

1890: Ghostly Procession

ON THE EVENING OF the 21st of August the passengers on one of the cars of the Mount Penn Gravity Road were startled by a sudden outcry from a man who had been gazing fixedly up the mountain. 'There is a hearse,' he cried, 'and a coffin!' Then, in awe-struck tones, he added: 'It is a ghostly funeral and means— death!' Within twenty-four hours the same train, in charge of the same conductor, was hurled from the track and a number of the passengers were killed. As a matter of fact, what seems at first to be a pronounced ghost story, has a most substantial basis of truth, and a number of people who were on the train on the night preceding the accident are willing to make affidavits that they saw the ghostly procession. Nothing was said of the matter at the time. Conductor Rettew, who was on both trains, and who was killed on the second night, seemed to think that there was something in the so-termed

warning, as he particularly requested a number of those who witnessed the incident to say nothing about it. Now it is the talk of the town."

Source:

"A Spectral Funeral Procession," 1890. *New York Times* (September 15). Reprinted from the *Philadelphia Times,* September 5.

1892: "Black Ghost"

SUPERSTITIOUS PEOPLE in this city [Carbondale] and neighborhood—and there are many among the large mining population—are greatly disturbed over the appearance in this city of what they call a black ghost. This mysterious apparition has been seen three times within the past fortnight, each time just after midnight, and in different parts of the town. It is in the form of a woman dressed in black from head to foot. A 'caller' in the employ of the Erie Railroad company, whose duty it was to awaken their men who go out on the trains, was the first to see it. The woman in black was standing in the street near the railway depot. The caller approached her and she moved slowly away toward the city.

"The caller and another railroad man, wondering what could have brought a woman alone to that part of the town at such an unusual hour, followed her. She seemed to be moving slowly along the street, but although the men walked as rapidly as they could, and then broke into a run, they could not overtake the figure in black, she keeping a few yards in advance, and finally suddenly disappearing from sight entirely.

"A few nights later the woman appeared again in another part of the city, led two citizens on a similarly weird chase and then disappeared in the same uncanny way. Early on Friday morning [February 5] she was seen and disappeared under the same mysterious circumstances, near the old Coal Brook mine entrance. Miners say that a short time before the disastrous cave-in at the Delaware and Hudson Canal company's old No. 1 mine in this city, fifty years ago, [the same ghost] that is prowling about the town now, appeared under the same circumstances three times. Twenty-eight years ago this winter, the same woman in black, or one with the same habits, appeared three times, just as this one has done, and the memorable plague of black fever, which carried away scores of men, women and children in Carbondale and vicinity, followed her appearance. Superstitious people hereabouts are greatly disturbed over the reappearance of this black ghost."

Source:
"A Black Ghost," 1892. *Bismarck* [North Dakota] *Daily Tribune* (February 12).

1892: A Monster with Heart-Shaped Scales

THREE YOUNG MEN set out one morning in November to pick chestnuts in Montgomery County. On their way, they walked along the tracks of the Port Richmond branch of the Reading Railroad. Their trek led them past the Old Oaks Cemetery. From there they heard a strange sound, something like a combination of ringing and hissing, apparently coming from trees a short distance away, on the north side of the track.

One of the men, Aaron Jobbings, climbed up the embankment and started toward the trees. His friends heard him shout something, then he collapsed. When the other two revived him some moments later, he asked, "Did you see it?" He went on to describe a snakelike creature of remarkable size. The press account, allegedly true, if hard to believe, seems to draw on a description in some ways reminiscent of that associated with another staple of period newspapers, the sea serpent—and in others of the medieval dragon:

"The thing they beheld was as long as a fence rail and was about a foot in diameter through the thickest part of its body. Its head had the resemblance of a horse's head, with nostrils more like those of an ox. From between the eyes and extending to the top of the head and for a short distance down the body was a growth of bristle-like hair that stood up like the clipped mane of a mule. Back from where the hair ended was on either side a web-like protuberance nearly a foot in length and much the shape of an elephant's ear.

"The monster's body was covered with heart-shaped scales, the point of the heart circling or pointing inward. At every moment they gave out a metallic ring which sounded like a small sleigh bell, though more silvery. The color of the monster was a beautiful bronze green, with a row of purple on either side which blended gradually to almost a pure white on its under parts. Its tail was like that of a fish."

When the men tossed sticks at it, the creature reared up its head in the fashion of a horse and turned toward the tall grass. The three followed it at a cautious distance as it went down the hill. Soon it entered a culvert under the Pennsylvania Railroad tracks. Half an hour's searching failed to produce a second sighting.

The witnesses related what they had seen to some railroad men at a depot at Schulkill Falls, a short distance away. One trainman claimed that sightings of the creature went back to at least 1875, and that once a naturalist named John H. Richards had tried to capture it for the Smithsonian Institution, searching through the rock where it had its lair. "Before his death in 1881," the unnamed man is said to have said, "he saw the rock blasted away [for dam construction], and deep down in the crevices were found dry sections of the molted skin, which were preserved among his collection of curios."

Source:
"A Strange Monster," 1892. *Atlanta Constitution* (November 21).

1895: Insect-Black Snow

S. S. W. HAMMER, justice of the peace, and J. B. Myers, ex-director of the poor, residing near Gettysburg, Pa., on Monday, January 7, walked several miles to the Western Maryland Railroad to take the train for Gettysburg, and on their return at 10:10 a.m., they followed their foottracks home again from the railroad. When about three quarters of the way they were astonished to see the snow covered with something black. The strip of the country covered was about 75 yards in width, the length as far as they could see. The snow looked as if it had been covered with fine rifle powder. On small ponds of water there was a stiff, black crust, and the old footprints made in the morning were one half inch deep with unknown insects. Mr. Hammer took several on his knife blade and they jumped off. They were about half the size of a common flea and on the snow they seemed to be powerless. They were not the snow louse as seen in the springtime. Singly, these black specks were barely large enough to see with the naked eye."

Source:
"Snow Black with Insects," 1895. *Frederick* [Maryland] *News* (January 12).

1897: Flying Cigar

A DISPATCH FROM DERRY, Westmoreland county, dated April 18, says: 'The air ship[,] which has been seen in Indiana, Wisconsin and Iowa, passed here last evening, according to the testimony of many prominent persons, who claim to have sighted it. It is cigar-shaped and has red and green lights and a very small center light, white and very brilliant. The ship was headed east, traveling very rapidly and about 500 feet high. A car is hanging about ten feet from the airship entirely enclosed.'"

Source:
"The Air Ship at Derry," 1897. *Indiana* [Pennsylvania] *Messenger* (April 21).

1897: "Swung by an Unseen Hand"

RAILROADERS IN PENNSYLVANIA on the Beach Creek line, between Williamsport and Lock Haven, have been frightened by a ghostly figure that flags the midnight express at a lonely gorge in the Alleghenies. The trainmen say that when the express is far down the road a red lantern, swung by an unseen hand, can be

seen swinging across the tracks. But before the train reaches the point[,] lantern and flagman disappear. Watchmen have been placed there, but the ghostly watchman swings his lantern just above or below the place where the guard is stationed. The railroaders regard the appearance of the ghostly lanterns as a token of a fatal wreck to the crew that sees it."

Source:
"A Railroad Ghost," 1897. *Olean* [New York] *Democrat* (May 7).

1902: Cross on the Moon

MARKET MEN DRIVING in from the country and motormen on suburban lines and others astir at 5:30 a.m. at Reading, Pa., the other day saw an unusual sight in the heavens. A huge illuminated cross seemed to be flashed across the face of the moon. The moon was unusually bright. From the top, bottom, and sides extended for some distance the projecting ends of the cross as if thrown upon the planet by some gigantic magic light. Those who looked upon the unusual sight associating it with Christmas-tide were more or less awe stricken."

Source:
"Giant Cross Seen in the Sky," 1902. *Emery County* [Utah] *Progress* (January 3).

1903: Raining Frogs

WHILE SPINNING IN from the suburbs of this city the trolley car on which Amos Kennard was motorman was caught in a thunder storm. The man at the comptroller was pelted with rain, then with hail, and next, he says, it began to rain frogs. The passengers on the train were skeptical, but Kennard exhibited a small frog which, he says, he caught in his hand."

Source:
"Rained Frogs at Chester," 1903. *Bluefield* [West Virginia] *Daily Telegraph* (October 1).

1906: Winged Water Serpent

W. A. TALBOT, PRESIDENT of the Piso Company, a Pennsylvania book publisher located in Warren, had "a summer villa opposite Grunderville." In October, the press reported that his daughter saw a sea serpent in the Allegheny River: "Miss Rachel Talbot . . . was first to see the creature as it came swimming up the middle of the [Allegheny] river, its head protruding several feet above the surface. She called to 'Hank' Jackson, ferryman for the Warren Lumber company, who ran for his rifle and opened fire. Immediately the reptile reared his head at least ten feet in the air, Jackson says, and charged for the shore, its eyes, as big as saucers, fixed on him and propelling itself like lightning, partly by lashing the water with the tail end of its body and partly by means of two enormous fin-like wings spreading out from either side of its neck, with which it fanned the air furiously."

Jackson fired on the creature, and a bullet pierced one of the wings, disabling the serpent and knocking it down. It then proceeded to move in a circle, the lower part of the body under water, the head in the air as it flicked a long, forked tongue at him, hissing all the while. On the other side of the river, barge-builders at work secured guns and blasted away. As all of this was going on, the serpent dived under water and was out of sight for short periods of time.

Finally, the serpent used its wings and tried to fly away. It was 20 feet in the air when Jackson managed to disable its other wing. It emitted a piercing shriek, then hit the water with a great splash and swam rapidly down the river.

The account said: "It is believed by those who saw the monster it is the same one reported as having been seen in Oil creek at Titusville on August 1, and it probably entered the Allegheny at Oil City. Two boys, Jackson Miller and Harold Boynton, saw the serpent while they were camping on the river near Irvington, but the story told by these lads was not believed until the experience of the Grunderville people."

Source:
"A Sea Serpent in Fresh Water," 1906. *Fairbanks* [Alaska] *Evening News* (October 6).

1907: Unidentified Flying Object

A PECULIAR BRILLIANT streak of red light about 3 degrees in length, which suddenly appeared in the heavens about 10 degrees southeast of the zenith and died away to a brilliant red point after some minutes duration, was observed here [Allentown] last night."

Source:
Untitled, 1907. *Indiana* [Pennsylvania] *Evening Gazette* (February 8).

1907: "Volumes of Flames from Its Mouth"

THE RESIDENTS OF Pennville . . . are alarmed over the repeated appearance of a 'spook' on the macadamized path leading to that quiet suburb. It is described as an object dressed in black and emitting volumes of flames from its mouth as it glides through the darkness. Several persons have seen it and a number of children were terribly frightened by its appearance. The parents of these children are on the lookout for the apparition with shot guns."

Source:
"'Spook' Alarms Hanover." *New Oxford* [Pennsylvania] *Item* (November 7).

1908: Apparitional Ball of Fire

ARBOR, YORK COUNTY, is excited over the nightly visitation of a ghost. When shot at[,] the apparition changes into a ball of fire, drops to the ground and vanishes. There isn't a saloon in the place, either."

Source:
"Here and There," 1908. *New Oxford* [Pennsylvania] *Item* (January 30).

1908: "The Ghost Answers the Telephone"

OLD ST. PAUL'S CHURCH, the headquarters of the Protestant Episcopal City mission in Philadelphia, has a 'ghost.' It is an eccentric shade which whisks up stairways and disappears into nothingness, but it is also up to date. When the office force has departed and the quaint old building on Third street below Walnut is secured against intruders with stout locks and bolts[,] the ghost answers the telephone. Such conduct has dumfounded the Rev. H. Cresson McHenry and his assistants.

"The ghost once informed a friend of Mr. McHenry's that he 'had just left the mission,' and to Mrs. George Sommerer, the wife of one of Mr. McHenry's assistants, it imparted the information that her husband 'would be home to supper.' Both persons who conversed with the unknown occupant of the City mission declare that the voice was modulated to the softest tones.

"Mr. McHenry a short time ago, when the office force had a holiday, visited the church to open his mail. As he was unlocking the iron gate at the entrance to the churchyard he glanced up and was astonished to see what appeared to be a man standing on the stairway inside the building. The figure glided rapidly up the stairway, disappearing from view. The minister entered the church, locked the door behind him and searched the entire church from cellar to roof. He failed to find the visitor or any trace of occupation of the building. Every door and window was locked securely, and the desks were untouched."

Source:

"Ghost Answers the Phone," 1908. *Indiana* [Pennsylvania] *Evening Gazette* (October 26).

1911: "A Black Snake between 12 and 15 Feet Long"

WHILE GEORGE B. SHAFFER, who resides near Mont Alto and supplies milk to the residents of that town, was out in the field looking after his cattle he saw a black snake between 12 and 15 feet long. Mr. Shaffer estimates its length by a lane across which it was stretched. It reached from one side to the other and was as thick as an ordinary fire hose. He didn't carry his investigation very far, fearing that his snakeship might resent too close inspection. This snake has been reported before. It was seen about a year ago and is believed to be the same snake for which John Robison's circus men hunted for a week about 30 years ago."

Source:

"Monster Snake Seen," 1911. *Adams County* [Pennsylvania] *News* (May 27).

Rhode Island

Circa 1875–Early 1880s: Ghosts on Trains

ABOUT TEN YEARS AGO the engine [of] Matt Morgan blew up while standing on the track of the Shore Line Road near the station in Providence, R.I., killing the engineer. The engine was subsequently rebuilt and put on the road. On the first trip that she made after being rebuilt she went tearing into Providence in the night[,] the train swinging behind and the sleeping town echoing to the shrill whistle. On approaching the station the engineer leaned forward to shut off the steam, but to his horror a ghostly form appeared at his side and a ghostly hand grasped his wrist and held him fast. When the station was reached the ghost disappeared and the engineer stopped the train some distance beyond. At least, that's what the engineer tells.

"Many people have not forgotten the terrible Richmond [Rhode Island] switch disaster several years ago on the Providence & Stonington Road. A little brook became swollen by the rain and carried away a railroad bridge. The train came rushing along that night and was hurled into the chasm. Giles, the engineer, when he saw the danger ahead, instead of leaping from the engine, as his fireman did, grasped the lever and reversed the engine. But it was too late. The train was going at such speed that the locomotive leaped clear across the stream, and they found Giles lying under his overturned engine with the lever driven through his body, and one hand clutching the throttle valve with the grasp of death. Giles, when he came into Providence, was accustomed to give two peculiar whistles as a signal to his wife, who lived near the railroad where it enters the suburbs of the city, that he was all right and would soon be home. The absence of those whistles was the first intimation which was received at Providence of the disaster. When the engine which made the terrible leap on that stormy night was rebuilt and put on the road again, there was at first great trouble in getting engineers for it, with such a superstitious horror was it regarded. To-day there are people ready to swear that they heard whistles, such as Giles used to blow as signals to his wife, sound through the suburbs of Providence, when no train was coming up the road."

Source:

"Haunted Engines," 1885. *Fort Wayne* [Indiana] *Daily Gazette* (November 18). Reprinted from the *New York Tribune*.

1887–1888: Entity and Anomalous Light

LONELY, ISOLATED BLOCK ISLAND . . . is troubled by a wild man this winter. He appears at the dead of night in a wild, untenanted hollow of the hills, to which the throbbing of the ocean comes as a witch-like murmur; he digs a hole in the brittle peat; the swinging of a spectral light is seen in the drifting mists; the islanders go out to find him; they stumble upon the hole, but the swinging light dances away over the hills with the illusions of an ignis fatuus, and no one is any wiser about the strange visitant. The spot that the wild visitor loves to haunt is locally known as Monet's valley, and the

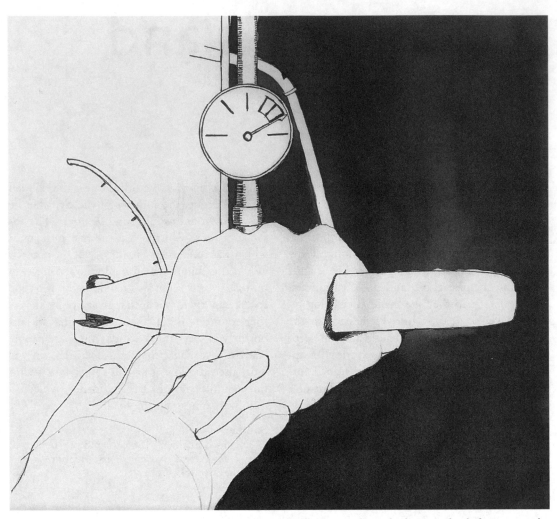

On approaching the station the engineer leaned forward to shut off the steam, but to his horror a ghostly form appeared at his side and a ghostly hand grasped his wrist and held him fast.

people are divided in opinion as to whether he is a wraith from the 'Phantom Ship Palatine,' that beached on the island two hundred years ago, or a ghost of Captain Kidd's crew, who came with the great pirate to bury an iron box of gold, with a Bible, at desolate Sand's Point."

Source:
"Wild Men," 1888. *Atlanta Constitution* (January 9).

1888: "As Big around as a Barrel"

Captain Delaney and crew, of the Maryland, saw the sea serpent, Saturday [August 4], off Port Judith. The creature was seventy feet long, as big around as a barrel, with eyes as big around as a crown of a bat, and its jaws five feet long and studded with six-inch teeth."

Source:
"The Sea Serpent," 1888. *Atchison* [Kansas] *Daily Globe* (August 7).

1896: Ghost "as Big as an Elephant"

It is from Rhode Island this time it comes and from a locality known euphoniously as Dandelion Hill. A man was going home from work one night and the ghost got after him. Its appearance tends to revive the belief in a red-hot place of punishment. Ghosts do not usually have breath. Being all breath themselves they do not need any. This one, however, was different. It had a breath that fairly scorched the affrighted man. The ghost was as big as an elephant, he said. He ran as fast as his feet could carry him, the ghost after him. It emitted a metallic sound as it ran, as of joints grown rusty from long lying around nights. Just before it got him it evidently changed its mind and had mercy on that poor mortal, for it suddenly plunged off into Gloucester woods, clanking joints, scorching breath and all."

Source:

"Jolly Old Ghost," 1896. *Coshocton* [Ohio] *Democratic Standard* (February 14).

South Carolina

1870: "Hideous Creature"

A PARTY COMING TO this city [Savannah, Georgia] from South Carolina through Wright river in a small sail boat manned by three Negro oarsmen, has furnished us with a decidedly sensational account of his adventures with one of the most [illegible] of creatures that ever crawled or floated under the firmament of heaven, and assures us positively that he has not been deceived by any freak of fancy or undue excitement of mind.

"Our informant, on the morning of the 28th [of February], toward noon, as he tells us, was about half a mile from Wright river, a stream merging into the Savannah, two miles above Fort Pulaski, with his Negro men pulling quietly along near the shore, when the slight built craft was suddenly and without any premonitory sign lifted up, as if by some immense roller, throwing the crew out of their seats and completely scaring the life out of them. The shock was so sudden that danger existed for a second of the boat turning over, but luckily it righted again and sank back into the water, which foamed like breakers.

"'But,' says the hero of the adventure, 'I did not heed the danger around me in this respect, nor the groveling fear of the men with me, for I could not, if my life was at stake, have taken my eyes away from the hideous creature that had caused all the commotion, and was making its way lazily out of the river into the long rushes on the bank. Never before had I anticipated such a monstrosity, nor do I ever wish to see another. A creature almost indescribable, though its general appearance is fixed in my mind's eye too indelibly for pleasant afterthought.'

"The beast, fish or reptile, whatever species of God's creation it might be classed under, was a tawny greenish color, growing more definite toward the head. The body of the creature was seal shaped, apparently twenty feet long, and as thick as the carcass of the largest sized elephant. From this trunk sprung the most remarkable feature of the phenomenon, a long, curved, swan-like neck, large enough, apparently, to have taken a man in whole, terminated by a head and jaws similar to that of an immense boa constrictor, the eyes fishy yet possessing ferocity enough in their expression to make a man tremble. The back of the beast was deeply ridged, running from the base of the neck to the extreme end of the tail, and several inches deep. An immense tail, shaped something like an alligator's, and three times longer, so it seemed, than the body, completed the *tout ensemble* of this wonderful anomaly. The creature navigated by feet resembling the fore feet of an alligator, and its progress on land was slow.

"'With all this combination of the terrible before me,' says our friend, 'it was not strange that I trembled, but before the frightened men had time to act, or I time to advise, the cause of our terror drew itself across the little island, out of sight, into the water beyond. It did not take us long to recover our senses, and as quickly leave the scene, though the shock to our nerves, and indeed to our belief in things possible and impossible, precluding anything like hard work.'

"The above statement we have from the lips of the gentleman himself, and being duly vouched for, we have every reason to believe in its truth."

Source:

"A Hideous Sea Monster," 1870. *Morning Oregonian* (May 13). Reprinted from the *Savannah* [Georgia] *Republican,* March 5.

1886: Falling Stones

A T 9:50 ON THE EVENING of July 31, an immense earthquake shook Charleston. Three other shocks followed, at 2, 4, and 8:30 a.m. August 1. Thousands of houses and other structures were destroyed or damaged by the quake and the fires that accompanied it, and 110 people lost their lives.

The quake, one of the most severe ever experienced on the North American continent, was felt as far away as Chicago. Even Boston suffered some minor effects. The city of Charleston was rendered nearly incommunicado with the rest of the world. A single telegraph line remained to alert potential rescuers to the disaster.

On August 4, as residents were still struggling to recover from the devastation, something else happened—fortunately not so destructive, but unexpected nonetheless. At 10:30 a.m., out of a lightly clouded sky, a rain of rocks—described in one press account as a "terrific spout of small blue and gray water-hardened stones"—fell on the printing rooms of the *Charleston News and Courier.*

It was later determined that two earlier falls had taken place, one at around 2:30 the same morning, a little to the south but still close to the newspaper office. The second one had occurred at 7:30. In each case, witnesses insisted that the fall had occurred *on a slant* from north to south. The stones were also warm to the touch. They were of various sizes, some the dimensions of grapes, others of eggs, most made of flint or a flinty-looking substance. One press account quoted an unnamed "expert" as saying "they looked as if they were part of a cabinet of mineralogical specimens" (quoted in Dwight and Mangiacopra 2002). Roofs all over the neighborhood were covered with the stones.

Whether the quakes and the falls were related was, of course, an issue much discussed, but no link could be established. In 1931, reviewing the episode, *Monthly Weather Review* could only surmise that "the public was being made the victim of a practical joke" (ibid.). If contemporary press accounts are reasonably accurate, however, that seems most unlikely.

Sources:

Smith, Dwight G., and Gary S. Mangiacopra, 2002. "Southern Falls—Flesh and Stones: A Reexamination of Charles Fort's Two Classic 19th Century Falls from the Skies." *The Anomalist* 10: 64–88.

1892: Alligator Man

The people residing along Palmetto creek . . . as well as those for miles back in the 'slashes,' are highly excited over the appearance of a strange and uncouth creature in that vicinity. The beast is described as being a creature that far outdoes the nightmare ideas of the mythologists. It is equally at home in the water, on the land or among the tall trees of the neighborhood, where it has been most frequently seen. The general contour of the head reminds one of a gigantic serpent with this exception: The 'snout' terminates in a bulbus [*sic*], monkey faced knot, which much resembles the physiognomy of some gigantic ape. From the neck down, with the exception of some fin shaped flippers, which extend from the arms to the waist, the creature resembles a man, only that the toes and fingers are armed with claws from two to six inches long.

"Tracks made by the beast in the soft mud around Hennis lake have been taken to Donner's Grove, where they are kept on exhibition in a druggist's showcase. Those who have seen the horn'd thing face to face say that it is a full nine feet in height, which could hardly be believed only for the fact that the tracks mentioned above are within a small fraction of fifteen inches in length. Fishermen who surprised the monster sitting silently on a mass of driftwood declared that its back looked like an alligator's, and that it had a caudal termination a yard long, which forked like the tail of a fish."

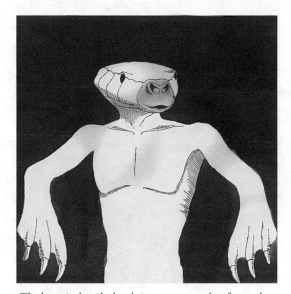

The beast is described as being a creature that far outdoes the nightmare ideas of the mythologists.

Source:
  "An Uncanny Monster," 1892. *Woodland* [California] *Daily Democrat* (August 10).

1897: Sky Serpent

THE FLYING SNAKE was seen near Newman swamp by a Mr. Odom on Sunday afternoon [probably July 11], say at 6 o'clock. This is about ten miles south of this place [Hartsville]. Later, say at 7 o'clock on the same day, it was seen by Mr. Henry Polson, in Chesterfield county, twelve miles north of here. Mr. Polson says: 'The monster was low down, just above the tree tops, had its head thrown back in a position to strike and was just floating through the atmosphere lengthwise.' He says it seemed to be 25 to 40 feet long and about 8 to 10 inches through at the largest part. In the language of Mr. Polson, 'he do [*sic*] not say it was a snake, but he do [*sic*] say it was the most like a snake of anything he ever saw, and he believes it was a snake.' There are all kinds of opinions as to what the monster could be. . . . William J. Johnson, a near neighbor of Mr. Polson, . . . says the snake was seen near Chesterfield court house and also in several towns in North Carolina."

Source:
  "The Flying Snake in South Carolina," 1897. *Statesville* [North Carolina] *Landmark* (July 20). Reprinted from the *Charleston News and Courier.*

1901: "Swimming between the Rows"

THE WEATHER BUREAU here [Columbia] announces the receipt of a report from Mr. J. W. Gardner, of Kershaw county, of 'an unusual phenomenon that occurred near Tiller's Ferry during a heavy rain near the close of June.' 'He states,' says the bureau official, 'that hundreds of little fishes were found swimming between the rows of a cotton field, and that among them were cat fish, perch and a few that looked like trout. There is no creek or pond near the field whence they might have come. And he asks where did they come from.'"

Source:
  Untitled, 1901. *Statesville* [North Carolina] *Landmark* (August 2). Reprinted from the *Columbia* [South Carolina] *Landmark.*

South Dakota

1889: Pillars of Fire

A PECULIAR HEAVENLY phenomenon was seen here [Sioux Falls] Wednesday evening [May 1]. The sky was perfectly clear, except for a long, black cloud in the western horizon from which protruded immense pillars of fire. They were of a bright red color. The pillars remained for about twenty minutes and disappeared."

Source:
"Peculiar Heavenly Phenomenon," 1889. *Marion* [Ohio] *Daily Star* (May 3).

1897: "Queer Lights Floating in the Air"

THE *ST. PAUL* [MINNESOTA] *Pioneer Press* reported an airship story on April 13: "People have gazed upon the airship and noted its flight in the skies. Last night one young man, while returning to his home from a social call at a late hour, distinctly saw queer lights floating in the air above him and the dim outline of a mysterious something sailing slowly over the city [Deadwood]. It moved in a leisurely manner and was in sight for at least twenty minutes. The same thing was observed by four other gentlemen, and each gives a different account of its appearance. The gentlemen are personally unacquainted with one another, which would seem to give their statements a semblance of truth."

Another story on April 19 in the *Sioux Falls* [South Dakota] *Argus-Leader* said: "C. A. Peckinpaugh[,] who was on shift at the Welcome hoist [in Deadwood] on Thursday night [April 15], and a companion, say they saw an air ship light on Terry Peak. It is said that it was also seen by fifteen or twenty others, some of whom started up to investigate[,] but before they reached it, it rose and flew away. It was cigar shaped with enormous wings."

Source:
"State Notes in General," 1897. *Sioux Falls* [South Dakota] *Argus-Leader* (April 19).
"Whole Fleet of Airships," 1897. *St. Paul* [Minnesota] *Pioneer Press* (April 13).

1903: "Some Gaseous Matter in the Air"

JUST PREVIOUS TO A heavy shower, Thursday [August 6], nearly every kitten less than six months old in the vicinity died, apparently from the effects of some gaseous matter in the air. A bunch of eighteen cattle were seen coming from the flats when eight of them fell to the ground. A farmer living nearby [to Oacoma] ran out to them and found six of the eight already dead. John Morris, a stockman living on White river, reports the loss of six head in the same way. It is generally believed that the copious rains which have fallen since have neutralized whatever poison to animal life may have been in the air. It was a phenomenon heretofore unknown here."

Source:
"Remarkable Phenomenon in South Dakota Thursday," 1903. *Marion* [Ohio] *Daily Star* (August 7).

Tennessee

1820s: American Pygmies?

ACCORDING TO TENNESSEE press accounts, small stone coffins containing the remains of "little people" were discovered in Sparta, in White County. They were contained in burial grounds, each half an acre to an acre long. Each grave was 2 feet deep, and each contained the remains of individuals no more than 2 feet tall. The heads were all aligned with the east. Each skeleton lay on its back, with the hands folded on the chest. "In the bend of the left arm," *Harper's New Monthly Magazine* reported, "was found a cruse or vessel that would hold nearly a pint, made of ground stone or shell of a gray color, in which [were] found two or three shells. One of these skeletons had about its neck 94 pearl beads."

Unlike other contemporary stories of fantastic archaeological discoveries in North America, this one is not a product of a hoaxer's imagination. Eventually, the Smithsonian Institution investigated the remains. In 1891, Cyrus Thomas, head of the Bureau of American Ethnology, however, declared that the skeletons' extraordinary features were more apparent than real. He said many of the bones were from small children born to local Indian tribes. The rest were from adults who, in line with custom, were stripped of flesh and their bones taken apart prior to burial, making it possible to place each body within a very small space.

In the late twentieth century, however, Filipino American anthropologist Virgilio R. Pilapil speculated that the remains really were of pygmies. Pilapil noted that according to measurements made by Barry Fell, author of controversial theories about pre-Columbian visitors to America, the brain capacity of the skulls was only about 950 cubic centimeters—that of a seven-year-old child—yet the teeth were developed and showed signs of wear of a kind associated with adults, and the skulls had projecting jaws. In Pilapil's view, which so far no mainstream archaeologist has endorsed, these features were consistent with the pygmy people of the Philippines, the Aetas.

Sources:

"American Pygmies," 1992. *Science Frontiers* 84 (November/December), http://www.science-frontiers.com/sf084/sf084a02.htm.

Miles, Jim, 1985. "Pygmies and Giants." *Fate* 38, 3 (March): 34–39.

Seaver, William A., 1869. "Giants and Dwarves." *Harper's New Monthly Magazine* 39 (July): 202–210.

1841: Blood and Flesh from a Red Cloud

ON TUESDAY [AUGUST 3] we heard from various persons that a *shower, apparently of flesh and blood,* had fallen in Wilson county, near Lebanon, in this State, and that the fields were covered to a considerable extent. The account staggered our belief; but, strange as it may appear, it has been confirmed by the statement of several gentlemen of high character, who have personally examined the scene of this phenomenon. They state that the space covered by this extraordinary shower is half a mile in length and about seventy-five yards in width.

"In addition to the information just received, we have been favored by Dr. Troost, professor of chemistry in the University of Nashville, with the following letter from a highly respectable physician in Lebanon. We have also seen the specimens sent to him for examination. *To us they appear to be animal matter, and the order is that of putrid flesh. We do not pretend to offer any theory to account for this phenomenon; we leave that to abler and more scientific hands. When the specimens have passed through the crucibles of Dr. Troost, we will furnish our readers with the result* [italics in original]:

"*Lebanon, August 8, 1841.*
"*Dr. G. Troost:*

I have sent you some matter, which appears, from an authentic source[,] to have fallen from the clouds.

"With me there can be no doubt of its being animal matter—blood, muscular fiber, abipous matter. Please account to us, if you can, on philosophical principles, for the cause of the phenomenon. The particles I send you I gathered with my own hands. From the extent of surface over which it has spread, and the regular manner it exhibited on some green tobacco leaves, leaves very little doubt of its having fallen like a shower of rain; and it is stated . . . to have fallen from a small red cloud; no other clouds visible in the heavens at the time. It took place on Friday last, between eleven and twelve o'clock, about five miles north east of Lebanon. I have sent what I think to be a drop of blood, the other particles composed of muscle and fat, although the proportions of the shower appeared to be a much larger quantity of blood than of other properties. Yours in haste,
W. P. SAYLE."(1)

☞

"Professor Troost has published a long article in the Nashville Banner with reference to the recent shower of blood which occurred at Lebanon, Tennessee, on the 17th [of September]. The explanation, which he gives of this remarkable phenomenon differs essentially from that of Professor Hallowell. He took especial pains to investigate all the facts in the case, and says there can be no doubt that the drops of blood found upon the tobacco leaves did really fall from a small red cloud passing swiftly from East to West. The extent of the shower was from forty to sixty yards in breadth, and six or eight hundred in length; it was thinly scattered, probably a drop every ten or fifteen feet, irregularly dispersed. Some of the pieces which fell were nearly two inches long. Professor Troost believes the fragments to be animal matter, but does not consider any part of them to be blood. There was about them an offensive smell of putrid matter, and the fibers were distinctly visible. On exposing the particles to the action of heat, they were affected precisely as a piece of beef. 'There is no doubt,' says the Professor, 'that this substance is animal matter, and belongs to our globe.' He then enumerates many instances in which remarkable showers have been observed from the year 472 to the present day. If the facts as stated by him are correct, there can be no ground for the explanation of Professor Hallowell that the matter was thrown off by insects in their chrysalis form. Professor Troost ascribes it to the action of a hurricane, which, he thinks, may have taken up part of an animal which was in a state of decomposition and have brought it in contact with an electric cloud, in which it was kept in a state approaching to a partial fluidity or viscosity.

"F. Castlenan of this city [New York] has addressed a letter to the Editors of the National Intelligencer, stating that the shower of blood recently observed in Tennessee, in his opinion was produced by insects flying in great numbers over the field. He says that insects of the order Homoptera have the property of producing a liquid of a red color having in all respects the appearance of blood or putrid matter."(2)

Source:

(1) "Atmospherical Phenomenon," 1841. *Ohio Repository* (September 9). Reprinted from the *Nashville Banner.*
(2) "The Shower of Blood," 1841. *Ohio Repository* (October 7). Reprinted from the *New York Tribune.*

1853: Strange Lights in the Night

WE HAVE RECEIVED A letter from Professor A. C. Carnes, of Burritt College, Tenn., with the following account of a singular phenomenon, that was seen by a number of the students, on June 1st, at 4 1/2 A.M., just as the sun was rising:

"'Two luminous spots were seen, one about 2 [degrees] north of the sun, and the other about 30 minutes further in the same direction. When seen, the first had the appearance of a small new moon; the other that of a large star. The small one soon diminished, and became invisible; the other assumed a globular shape, and then elongated parallel with the horizon. The first then became visible again, and increased rapidly in size, while the other diminished, and the two spots kept changing thus for about half an hour. There was considerable wind at the time, and light fleecy clouds passed by, showing the lights to be confined to one place.'

"The students have asked for an explanation, but neither the President nor the Professors are satisfied as to the character of the lights, but think that electricity has something to do with it. The phenomenon was certainly not an electrical one, so far as we can judge, and possibly was produced by distant clouds of moisture."

Source:

"Singular Phenomenon," 1853. *Scientific American* 8, 42 (July 2): 333.

1869: Pillar of Fire

AROUND NOON ONE day in Cheatham County, farmer Ed Sharp looked out at some nearby woods to see a whirlwind overhead. The whirlwind sucked up branches and leaves as it passed. Weirdly, it also burned them. The effect was a moving pillar of fire, its size expanding the more it took in. When it passed over a team of horses, it singed their manes and tails. As it moved toward Sharp's house, it consumed a stack of hay.

Hotter than ever, it moved over the roof. The shingles ignited, and within ten minutes the structure was consumed in flames. It crossed a field where wheat had recently been stacked, burning up everything in its path. It crossed into a wooded area leading to the Cumberland River and incinerated the leaves for 20 yards. Reaching the water, it changed direction and followed the river south, sending clouds of steam upward. After half a mile it finally died out.

A correspondent reported "this strangest of strange phenomena" to *Symons's Monthly Meteorological Magazine* not long afterward. He claimed that no fewer than 200 witnesses had observed the curious manifestation, "and all of them tell substantially the same story about it."

Sources:

"A Fiery Wind," 1869. *Symons's Monthly Meteorological Magazine* 4: 123–124. Cited in William R. Corliss, ed., 2001. *Remarkable Luminous Phenomena in Nature: A Catalog of Geophysical Anomalies.* Glen Arm, MD: Sourcebook Project, 209.

"A Pillar of Fire," 1869. *Grand Traverse* [Michigan] *Herald* (August 26). Reprinted from the *Nashville Press.*

1876: Pygmy Skeletons

ACCORDING TO AN 1876 issue of the *Anthropological Institute Journal,* an "ancient graveyard of vast proportions" found in Cheatham County contained the bodies of many members of a pygmy race. They were discovered when a farmer plowing a field noticed a human skull and bones brought to the surface in his labors. He notified others, and they began digging. They found that the bones were from a burying ground some 6 acres in diameter.

The dead, who were 3 feet tall, had been buried in sitting or standing positions. "It is estimated," the correspondent wrote, "that there were about 75,000 to 100,000 buried there."

Source:

"A Pigmy Graveyard in Tennessee," 1876. *Anthropological Institute Journal* 6: 100. Cited in William R. Corliss, ed., 1978. *Ancient Man: A Handbook of Puzzling Artifacts.* Glen Arm, MD: Sourcebook Project, 681.

1877: A Rain of Somethings

A VIOLENT RAINSTORM SWEPT through Memphis starting at 10:20 on the morning of January 15, 1877. Coming in from the southwest, it lasted for fifteen or twenty minutes. When it stopped, residents of a two-block area in the southern part of the city stepped outside to make a fantastic discovery. Many thousands of writhing, clumped-together creatures covered yards, sidewalks, and streets. Dark in color, with black spots and small black heads, they were 12 to 18 inches in length. Some remained as much as two days later. They were characterized as snakes of some unknown kind.

A neighborhood man put a dozen of the "snakes" into a jar and carried it to the office of the *Weekly Public Ledger*. The *Ledger* immediately put the story on the press wires. On the eighteenth, the *New York Times* covered the story under the title "Thousands of Snakes in Memphis," and other newspapers picked it up. The *New York Herald Tribune* sneered but still tried, unsuccessfully, to obtain specimens for scientists to examine. It did, however, contact a Memphis man identified only as Dr. Scherer. Scherer conducted inquiries for the paper, interviewing a Sgt. McElroy of the U.S. Signal Corps. McElroy said he had collected specimens two days after the fall and forwarded them to Washington. The creatures, he related, were as thick (or as thin) as knitting needles. Their movement was unsnake-like; they "would shove the fore part of the body ahead and draw the balance up in a hoop shape. They would also raise themselves five or six inches in an upright position as though they were seeking some support."

The *Herald Tribune* agreed that these could not be snakes. More likely, it stated, they were from a "wormy colony," and the storm had picked them up and deposited them miles later on the streets of Memphis. That explanation may be correct—it certainly has a surface plausibility—but as in other cases of anomalous falls (witnessed or, as here, inferred), the curious selectivity remains unaddressed. Why worms only? Why not other creatures, leaves, mud, branches, and other phenomena that storm winds should have carried as well?

Both *Scientific American* (February 10) and *Monthly Weather Review* (January) took note of the event, passing on the dubious assertion that snakes had fallen. After that, the incident was forgotten until anomalies chronicler Charles Fort discussed it in two pages of his *Book of the Damned* (1919) and in a letter to the *Philadelphia Public Ledger* (1924). In the latter, Fort proposed a bizarre theory: "Between this earth and other worlds there may be definite currents to which living things in other worlds respond migratorily. If living things can come to this earth from other worlds, we have the material for visions such as have not been excited since the year 1492."

In the early 1980s, Gregory L. Little, a Memphis psychologist interested in the unexplained, conducted his own inquiries. Finding no record of the occurrence in Signal Corps archives, he could only draw on contemporary press accounts to reconstruct what may have happened. He took them to scientists in the biology department of Memphis State University. In the view of most, the creatures were most likely horsehair worms, which sometimes show up after rainstorms, but not because they have plummeted from the heavens. They are parasites that grow inside grasshoppers, beetles, and other large insects. In their later stages of growth, if they are exposed to water, they will exit their hosts as thin worms that observers may mistakenly think are snakes. They are the same color and size as the animals reported in south Memphis.

Little wrote, however, that the match was hardly perfect, since horsehair worms "have few external features" such as heads or spots. Nor do they "move in the hoop-push fashion." Moreover, Little thinks it is "improbable that thousands of dead insects infested with the worms would be so localized in a two-block area." More probably, in Little's judgment, the animals did fall from the sky, having been swept by waterspout from a lake. "Northern freshwater leeches reach their numerical peak in midwinter. Aquatic leeches," he added, "congregate in masses numbering up to 700 per cubic meter."

Sources:
Fort, Charles, 1924. Letter to *Philadelphia Public Ledger* (July 27).

———, 1941. *The Books of Charles Fort.* New York: Henry Holt, 93–94.

Little, Gregory L., 1985. "Snakes Fell on Memphis." *Fate* 38, 2 (February): 74–77.

"A Forgotten Fort Letter." 1996. *INFO Journal* 74 (Winter): 22–23.

1880: Into Thin Air

ACCORDING TO A STORY widely told as true, a man vanished into thin air in full view of his family and two other witnesses near Gallatin. The man's name was David Lang, and the incident took place on September 23, 1880.

He was crossing a field that afternoon while his two children, George and Sarah, were playing in the front yard. His wife was watching him from the porch. At the same moment, a buggy was pulling up to the farmhouse. Aboard it were Gallatin lawyer August Peck and his brother-in-law, identified as a Mr. Wade from Akron, Ohio. Noting their arrival, Lang turned around, waved, and started back to the house. At that moment, he suddenly faded from sight. A search of the spot, and then the entire field, produced no clue to Lang's fate. Before the day was over, neighbors and authorities had joined the search.

Mrs. Lang was so upset that she lay bedridden for weeks. The county surveyor determined that the ground beneath the field contained no sinkholes into which anyone could fall. Some months later, the grass at the site of the disappearance grew a strange yellow color within a 15-foot circle. For some reason, Sarah called out to her father, and in response she heard what sounded like his voice, but later she could not remember what it had said. After a short time the voice faded away.

In 1953, *Fate,* a popular digest specializing in stories of unexplained occurrences, published a piece by Stuart Palmer. Palmer claimed that in 1929 he had interviewed Sarah Lang, by then an old woman. Sarah provided some new details, the most dramatic of which concerned her receipt of psychic messages from her deceased mother. The messages indicated that Mrs. Lang had searched for her husband in the afterlife. In April 1929, through automatic writing, Sarah heard from her father again: "Together now. To-

gether now and forever . . . after many years. . . . God bless you."

Palmer quoted Sarah as saying that the famous short-story writer and essayist Ambrose Bierce visited the farm a month after her father's vanishing. Subsequently, he wrote three science-fiction tales based on the incident. She did not mention them by name, but the reference clearly is to "The Difficulty of Crossing a Field," "An Unfinished Race," and "Charles Ashmore's Trail," all of which were published in *Can Such Things Be?* (1893).

After the *Fate* article appeared, the story circulated through popular "true mysteries" literature, for example Harold T. Wilkins's *Strange Mysteries of Time and Space* (1958) and Frank Edwards's *Stranger Than Science* (1959). It was only in the 1970s that the story was exposed as fantasy. None of the principals had ever existed, and an expert who examined the alleged handwriting samples Palmer printed with his article (they were supposed to be from the automatic writing Sarah received) found they were all done by one person, presumably Palmer.

One theory holds that the tale is an authentic legend, planted by a late nineteenth-century traveling salesman and yarn-spinner variously called Joe Mulhatten or Joe Mulholland. Mulhatten/Mulholland may himself be no more than a legend, however; no one has been able to find any evidence of his physical existence. Since no written account of the disappearance is known from before 1953, it seems more probable that Palmer himself is responsible, patching the tale together from the three Bierce short stories he himself mentions.

The elements Palmer (if he was indeed the culprit) drew from the Bierce stories are clear enough. Bierce himself, using a popular literary conceit of the period, prefaced his tales with words to the effect that he had heard of them from a friend who collected reports of mysterious events. In that spirit, Bierce gave these unreal events real places and dates.

In "The Difficulty of Crossing a Field," "a planter named Williamson" near Selma, Alabama, vanishes one morning in 1854 while walking in a pasture across the road from his house, as several persons, including his wife, look on. An intensive search fails to uncover any trace of him. "An Unfinished Race" is set in Warwickshire, England, on September 3, 1873. A runner engaged in a race is tailed by three friends in a wagon. Bierce wrote, "Suddenly—in the very middle of the roadway, not a dozen yards from them, and with their eyes full upon him—the man seemed to stumble, pitched headlong forward, uttered a terrible cry and vanished!" The Charles Ashmore of "Charles Ashmore's Trail" is a young man from a well-to-do Quincy, Illinois, farm family. On the evening of November 9, 1878, he walks to a nearby spring to get water. When he does not return, his family goes looking for him, to discover that his footprints in the snow disappear halfway to the spring. Four days later, "the grief-stricken mother herself went to the spring for water. She came back and related that in passing the spot where the footprints had ended she had heard the voice of her son and had been eagerly calling him, wandering about the place, as she had fancied the voice to be now in one direction, now in another, until she was exhausted with fatigue and emotion. Questioned as to what the voice had said, she was unable to tell, yet averred that the words were perfectly distinct. . . . For months afterwards, the voice was heard by the several members of the family." Strangely, none could ever recall what it said.

The Charles Ashmore story is apparently the inspiration for another classic vanishing hoax, that associated with a young Indiana man named Oliver Lerch.

Sources:
Bierce, Ambrose, 1893. *Can Such Things Be?* New York: Cassell.
"Fortean Corrigenda: The Disappearance of David Lang," 1976. *Fortean Times* 18 (October): 6–7.
Haslam, Garth, 2000. "The Mystery of David Lang," http://anomalyinfo.com/articles/sa00010.shtml.
Nickell, Joe, with John F. Fischer, 1988. *Secrets of the Supernatural: Investigating the World's Occult Mysteries.* Buffalo, NY: Prometheus Books, 65–73.
Palmer, Stuart, 1953. "How Lost Was My Father?" *Fate* 6, 7 (July): 75–85.

Park, T. Peter, 1998/1999. "Vanishing Vanishings." *The Anomalist* (Winter): 158–178.
Schadewald, Robert, 1977. "David Lang Vanishes . . . Forever." *Fate* 30, 12 (December): 54–60.

1891: Fishes with the Snow

A SINGULAR PHENOMENON was observed in connection with a heavy fall of snow which occurred in this city. Large quantities of small fishes came down with the snow, and in many localities the ground was literally covered with them. The largest were nearly two inches in length. As such a downpour from the heavens has heretofore been unheard of, it has created much comment."

Source:
"Fishes Fall from Heaven," 1891. *Indiana [Pennsylvania] County Gazette* (January 29).

1891: Yellow Substance

A SEVERE ELECTRIC storm passed over this section today [April 18], and inside the corporation quite a heavy rain fell for a short time. The pyrotechnic display in the heavens was beautiful. After the storm it was noticed that on the housetops, and everywhere that rain had fallen, was a thin coating of yellow powder, resembling sulphur. Along the gutters large deposits of the powder are to be seen, left there by the receding waters, and the Market street asphalt paving is in many places streaked with it. Alleged sulphur showers have been heard of recently, notably the one in the vicinity of Nashville last February, in which the yellow deposit was similar to the present case. The explanation of the phenomenon in that case was that the so-called powder was pollen from the pine forests, driven here by the wind for, perhaps, hundreds of miles and held in the air until driven to the ground by rain."

Sources:
"Sulphur on the Housetops," 1891. *Atlanta Constitution* (April 19).

Now he could see, in the words of a newspaper account, "a huge balloon, but of a pattern and appearance he had never in his life before seen."

1907: "More Like Angels Than Mortals"

ONE DAY IN LATE SPRING, Walter Stephenson was training his two bloodhounds "near the Dikeman springs" (presumably in central Tennessee). When he sat down on a log to take a break, he noticed a speck just above the eastern horizon. Assuming it to be a large kite, he turned his attention elsewhere. Soon afterward, hearing a whirring noise, he looked up and was shocked to see that the "speck" was rapidly approaching him, and in moments it was directly overhead. Now he could see, in the words of a newspaper account, "a huge balloon, but of a pattern and appearance he had never in his life before seen."

As it descended toward the ground, "strains of music calculated to charm the spheres" emanated from the object. While the music played, the balloon circled repeatedly before finally landing. "Strange people" stepped out of the ship, "which was closely curtained with a substance that fairly glistened in the sunshine that temporarily burst through the obscuring clouds." The crew members (apparently masked) walked to the nearby spring and fell to their knees in a worshipful pose. Stephenson watched from a short distance and said nothing until the visitors rose to their feet. Then he asked politely who they were and what their mission might be. One of them lifted the mask, revealing the "benign face of a lady." Speaking in German, she asked him if he had prayed.

Then "instantly all were aboard, the airship rose, circled about for a minute or more, and was gone in a westerly direction," in the words of a press account. Mr. Stephenson says that the incident left an impression upon him that he can never forget, and while he knows that it was some human invention, it looked and the music sounded more like that of angels than of mortals."

Source:

"Here's a Weird Tale," 1907. *Ada* [Oklahoma] *Evening News* (July 1). Reprinted from *Nashville American.*

1908: Headless-Ghost Omen

HOUSTON ERWIN, A farmer, living near Madisonville, appeared at a neighbor's last night, greatly agitated. He said that he had seen a ghost in the form of a headless man. Soon after starting home, some one from the darkness shot Erwin to death."

Source:

"Houston Erwin Saw a Ghost," 1908. *Fort Wayne* [Indiana] *News* (March 14).

Texas

1852 and After: Rock Wall

AS HE WAS DIGGING a well on the east side of the Trinity River Valley's east fork, T. U. Wade came upon a stone formation. Further digging uncovered what he thought of as a "rock wall" running for some undetermined, though apparently considerable, distance and depth. The discovery had the fortuitous effect of resolving a dispute among three men, each of whom wanted the town-in-the-making named after himself, and caused the community to be called Rockwall.

In the years to come, there would be much speculation, most of it only locally circulated, about the wall's nature and significance. Wade and his neighbors did some excavating. According to Rockwall lore, the men found "rooms," which, when walked through, led to a "corridor," apparently heading toward the hill on which the rest of the town sits. In 1906, two men explored the corridor, which they had to shovel out because it had become filled up with soil and stones, into the hill, where they observed steep slopes like an arched ceiling. The farther into the corridor they went, the steeper the ceiling, until finally the men, who were treasure-hunting, feared that it would collapse on them if they disturbed it even a little.

A Rockwall man allegedly dug 40 feet down into the wall, noting that it curved inward, as though the unknown builders had made an intentional effort to thicken the walls. Other early Rockwall residents claimed to have seen a window and a doorway. In 1949, a Fort Worth man attempting an excavation of the wall unearthed four rocks, the heaviest weighing two tons. They contained what onlookers took to be "inscriptions" (called "pictographs" by a more recent theorist). Photographs survive, but the rocks themselves disappeared long ago. They were the only alleged artifacts associated with the wall.

On October 21, 1999, Texas architect John Lindsey lectured to a Dallas-based New Age group, Eclectic Viewpoint, to report his ongoing excavations at Rockwall. He said that by the time he got to the (literal and figurative) bottom of the wall, he was certain he would find artifacts conclusively linking it to an unknown civilization 30,000 to 100,000 years old. Among those attending was geologist James Cunliffe, who subsequently wrote a scathing account in the newsletter of the North Texas Skeptics. "Anyone with a basic knowledge of geology and archeology would be stunned by Lindsey's intepretation," he wrote.

Professional archaeologists and geologists hold that the so-called wall is in fact a sandstone dike, an upright sheet of sandstone formed when sand is driven into rock fissures. In the absence of any evidence beyond a handful of disputed photographs of alleged pictographs, there can be no reason to conclude otherwise. As anomalist William R. Corliss wrote, "The geologists have won this argument."

Nonetheless, the Rockwall rock wall is now a part of the lore of New Age archaeology. In early 2001, a channeled entity named Lady Kadjina was asked about its origins. She declared the wall to be an extraterrestrial creation, built sometime after the fall of Atlantis to store technological devices the space people had given the Atlanteans.

Sources:

Corliss, William R., ed., 1999. *Ancient Infrastructure: Remarkable Roads, Mines, Walls, Mounds, Stone Circles.* Glen Arm, MD: Sourcebook Project, 380–381.

Cunliffe, James, 1999. "Cult Archaeology in Rockwall" (Part One, December), http://www.ntskeptics.org/1999/1999december/december1999.htm.

———, 2000. "Cult Archaeology in Rockwall" (Part Two, January), http://www.ntskeptics.org/2000/2000january/january2000.htm.

"The Discovery of the 'Rock Wall,'" n.d., http://www.rockwallfoundation.org/background.htm.

Howell, Gretchen, 1939. "The Mystery of the 'Wall.'" *Nature Magazine* 32 (January): 23.

"Lady Kadjina Speaks," 2001, http://groups.com/group/RMNEWS_DAILY_EMAILS/message/13749.

1873: A Dragon in the Clouds

NEAR BONHAM, THE local newspaper, the *Fort Scott Monitor* reported in June, a farmer identified only as a Mr. Hardin, along with several groups of men and boys, witnessed the passage of a serpent, floating in a cloud, across the sky 5 or 6 miles east of town. The length and diameter of a telephone pole, the creature had yellow stripes. It floated without apparent effort. Observers allegedly saw it "coil itself up, turn over, and thrust forward its huge head as if striking at something." As the serpent and cloud continued on their eastward course, the press account stated, other rural people saw it, too.

Noting the latter story in its July 6 issue, the *New York Times* called it "the very worst case of delirium tremens on record."

Source:

"Signs and Wonders," 1873. *Fort Scott* [Texas] *Monitor* (June 24).

The length and diameter of a telephone pole, the creature had yellow stripes. It floated without apparent effort.

1877: "Aerial Fireworks"

LAST EVENING JUST before the sudden change in the weather, an unusual display of aerial fireworks was visible in the heavens in the direction of the Indian Territory [later, Oklahoma]. Moving from east to west, they looked like globes of fire followed by a luminous trail. We suppose they were aerolites, but are unable to account for the phenomena which preluded the sudden change of weather."

Source:

Untitled, 1877. *Sedalia* [Missouri] *Daily Democrat* (January 11). Reprinted from the *Denison* [Texas] *Cresset.*

1878: Falling Leaves

ROCKDALE, ON MONDAY [probably March 11], witnessed a strange phenomenon. It was a shower of leaves. As far as the eye could reach the heavens above were filled with them, reminding one of the days of the grasshoppers. There was no commotion of the atmosphere to be observed in any direction on the horizon. The supposition is that they were lifted up by an immense whirlwind and carried a great distance. They were falling all day long."

Source:

Untitled, 1878. *Atlanta Daily Constitution* (March 19). Reprinted from the *Rockdale* [Texas] *Messenger.*

1880s: Misty Ghost across the Track

ANOTHER TRAGIC POINT on the Houston and Texas railroad is Elm creek bottom, a few miles south of Corsicana, where a convict employed in the construction of the road was killed while attempting to escape. The tradition is that the unfortunate man was crammed in a water barrel, head down, and buried in scandalous haste. The superstitious think they can see his misty ghost as it flits across the track in front of the engineer, ominous of evil. Here accident after accident has occurred; many engineers and firemen have lost their lives or been crippled for good, and thousands of dollars' worth of rolling stock and merchandise has been destroyed. The train men have therefore changed the name of the place to Bad Medicine, and declare that these calamities will be perpetual until the convict is exhumed and buried horizontally, so as to destroy his chances to kick at the road which cost him his life here below."

Source:

"Superstitious Railroad Men," 1890. *Marion* [Ohio] *Daily Star* (March 29). Reprinted from the *St. Louis Dispatch.*

1885: A Plague of Frogs

A CURIOUS PHENOMENON that is mystifying the intelligent and arousing the fears of the ignorant and superstitious has occurred here [Eagle Pass] in the last few days. Three days ago [July 5] an unusual number of small frogs were first noticed hopping about, which have so rapidly multiplied that yesterday the earth was so thickly covered with them that it was almost impossible to step on the ground without crushing them. They are of various sizes, from the size of a pea to as large as a coal [illegible], and in color resembling the different shades of brown and green. They are rapidly increasing in number, and to-day the merchants in town were kept busy sweeping them out of their stores, as they hopped in through the doorways. They are moving in a steady, hopping stream in a southerly direction, and as far as the eye can reach frogs can be seen moving, and millions cover the ground in every direction. Ranchmen living in the country some distance from here report them as bad, if not worse, on the prairie, and say that some of the steep-banked arroyas are a swimming mass of small frogs [which] have fallen in and are unable to get out. None are reported on the other side of the Rio Grande in Mexico. No satisfactory explanation can be given for this remarkable phenomenon."

Source:

"A Great Frog Story," 1885. *Waukesha* [Wisconsin] *Freeman* (July 9).

1887: Strange Storm

AT ABOUT 5:20 O'CLOCK last evening [May 18] a sudden splash and rushing of water on the southeast corner of the square [at Pilot Point] was heard and seen, and teams hitched to carriages standing close by dashed off in fright, but fortunately the drivers had their lines in hand or serious damage would have been done. Upon the first reaction from alarm thus occasioned[,] the cause of this sudden precipitation of such a small body of water, less than five barrels, with a force and accompanied by such a peculiar detonation, as it could have only descended from the clouds, was observed to be two clouds immediately above, one a bronze, angry-looking and the other a purplish, smoke colored cloud with silver streaks through it, repelling each other with great violence."

Source:
"A Queer Little Cloudburst," 1887. *New York Sun* (May 29). Reprinted from the *Galveston News*.

1888: "This Weird Visitor from the Great Unknown"

HORSE THIEF RED PAGE met his end at the hand of an outraged Madison County lynch mob one bloody day in February. He was strung up, in venerable vigilante fashion, on the first available tree.

Soon afterward, a strange story began to circulate. Supposedly, riders who passed the tree, in a remote, little populated area of the county, were seeing a ghostly form, 8 feet long, emitting a pale, silvery light and hanging in mid-air. When word got to Madisonville, the reaction was laughter—though tinged with curiosity and a little uneasiness. Finally, three county residents of good reputation investigated the matter for themselves. They returned to report that the stories were true; they had seen it with their own eyes.

A party of twenty-five quickly assembled and rode to within 100 yards of the site. At that point, apparently, nerves started to buckle, and the group stopped. After discussing their options, they decided to send six of the braver

members up to the tree. They did so but, seeing nothing, these six summoned the rest. As the band sat on their horses and talked, one member glanced toward the tree and gasped in horror. Just 20 feet away, there it was. One witness provided this account:

"The blood ran chill and cold in my veins. An indescribable sensation—a sickening, terrifying feeling—crept over me that I shall never forget until my dying day. . . . It seemed to be composed of a grayish white substance, and was surrounded by a faint, pale, mysterious light, and appeared to be about eight feet in length, with long, shining, shriveled neck. Its face could not be seen distinctly. After viewing the unearthly object for some minutes the crowd rode off a few hundred yards and a consultation was held, but of all the brave men in that crowd there was not one who would approach it. We were all satisfied that it was not of this earth, and no one present was particularly anxious of forming more intimate acquaintance with this weird visitor from the great unknown. We returned to town without investigating further. What it is, where it came from, or what its object in coming is I do not know, but as sure as faith I saw it. I looked at it, I scrutinized it carefully, and from that moment I will be a firm believer in ghosts."

Source:
"A Texas Specter," 1888. *Atlanta Constitution* (March 10). Reprinted from the *Fort Worth Gazette*.

1897: The Landing at Beaumont

MR. J. F. LIGON, local agent for the Magnolia brewery of Houston, this afternoon [April 20] informs The Post correspondent that the airship visited Beaumont Monday night [the 19th], and that he not only saw the flying machine, but he had a chance to inspect it from the outside and conversed with one of the men who was traveling in it.

"'I and my son Charley drove home at 11 o'clock Monday night,' said Mr. Ligon, 'and were unhitching the horse when we saw lights in the Johnson pasture, a few hundred yards distant. We went over to investigate and discovered four men moving around a large dark object, and when I hailed them they answered and asked if they could get some water. I told them that they could. They went to my house, each bringing two buckets, and after filling them, started to return. I escorted one of the men and he told me his name was Wilson and that he and his companions were traveling in a flying machine. They were returning from a trip out on the gulf and were now headed toward Iowa, where the airship was built.'

"Mr. Ligon said he accompanied the men to the ship, and describes it as being 180 feet long and 20 feet wide. It was propelled by four large wings, two on either side, and steered by propellers attached to the bow and stern, electricity being the power used. The hull of the ship is made of steel and contains apartments into which compressed air is pumped when the ship is in action. The ship contains a water ballast that is pumped to the bow where the course is skyward. Mr. Ligon says the workings of the ship were thus described to him by the man Wilson, who also stated that this ship was only one of five that had been built in a quiet Iowa village."

Source:
"Inspected the Air Ship," 1897. *Houston Daily Post* (April 21).

1897: The Beaumont Airship

THERE IS A MAN in New Orleans who has seen the airship and conversed with one of the occupants. He is Rabbi A. Levy, of Beaumont, Tex., and his clerical position entitles him to credence. Dr. Levy was seen by a Picayune man last night at the residence of Mrs. G. Levy, No. 48 St. Peter street, where he is stopping, having come to this city to attend the wedding of Mrs. Levy's daughter this afternoon, the young lady being his niece.

"'You can take my word for it,' said he, 'that the airship is no myth. I had heard a good deal about it, but placed little reliance in the stories that were circulated, and doubted until the mo-

ment I saw it. It was about ten days ago on a farm about two miles from Beaumont, which is my home. About 10 o'clock that night the whole country around was aroused by a report that the airship had been seen and that it had alighted on a farm near by. My curiosity was aroused and I went to see it. I learned that they had stopped to lay in a fresh supply of water.

"'It was dark as pitch then, and I could see very little except the outlines of the ship. It was about 150 feet long, the body being shaped something like the shuttles used in an ordinary sewing machine. On either side were immense wings, about 100 feet long. It seemed to be made of some light material, what, I could not say. I spoke to one of the men when he went into the farmer's house, and shook hands with him. It is run by electricity, but how it is applied I do not know.

"'Yes, I did hear him say where it was built, but I can't remember the name of the place, or the name of the inventor. He said that they had been traveling a great deal, and were testing the machine. I was so dumbfounded that I could not frame an intelligent question to ask, so you see I can give you but very meager details. One thing I do know, and that is that an airship is an accomplished fact, for I have seen it, and many of my friends have seen it flying in the air. It went to Dallas, Austin, Fort Worth and hovered all around Texas for some time.'"

Source:
"The Airship," 1897. *New Orleans Daily Picayune* (April 25).

1899: Dellschau and the Aeros

CHARLES AUGUST ALBERT Dellschau was born on June 4, 1830, in Germany and died on April 20, 1923, leaving behind both a legacy that would not be appreciated for decades and a mystery that may never be solved.

Biographical information about Dellschau is sketchy. He is known to have come to the United States in 1853. In 1856, he lived in Harris County, Texas; four years later in Fort Bend

County. The great enigma of Dellschau's life concerns where he was in the years between.

In 1899 (or 1903 or 1908, according to other sources), living in Houston with his stepdaughter and her husband, he began to fill notebooks with writings and ink-and-watercolor drawings of exotic flying machines, described by a modern art critic as "a sort of febrile hybrid of Jules Verne and Leonardo da Vinci. [They] sometimes look like dirigibles or balloons. At other times they resemble pneumatic UFOs or sailing ships from some unknown world of water" (Johnson 1998). After his death, the materials were stored in the attic, where they lay until the 1960s, when the city declared the house a fire hazard and demanded that it be cleared. The notebooks ended up in the hands of a trashman, who sold them to Houston junk collector Fred Washington, owner of the O.K. Trading Center. In 1969, Mary Jane Victor, an art-history major at Houston's University of St. Thomas, found them there. She worked for Dominique de Menil, a wealthy heiress who collected folk art, and when Victor told her employer about the scrapbooks, de Menil gave her $1,500 to purchase the first four.

Soon afterward, at a St. Thomas exhibit on the theme of flight, Dellschau's drawings were displayed. A Houston commercial artist, P. G. Navarro, read of the exhibit in a newspaper and was intrigued. Through his interest in UFOs, he knew of late nineteenth-century reports of mysterious airships, and he wondered if there might be a connection. After seeing the drawings, he grew even more interested. In 1972, he learned that other Dellschau notebooks survived at Washington's junk shop. He purchased seven of them. By now he was in the grips of something of an obsession.

The notebooks contain, as one writer described them, "intricate collages [which] show shiplike decks supported by striped balloon pontoons; they show bright-colored helicopters and evil-looking striped dirigibles outfitted for war; they show crews of dapper little gentlemen accompanied by the occasional cat. Many pages are bedecked with little newspaper clippings about aviation, and text in his weird Germanic lettering celebrates the pure, unexcelled marvelousness of the flying machines" (Greenwood 1998).

The devices, according to Dellschau, were the creations of the Aero Club, later (eight years into its initiation) the Sonora Aero Club, formed in 1850 at a meeting in a hotel in Sonora, California, a Gold Rush boom town in the Sierra Nevada foothills. The club, a sixty-two-member secret society consisting mostly of German immigrants, with some of English, Spanish or Mexican, and (in one instance) French background, got its funding from a mysterious group "back east." Dellschau referred to that group only by its initials, NYMZA, and provided no further details about it, except to indicate that it was not a government agency and that it did not want to use the aeros for military purposes. The Sonora Aero Club, however, was headed by a brilliant man named Peter Mennis, inventor of a gas known only as "NB," made from a substance Dellschau rendered as "suppe" (possibly "soup" in Dellschau's eccentric spelling). NB, with what seem to be antigravitic qualities, helped power the airships, called aeros, that the members built and flew, apparently beginning in 1856 or 1857. The ships bore varied names: Aero Trump, Aero Mio, Aero Dora, Aero Goosey, and the like. Dellschau's drawings depict about 100 of them. Dellschau himself was neither an inventor nor a pilot; his role in the club apparently was to record its activities, though not for public consumption.

There is a sinister undertone to Dellschau's account, some of it written in code as if there were secrets that needed to be kept long after the club's demise. He intimates that one Jacob Mischer, who died in the crash of his Aero Gander, may have been the victim of sabotage by fellow club members who objected to Mischer's using his aero for a commercial enterprise, cargo transportation. Another member who spoke too openly with outsiders was forbidden to build his own machine. Mennis himself would not share the secret behind his NB; thus, when he died accidentally in a fire in the 1860s, the club collapsed in disarray. In one of the scrapbooks, Dellschau would write, "Peter Mennis, you are not forgotten."

Elsewhere, he addressed a future someone who he thought would find the scrapbooks and attempt to unlock their secrets, "You, Wonder Weaver, will unriddle these writings . . . which are my stock of open knowledge." Navarro, who came to think of himself as that Wonder Weaver, devoted years to deciphering Dellschau's coded messages. Eventually, he satisfied himself that he had succeeded, but he has shared his findings with few other researchers, and these he has sworn to confidentiality in the hope of producing a book on the subject. In recent years, Dellschau's works have attracted wide attention and big prices from art dealers. A single page from a notebook has brought as much as $15,000.

Did the Aero Club and the aeros really exist? Those who have studied the scrapbooks tend to think so, impressed not only by the extensive detail of the drawings but by Dellschau's recounting of the personalities and events associated with the organization. And yet, Navarro and others who have gone to Sonora seeking evidence have returned with little to show for their efforts. None of the named individuals left any trace of their existence, if they ever lived at all. Area newspapers from the 1850s, as far as anyone has been able to determine so far, do not mention sightings of strange airships, whose passages through the sky would have been all but impossible to conceal and which surely would have been remarked on. There is no compelling reason to connect the Aero Club with unexplained airships allegedly sighted in California in late 1896 and elsewhere in the United States between February and May of the next year, though some theorists have tried. Dellschau's scrapbooks contain numerous clippings documenting early aviation experiments and disasters, but none concern reports of the UFO-like airships that filled newspapers for a few months in the late nineteenth century.

The only evidence for the club's possible reality is broadly circumstantial. Near Sonora, in the small town of Columbia, is a desert airport on flat ground surrounded by high hills—a place that certainly *could* have served as a location for aero-testing. No one has been able to establish that Dellschau lived in the area in the late 1850s, however, though there is no evidence that he didn't, either. As remarked above, his place of residence between 1856 and 1860 is uncertain.

Sources:

Greenwood, Cynthia, 1998. "Secrets of the Sonora Aero Club." *Houston Post* (December 10).

Johnson, Patricia C., 1998. "Folk Art in Flight." *Houston Chronicle* (June 25).

"Lauren Redniss Uncovers the Fantastical Flying Machines Inside the Aeronautical Notebooks of Charles Dellschau," n.d., http://www.rawvision.com/back/dellschau;dellsc.html.

Paijmans, Theo, 1998. *Free Energy Pioneer: John Worrell Keely.* Lilburn, GA: IllumiNet Press, 373–381.

1908: Man in a Mirror

OF ALL THE PECULIAR stories that have been told in this city [Gainesville] recently, the following takes the prize, but that it is absolutely true is vouched for by almost half a hundred people who saw the weird image which caused the excitement.

"Last Friday afternoon [January 24] Mrs. J. N. Shacklett was sitting in her bedroom at her home on South Clements street when she glanced in the mirror and noticed a dark spot on it. Watching the dark place, she saw it take the shape of a man's head and after a few minutes, she beheld the unmistakable features of her brother, Lester Harrison, who lives in Vernon.

"Mrs. Shacklett was frightened and called to her sister. Her sister immediately recognized the features of her brother, and called for other people. Within the space of an hour, probably fifty people had been called in to witness the phenomenon, and everyone acquainted with Harrison recognized his features.

"About 8 o'clock the next morning the picture began to fade and within a few minutes it had entirely disappeared.

"The peculiar occurrence aroused the curiosity of all who saw it and thus far no one has offered any explanation. At first it was thought that the heat had affected the quicksilver on the back of the mirror, but when the image disappeared, the mirror was as clear as ever before; others offered the explanation that it was the reflection of a photograph, but this was upset for the reason that there was no picture of him in the room; in fact, the only picture of him in the house was a crayon drawing which was very different from the one on the mirror.

"In one room of the house was his picture in which he had posed for the crayon artist and in this he wore a hat. The doors between the rooms were closed and yet in the mirror could be seen his bust, but he wore no hat, his forehead being plainly visible.

"Soon after the image was discovered in the mirror, a phone call was put in for Harrison in Vernon and he responded promptly. He was told of the phenomenon and on the next morning he rang up to see how his picture was getting along. . . .

"There can be no question of the fact that the picture was on the mirror; it was seen by too many reputable witnesses to deny that; so it is now up to the scientist to explain it. For a time Mrs. Shacklett thought it was some trick, and she took soap and water . . . and tried to wash it off, but this made not the least impression on the picture. It was as clear after each bath as it had been before."

Source:
"True Ghost Story," 1908. *Wichita* [Texas] *Daily Times* (January 29). Reprinted from the *Gainesville Messenger*.

1908: "Not Less Than 200 Feet Long"

THE OFFICERS, CREW and passengers of the steamship Livingstone, which arrived today [July 1], en route from Galveston to Frontera, Mexico, saw a sea serpent and got within sixty feet of the monster. Much has been written and published about sea serpents, but rarely, if ever before, has anybody been ready to make an affidavit setting forth that they have seen and in-

spected one of these terrible monsters of the deep. But those on board the steamship Livingstone had the opportunity of seeing and inspecting the sea serpent and accepted it. Those who saw the reptile are willing to make oath to what they saw and did make affidavit, a copy of which is attached. The witnesses swear to only what they actually saw, and that was enough to establish the fact that a monster of great power and size exists in the sea.

"Captain G. A. Olsen stated that they didn't see the head of the sea serpent, but could distinctly see its body, which was not less than 200 feet long.

"'I ran the ship to about sixty feet of the monster,' said Captain Olsen, 'and some of the passengers were afraid even to approach that close. The monster appeared to be sleeping and we had no desire to arouse him. It would not be safe to molest an animal or fish or whatever it was of that size and especially with passengers on board. No doubt with a swish of his tail he could have wrecked the ship, so we were content with studying him through the glasses and we are positive of what we saw. It was alive all right, for shortly after we pulled away a squall came up and the reptile was lost to sight.'"

Source:
"Great Sea Serpent," 1908. *Wichita* [Texas] *Daily Times* (July 1).

1909 and After: Paluxy Tracks

IN 1909, NEAR GLEN Rose, southwest of Fort Worth, a teenager noticed large, three-toed tracks embedded in limestone in a tributary of the Paluxy River. The next year, two young men fishing near the site noticed another kind of fossil track alongside the familiar three-toed ones. They were 15 to 18 inches long and looked like something a giant man could have left. Both the three-toed tracks (subsequently attributed to a carnivorous, bipedal dinosaur called the theropod) and the "giant man tracks" (as locals referred to them) were used to attract tourists. By the 1930s, some were being removed for sale, and one Glen Rose man, George Adams, was manufacturing fake tracks, some still in circulation.

That same decade, Roland Bird, who conducted field work for the American Museum of Natural History, came to the Paluxy site and studied the theropod and sauropod tracks in evidence there. At one point a local man showed Bird an alleged man track. It was unlike anything Bird had seen before, and he could only speculate that it was from a heretofore unknown species of dinosaur.

The "giant man tracks" soon would play a significant role in the growing "creation science" movement, the effort of Christian fundamentalists to prove the literal truth of the Genesis creation story. Creationists hold that human beings came into existence in the Garden of Eden as a result of God's intervention; evolution never occurred, and thus human beings and dinosaurs coexisted. That was the view of an organization called the Deluge Society. The society's founder, Clifford Burdick, read Bird's remarks about the mysterious track and was sufficiently intrigued to take a quick trip to Paluxy. The July 25, 1950, issue of the Seventh Day Adventist publication *Signs of the Times* carried Burdick's article about the trip. In it Burdick claimed to have conclusively refuted evolution. After that, the Paluxy tracks became a staple of creationist literature, featured in such standard texts as John Whitcomb and Henry M. Morris's *The Genesis Flood* (1961) and A. E. Wilder-Smith's *Man's Origin, Man's Destiny* (1965). In 1975, Burdick devoted an entire book, *Footprints in the Sands of Time,* to the subject.

Even so, not all creationists were convinced. A research team from the fundamentalist Loma Linda University did work at the site in 1970 and left persuaded that these were dinosaur, not human, tracks. In fact, some "tracks" were not tracks at all, just eroded rock. In 1980, two young men with creationist sympathies, Tim Bartholomew and Glen J. Kuban, came to the same conclusions following their own trek to Paluxy.

Still, both creationists and secular theorists attracted to heterodox ideas about human origins continued to cite the tracks as something truly extraordinary. Orthodox scientists, who had largely ignored the issue, finally took notice, having grown alarmed at the growing power of the creationist movement. Four of them journeyed to Paluxy and nearby locations to examine all the prints cited by creationists. Some, they determined, were fakes, others the products of erosion; the rest seemed to be of dinosaurs whose toes for some reason had left no impression. Kuban's continuing investigations confirmed this last; he found that sediment had filled in the toe marks, then hardened into rock. The sediment was different (and differently colored) from that in which the rest of the track was set. Subsequent studies by other scientists backed up Kuban's observation.

Kuban, who had evolved from creationist to creationist critic, debated those who insisted that the Paluxy tracks were human. Eventually, he and colleague Ronnie Hastings brought prominent creation scientists to Paluxy and convinced them to abandon the tracks as a lost cause. A creationist production company even withdrew a documentary, *Footprints in Stone,* popular in fundamentalist circles.

The "giant man tracks" were not yet dead, however. A man named Carl Baugh took them up, going so far as to assert that he had found fossil human artifacts—a tooth, a finger, and a hammer—at the site. The first proved to be from a prehistoric fish, the second no more than a stone, and the third of no demonstrated age. As late as February 1996, NBC television broadcast a much-criticized documentary, *The Mysterious Origins of Man,* which used the Paluxy tracks and other "evidence" to argue that conventional scientific and archaeological ideas about the age of *Homo sapiens* are radically wrong.

Sources:

Kuban, Glen J., n.d. "On the Heels of Dinosaurs: An Informal History of the Texas 'Man Track' Controversy," http://members .aol.com/Paluxy2/onheel.htm.
———, n.d. "A Review of NBC's 'The Mysterious Origins of Man,'" http://members .aol.com/Paluxy2/nbc.htm.

Patterson, John W., 1985. "Dinosaurs and Men: The Case for Coexistence." *Pursuit* 18, 3 (Third Quarter): 98–102.

Schafersman, Steven, 1983. "Raiders of the Lost Tracks: The Best Little Footprints in Texas." *The Skeptical Inquirer* 7, 3 (Spring): 2–6.

Steiger, Brad, 1978. *Worlds before Our Own.* New York: Berkley-Putnam, 45–50.

Wilford, John Noble, 1986. "Fossils of 'Man Tracks' Shown to be Dinosaurian." *New York Times* (June 17).

1913: Death of a Little Green Man

IN 1978, A YEAR AND a half before his death, an old man would recall an experience from his childhood, when he claimed to have witnessed the killing of an otherworldly being. The incident happened, he said, in early May 1913, about 2.5 miles west of Farmersville, a Collin County town in the northeastern part of Texas.

Silbie Latham was twelve years old, living on a cotton farm, at the time. That morning, he was chopping cotton, along with his brothers Sid and Clyde, when suddenly the family dogs, Bob and Fox, let loose with a strange bark as if they were in "terrible distress." When the "deathly howl" persisted, the boys finally put down their hoes and went over to investigate.

The dogs were some 75 feet away, on the other side of a picket fence. Clyde, the oldest boy, got there first. When he saw what had so excited the animals, a look of astonishment crossed his face. He shouted, "It's a little man!"

Speaking to Larry Sessions, then of the Fort Worth Museum of Science and History, Silbie recalled that the figure "looked like he was resting on something. He was looking toward the north. He was no more than eighteen inches tall and kind of a dark green in color. . . . He didn't seem to have on any shoes, but I don't really remember his feet. His arms were hanging down just beside him, like they was growed down the side of him. He had on a kind of hat that reminded me of a Mexican hat. It was a little round hat that looked like it was built onto him. He didn't have on any clothes. Everything looked like a rubber suit including the hat. . . . He just stood still. I guess he was just scared to death."

The dogs jumped on the little creature and ripped him apart, leaving blood and organs in their wake. The boys stood by and watched, not knowing what else to do, then returned to their hoeing. On two or three occasions they went back to the site and observed the remains, now rotting in the hot sunlight. The dogs now appeared oddly frightened and stayed close to the boys at all times.

By the next day, all traces of the event were gone.

Two years later, however, Silbie and a brother saw something else out of the ordinary. While sitting on the porch of their uncle's farmhouse near Celeste, Texas, they watched a mysterious object carrying two lights—one in the front, one in the back—sail silently overhead. In the early darkness they could make out a large cylindrical shape "like an airplane without wings" between the lights.

Silbie Latham's experiences came to light when his grandson Lawrence Jones wrote the Illinois-based Center for UFO Studies (CUFOS) to report them. Referring to the green-man story, he said, "My grandfather has a most solid reputation for truth and honesty and has never told of this because of fear of ridicule. . . . He has agreed to tell this only after much prompting and encouragement from me, his history-oriented grandson. . . . There is no question in my mind that he is telling the truth." He added that the incident "has been discussed in my family for years." Sessions, who met personally with Latham, was impressed and found him credible, at least to the extent that "there's no doubt he *believes* it happened."

Source:
Evans, Alex, 1978. "Encounters with Little Men." *Fate* (November): 83–86.

Utah

1858: Sword Shape in the Sky

AT 1 A.M. ON MAY 26, residents of Salt Lake City reported seeing a large sword, its blade pointed eastward, in the sky. They took it to be an omen of God's support for their faith.

Source:

"News of the Day," 1858. *New York Times* (July 26).

1868 and Beyond: Monster Mammals

WRITING IN *HARPER'S New Monthly Magazine* in 1883, Phil Robinson, a British traveler in the American West, remarked, "There is abundant testimony on record—or the formally registered oath, moreover, of men whom I know from personal acquaintance to be incapable of willful untruth—of the actual existence at the present day of an immense aquatic animal of some species as yet unknown to science. . . . The men whose testimony is on record . . . agree as to their facts, [which] point to a very possible monster—in fact, a freshwater seal or manatee." A Utah writer, 120 years later, attributed the sightings to a large insect "known as humbug" (Bagley 2003).

Real or imagined, the Bear Lake monster survives only in late-nineteenth-century newspaper clippings. Those who could have shed light on the matter, the witnesses (or alleged witnesses), are long dead. All that can be stated with some reasonable certainty is that—unlike many other period lake-monster stories—this one was not at its core a prank or hoax. Most who claimed to have seen the beast were manifestly serious, and

no one has charged that the reports were mere journalistic fictions. Beyond that, it's all a matter of intriguing anecdote and not much, if anything, else.

Belief in the existence of a huge animal or animals in the lake goes back to the area's original inhabitants, the Shoshones. Interestingly, though, they thought the creatures had gone extinct along with the buffalo destroyed in devastating blizzards in the winter of 1830. They told white settlers that the animals, which on occasion had attacked and dragged away swimmers, were long and serpentine but also had legs about eighteen inches long, which enabled them to crawl out of the water a short distance to the shore. Water would sometimes spout upward out of their mouths.

Then in early July 1868 a man named S. M. Johnson, passing on a road not far from the lakeshore, happened to notice what he took to be a drowned person in the water. He turned off the road and drove to the shore, on the assumption that the high waves would soon wash the body to the sand. By the time he got there, the presumed body was no longer visible, but then at the spot it had disappeared, the head and part of the neck of some enormous animal rose out of the water. According to the Salt Lake City paper *Deseret News* (July 27), "It had ears or bunches on the side of its head nearly as big as a pint cup. The waves at times would dash over its head, when it would throw water from its mouth or nose. It did not drift landward, but appeared stationary, with the exception of turning its head" (quoted in Fife 1948). Johnson thought it must be sitting on the lake bottom, or it could not have been unaffected by the rolling waves.

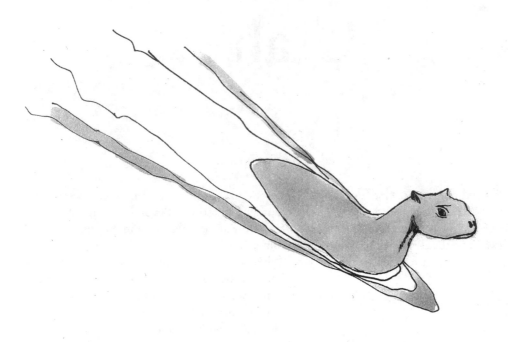

By the time he got there, the presumed body was no longer visible, but then at the spot it had disappeared, the head and part of the neck of some enormous animal rose out of the water.

A day later, at the same location, a party of four claimed to have seen the same animal, described as "very large" and swimming "much faster than a horse could run on land" (ibid.).

On July 22, a Sunday, ten persons—four men and six women—reported sighting a number of strange animals in the lake. The *News* reporter interviewed them and put together this not altogether lucid account:

"[T]heir attention was suddenly attracted to a peculiar motion or wave in the water, about three miles distant. The lake was not rough, only a little disturbed by a light wind. Mr. Slight says he distinctly saw the side of a very large animal that he supposed to be not less than ninety feet in length. Mr. Davis doesn't think he saw any part of the body, but is positive it must have been not less than forty feet in length, judging by the wake it left in the rear. It was going south, and all agreed that it swam with a speed almost incredible to their senses. Mr. Davis says he never saw a locomotive travel faster, and thinks it made a mile a minute easily. In a few minutes after the discovery of the first, a second one followed in its wake; but it seemed to be much smaller, appearing to Mr. Slight about the size of a horse. A larger one, in all, and six small ones hied southward out of sight.

"One of the large ones, before disappearing, made a sudden turn to the west, a short distance; then back to its former track. At this turn Mr. Slight says he could distinctly see it was a brownish color. They could judge somewhat of their speed by observing known distances on the other side of the lake, and all agreed that the velocity with which they propelled themselves through the water was astonishing. They represent the waves that rolled up in front and on each side of them as being three feet high from where they stood. . . . Messrs. Davis and Slight are prominent men, well known in this country [Utah], and all of them are reliable persons whose veracity is undoubted" (ibid.).

One day later in 1868, four men fishing at Swan Creek, a tributary immediately west of the lake, reported seeing one of the creatures. It had the head of a serpent, and a portion of its body about 20 feet long was visible above the water. Covered with light brown fur like an otter's, it "had two flippers extended upwards from the body," according to a press account. At one point it swam so close that the witnesses judged it within rifle range. The report does not make clear whether they refrained from firing out of choice or because they lacked firearms.

Other sightings attracted press attention that year. Incensed that his report had been greeted with ridicule, witness Phineas H. Cook vowed to capture the beast and prove its existence to the mockers. Unfortunately, his effort sparked only further ridicule. The *Semi-Weekly Telegraph,* a Mormon periodical, scoffed that the man "anticipated a big haul not only of his snakeship but of spondulicks [money], from Barnum or some other genius, for the exhibition of the eighth wonder of the world" (ibid.).

In August 1870, however, another Mormon newspaper, the *Salt Lake City Semi-Weekly Herald,* cited the highly respected Mormon figure William Budge, "who informs us that the Bear Lake Monster has been seen very frequently of late. Even the most skeptical are giving way. One reliable gentleman saw three of them together recently; and a large number have been seen by different persons at different times within a short period" (ibid.). In the face of sightings by Mormon elders, the *Telegraph* would soon find that the Bear Lake creatures were respectable, after all. Budge would become a witness himself on May 15, 1874, when he and three companions observed something out in the water about 60 feet from the shore. "At first sight we thought it might be a very large duck," he wrote Mormon patriarch Brigham Young three days later. "Its face and part of its head were distinctly seen, covered with fur, or short hair of a light snuff color." The flat-faced animal had "very full large eyes, and prominent ears" and a neck 4 to 5 feet long. Budge couldn't judge its overall size but thought "it did not look ferocious." It swam at a leisurely pace, then dived and reemerged to move "off into the Lake as fast as a man could walk" (ibid.).

On July 19, 1871, two men noticed a commotion in the water. As they turned to look, they saw the head and part of the body of an unusual animal. The body was "like that of a man," the head like that of a tuskless walrus. The men fired on it, and it swam off with a "wavy serpentine motion." Another story from around the same time made the even more fantastic claim that one of the monsters had been captured. It was "some twenty feet long" and "propelled itself through the water by the action of its tail and legs" (ibid.). It had a large mouth. If this story is in any way true, the remains have long since been lost.

The Bear Lake beasts had faded from history by the time the new century arrived. In fact, the 1883 *Harper's New Monthly Magazine* reference is perhaps the last printed account alluding to recent sightings. It is striking that—as is not the case with most contemporary claimed lake-monster reports—the animals are described as mammals. Perhaps that detail gives them a ring of authenticity, but nothing about this curious little episode can be stated with any certainty. Looking back on the matter in 1948, Austin Fife reported that "there has been a conscious effort made by resort owners of Bear Lake to squelch the legend because they feared it might intimidate wary bathers and thus depopulate their establishments. If this be true then one of Utah's finest legends is to be murdered in the guise of a magnificent who never yet seems to have swallowed a single person since the arrival of the white man."

Sources:

Bagley, Will, 2001. "Monster Tale Haunts Shore of Bear Lake." *Salt Lake Tribune* (April 1).

———, 2002. "Bear Lake Monster Season Is Over for This Year, but the Legend Lives On." *Salt Lake Tribune* (December 8).

———, 2003. "It's Monster Season Again, So Take Care on Shores of Bear Lake." *Salt Lake Tribune* (April 20).

Fife, Austin E., 1948. "The Bear Lake Monsters." *Utah Humanities Review* 2, 2 (April): 99–106.

Robinson, Phil, 1883. "Saunterings in Utah." *Harper's New Monthly Magazine* 67, 401 (October): 705–715.

1873: Monster in Pursuit

ON THE 16TH [OF August] Mr. Walker, his brother Sylvester, and their cousin John Coon, were felling timber for Mr. Standish's mill, in the right fork of Coon's Canyon, about three quarters of a mile from the point of the mountain west from this city, on the high peak south of Black Rock. Between 10 and 11 o'clock in the morning a deer ran by where they were working, and our informant snatched up a Sharp's rifle and started in pursuit. When he had continued the chase for about a mile, due north, he was startled by a loud, shrill whistle and hiss, which he at first thought might be a signal from an Indian.

"He came to a halt and looked about him, and heard the noise of rocks rattling southeast from where he stood. He turned, when, to his horror, he saw approaching him at a very rapid rate a serpent which he judged was between thirty and forty feet long and about ten inches through the body. The reptile's head was raised fully six feet from the ground, and his jaws were open fifteen or eighteen inches wide, with large fangs growing from both lower and upper jaws. Walker was almost petrified with fear, but the hope of saving his own life made him start to run.

"The serpent, however, was too quick for him, and jumped at and knocked him down, striking him on the left shoulder just below the shoulder blade, going over him and down the mountain in the southwest; for a short distance, he turned and pursued Walker, who had risen, and with a speed inspired by the deadliest fear was making his way to the top of the ridge. Unfortunately for our informant, he stumbled, and immediately felt the weight of his monstrous pursuer gliding over his body. He gave himself up for lost just then, but it seems hard to tell which was the most frightened, the man or the snake, for the latter did not seem disposed to run the risk of a contest, but gliding across the body of the prostrate man, he slid off at a tremendous rate toward the ridge of the mountain and across it to the east side.

"Walker rose and watched his movements, and says that after crossing to the east side and down the mountain a few yards, he then twined himself around a large mahogany tree, where he remained waving his head to and fro, flapping his tail on the rocks, and whistling and hissing defiance. That was the last he saw of him, for he made his way back to his companions as quickly as he could. They wanted to return and hunt for the serpent, but the hero of the adventure was too weak from fright and excitement to do so, but says they intend to go on an expedition to hunt the snake in a very short time.

"The color of the reptile was yellow, with a black mark like a half-moon on each side of his eyes; he had a beard or fuzz round his mouth, and what appeared to our informant to be a crown-shaped mass on the top of his head. The latter was about six inches high and varied in color, being green, blue, white, yellow and red. The head of the creature was about as large as that of a full-grown bull dog, and in shape between that of a bull dog and a monkey. His body was covered with hard scales six or eight inches long. Mr. Walker says he has been used to mountain life for years, and never was afraid of anything, but nothing could persuade him to go alone again into the right hand of Coon's Canyon."

Source:

"A Mountain Monster," 1873. *Hagerstown [Maryland] Herald and Torch Light* (September 3). Reprinted from the *Deseret News.*

1894: A "Uniform Hat" Cloud

LAST FRIDAY EVENING [July 20], Bernard Parry and wife are reported having seen a most unusual phenomenon. As they were nearing their home in Marriotts settlement, about sundown, they observed a small black cloud in the direction of North Ogden, coming swiftly toward them. The cloud seemed to enlarge, then collapse again, all the while rapidly moving toward the southwest. When it approached within 150 yards of the road it seemed to stop and remain motionless in the air. The sides appeared to be folded toward the center and pinned—something like the uniform hat of the old Continentals, and resembled the meshes of a net. In the center appeared to be many small birds[,] part black and part white. All at once the cloud wheeled toward Ogden and in a moment was lost to sight. Other parties living north of Marriotts claim to have seen the same startling phenomenon."

Source:

"Strange Phenomenon," 1894. *Ogden Standard* (July 26).

1894: Serpent Comes to Eden

ACCORDING TO REPORTS received at this office, there was seen last Friday [July 20], about sundown, coming over the mountains from the north, a monster snake, 60 feet in length and 18 inches in diameter, not wriggling along in the ordinary path of the snake—but floating gently through the air at thirty-five or forty miles per hour. It came swooping down in the direction of Wilbur's store, at the corner of Independence park, and when almost there came within twenty or thirty feet of the ground. Suddenly swerving to the left it disappeared up over the mountains in the direction of Middle Fork canyon. The movement of the monster was like a snake in water and it seemed to acquire speed without any effort whatever. Its skin seemed to be formed of scales like an alligator.

"The above story is vouched for by a number of Eden's reliable men who saw the grim specimen. It was seen by half the inhabitants and created great excitement. Mr. Fuller, one of the oldest inhabitants of Eden, says that a great many years ago a similar dragon appeared in the south of France."

Source:

"A Veritable Eden," 1894. *Ogden* [Utah] *Standard* (July 23).

Vermont

1843: Killing a Monster Snake

Mr. Joshua Buddington, of North Attleboro, has furnished . . . the particulars of a snake capture on his farm on the 10th inst. [July]. A cow had been missing for several days, and his son, while searching for her, found her laying [sic] dead, with an enormous snake entwined [in] her hind legs. The lad ran home, and a party sallied forth to slay the monster. They found that the snake had made an incision into the left side of the udder, through which he had inserted his head about four inches, and was in the act of extracting the milk at the time the party arrived. He was immediately dispatched by a tremendous blow from a club given in the region of the neck. On being struck he at once uncoiled from the limbs of the cow, drew forth his head, and after gasping three or four times, expired.

"His back is zebra striped, and the belly of a dark green, with small spots thickly interspersed. Around the neck and directly back of the jaws, are four stripes or rings of a bright yellow color, and just under the throat a small bag or hollow membrane is suspended, filled with a thin liquid substance. This membrane is perfectly transparent and through it the appearance of the contents is dark green. The length of the snake is 14 feet 3 inches—circumference around the largest part of the body, 1 feet [sic] 10 1/2 inches—from the end of the upper jaw to the eye, five inches—width of the head, which is very flat, 7 1/4 inches. The species to which the snake belongs has not been ascertained."

Source:
"A Terrific Snake Story," 1843. *Adams* [Pennsylvania] *Sentinel* (July 31).

1907: Fiery Torpedo

At 2 o'clock on the afternoon of July 2, a sudden explosion rocked downtown Burlington. According to a letter he wrote to *Monthly Weather Review,* Bishop John S. Michaud, looking to its source, spotted what he described as "a torpedo-shaped body, some three-hundred feet away, stationary in appearance, and suspended in the air, about fifty feet above the tops of the buildings" (quoted in Fort 1941).

Michaud estimated it to be 6 feet long by 8 inches in diameter. It had a dark shell, through which red-hot "tongues of fire"—apparently emanating through openings—issued. The object began to move slowly, and as it did so, the witness thought the shell seemed to be rupturing as more flames broke through it. It vanished to the south over the roof of a nearby store.

Source:
Fort, Charles, 1941. *The Books of Charles Fort.* New York: Henry Holt, 292–293.

☞

It had a dark shell, through which red-hot "tongues of fire"—apparently emanating through openings—issued.

Virginia

1850: Falling Flesh

ONE AFTERNOON IN the spring, along the south bank of the Pamunkey River in lower Hanover County, several men observed a small cloud passing overhead from northeast to southwest. As it floated overhead, pieces of some material dropped out of it and fell all around the witnesses. When examined, the material, scattered around a 5-yard area, proved to be "various pieces of flesh and liver, too well to be defined in each sort to allow any mistake in their character," in the words of physician G. W. Bassett, on whose estate the curious fall occurred (quoted in Splitter 1953).

The next morning, Dr. Bassett and an associate collected fifteen to twenty pieces, some weighing as much as an ounce. Bassett sent one sample to a doctor in Richmond and preserved the rest in alcohol "for the future inspection of the curious."

Source:
Splitter, Henry Winfred, 1953. "Wonders from the Sky." *Fate* 6, 10 (October): 33–40.

1853: Cod Fall

A CURIOUS PHENOMENON attended a recent hail storm in the city of Norfolk. Quantities of codfish, some measuring a foot in length, fell in different sections of the city, and some of the fields were literally strewed with them. Hundreds were picked up in the morning. The Norfolk *Argus* says, this is no piscatorial fabrication, but a fact which is attested by hundreds of citizens of that place."

Source:
"A Rain of Codfish," 1853. *Daily Alton* [Illinois] *Telegraph* (May 30).

1856: Black Bugs on White Snow

WE WERE SHOWN yesterday . . . by a gentleman living near Fairfax, a species of bug that fell at that place during the rain of Saturday night last [February 14]. The snow for several miles was robbed of its whiteness, and made to resemble a vast field of colored velvet. What is still more surprising, the intense coldness of the weather cannot kill them. They apparently seem to be stiffened by the raw atmosphere; but, if placed near the fire, will relax and exhibit signs of life. They are very black, and are but little larger than a grain of coarse powder. Who can enlighten us on the 'bug question?'"

Source:
"A Singular Phenomenon," 1856. *Ohio Repository* (February 20). Reprinted from the *Alexandria* [Virginia] *Gazette.*

1856: Invisible Train

WE UNDERSTAND THAT numbers of our citizens and persons living in the country on the line of the railroad, have been considerably mystified and no little alarmed, by a singular fact recently noticed on repeated occasions. Between the hours of 11 and 12 o'clock at night, the approach of a train of cars has been plainly heard, the shriek of the whistle and the rumbling of the train increasing in distinctness until the cars reached the Staunton depot and stopped. Persons have gone to the depot to find out the cause of an arrival at so unusual an hour, and when they got there found no train. The depot agents say that no train is on the road at that hour of the night, yet the approach of one is unmistakably heralded by the rumbling, and its arrival by the whistle."

Source:

"Phantom Train," 1856. *Liberty* [Missouri] *Weekly Tribune* (August 8). Reprinted from the *Staunton* [Virginia] *Spectator*.

1862: Sounds of Silence

THE BLOODY CIVIL WAR battle fought at Gaines's Mill on June 26, 1862, was only one episode of a larger struggle known as the Seven Days' Campaign. By the time it had run its course, the two sides together had suffered more than 15,000 casualties. The Confederate side prevailed, but the Union troops were able to retreat in orderly fashion to fight again another day.

Aside from its military aspects, the battle of Gaines's Mill is remembered for a strange, never entirely explained event sometimes referred to as the "acoustic shadow" or the "silent battle." Observers on both sides remarked on it. Major General William B. Franklin of the U.S. Army was no more than a mile and a half from the action, but he and his troops could not hear a sound. On the opposite field was the Confederate secretary of war, General G. W. Randolph. As he and R. G. H. Kean watched from a hilltop a mile away, they were unable to reconcile what they were seeing with what they were not hearing. Kean would write:

"Looking across the valley, I saw a great deal of the battle. [General Robert E.] Lee's right [flank] resting in the valley, the Federal left wing the same. My line of vision was nearly in the line of battle. I saw the advance of the Confederates, their repulse two or three times, and in the gray of the evening the final retreat of the Federal forces. I distinctly saw the musket fire of both

lines, and the smoke, individual discharges, the flash of the guns. I saw batteries of artillery on both sides come into action and fire rapidly. Several field batteries on each side were plainly in sight. Many more were hid by the timber which bounded the range of vision.

"Yet looking for nearly two hours, from about five to seven on a midsummer afternoon, at a battle in which at least 50,000 men were actually engaged and doubtless at least 100 pieces of field artillery, through an atmosphere optically as limpid as possible, *not a single sound of the battle* was audible to General Randolph or myself. I remarked it to him at the time as astonishing" (De Motte, n.d.).

Kean attributed the phenomenon to differences in air density associated with "varying . . . amounts of water vapor arranged like laminae at right angles to the acoustic waves as they came from the battlefield to me." Oddly, however, this type of anomaly was never reported before the Civil War, and it has not been reported since.

Another Confederate general, E. M. Law, compared the effect to a "pantomime." "Neither the tremendous roar of the musketry nor even the reports of artillery" could be heard.

Sources:

De Motte, John K., n.d. "The Cause of a Silent Battle," http://www.ehistory.com/uscw/library/books/battles/vol2/365.cfm.

Miles, Jim, 1982. "The Civil War's 'Silent Battles.'" *Fate* 35, 12 (December): 83–85.

1863: Cloud Rolls and Mystery Soldiers

A WRITER IN THE Staunton Spectator, dating at Lewisburg, Greenbrier county, Va., Sept. 15th, writes to that paper a description of a remarkable atmospheric phenomenon witnessed in that town. It was seen by our [Confederate] pickets a few miles from the town. The same scene has been described in several respectable papers, the editors of which all vouch for the reliability of their informants. The writer says:

"'A remarkable phenomenon was witnessed a few miles west of this place, at the house of Mrs.

Pearcy, on the 1st day of this month, at about 3 o'clock, P.M. Mr. Moses Dwyer, her neighbor, who happened to be seated in her porch at the time, as well as by others at or near the house.

"'The weather was quite hot and dry, not a cloud could be seen, no wind even ruffled the foliage on the surrounding trees. All things being propitious, the grand panorama began to move. Just over and through the tops of the trees on the adjacent hills on the South, immense numbers of rolls resembling cotton or smoke, apparently of the size and shape of doors, seemed to be passing rapidly through the air, yet in beautiful order and regularity. The rolls seemed to be tinged on the edge with light green, so as to resemble a border or deep fringe. There were apparently thousands of them, and were, perhaps, an hour in getting by. After these had passed over and out of the sight the scene was changed from the air above to the earth beneath, and became more intensely interesting to the spectators who were witnessing the panorama from different stand points. In the deep valley beneath, thousands upon thousands of (apparently) human beings (men) came in view traveling in the same direction of the rolls, marching in good order, some thirty or forty in depth, moving rapidly—"double quick"—and commenced ascending the sides of the almost insurmountable hills opposite and had the stoop peculiar to men when they ascend a steep mountain. There seemed to be great variety in the size of the men, some were very large whilst others were quite small. Their arms, legs, and heads could be distinctly seen in motion. They seemed to observe strict military discipline, and there were no stragglers.

"'There was uniformity of dress, loose white blouses or shirts, with white pants, [and the figures] wore hats, and were without guns, swords, or anything that indicated "men of war." On they came through the valley and over the steep hill crossing the road, and finally passing out of sight, in a direction due north from those who were looking on.

"'The gentlemen [sic] who witnessed this is a man with whom you were once acquainted, Mr. Editor, and as truthful a man as we have in this country, and as little liable to be carried away by

"fanciful speculations" as any man living. Four others (respectable ladies) and a servant girl witnessed this strange phenomenon. W.

"'P.S. On the 14th inst. [September] the same scene, almost identical, was seen by eight or ten of our pickets at Bunger's Mill, and by many of the citizens in that neighborhood; this is about four miles east of Pearcy's [and] it was about one hour passing.'"

Source:

"A Remarkable Phenomenon," 1864. *Raleigh* [North Carolina] *Weekly Standard* (October 7).

1863: "Beautiful Effects of Light and Electricity"

THE WRITER OF THIS was in 1863 near Warrenton, Virginia. In the afternoon he saw a most wonderful appearance of the heavens in which clouds were lighted up by the setting sun with all the colors of the rainbow, scenes were shifted with the rapidity of magic, bombs and shells were bursting in the air, steamboats moving in every direction and needing a little noise to form a realistic picture. We listened for the sound of guns, and yet knowing there was no enemy within many miles. That beautiful vision has never vanished, and it is easy to understand how people can be deceived in their interpretation of the beautiful effects of light and electricity upon moving vapor. Still, there may be 'things not dreamed of in our philosophy.'"

Source:

"Excited Imaginations," 1881. *Oshkosh* [Wisconsin] *Daily Northwestern* (October 12).

1868: The Minotaur of Prince William County

IT BECOMES MY duty to chronicle a most singular and extraordinary series of nocturnal visitations on the part of some ghostly apparition, to the farm of one whom I shall call Silas Brown, Esq., a peaceable and intelligent citizen of this county. Mr. Brown lives in what is known as the forest of Prince William, near the village of Independent Hill, and his residence is completely surrounded with the growth indigenous to that section of the county.

"For the past few weeks visions of an alarming character have been seen in the neighboring forest, but more particularly in the copse adjacent to Mr. Brown's barn and stable. At numbers of times has an immense figure been seen passing to and fro near the barn, with large horns and terrible claws, which it contracts to a sort of hoof, and has assaulted Mr. Brown, when he attempted after dark to feed his horses and stock, in such a manner and with such violence that he has been compelled to flee to his house for safety. The figure, to the best of Mr. Brown's recollection, seemed about three times as large as a man in its front, and having a back converging from its neck, and shoulders horizontally to the distance of some six to eight feet, and supplied on each side with huge and tremendous arms. It is of a pale blueish [sic] color when first seen, but upon being irritated by the near approach of any person becomes a deadly white, and issues from its surface a small volume of smoke, accompanied with a sickening smell.

"This ghoul, or unnatural and horrible animal or demon, has been seen as often as four times near Mr. Brown's stable; and when seen, it has lingered till its deadly effluvia has completely impregnated the surrounding atmosphere. One evening Mr. Brown, desiring to have another beside himself see this terrible visitant, induced a courageous gentleman whom I shall call Siger, who happened with his wife to spend the evening at Mr. Brown's, to go to the stable to feed his horses. Mr. Siger, not believing the story, went without hesitation, when, upon entering the stable, he was alarmed by the fall at or near his feet, with a deep rumbling sound, of a tremendous stone. Mr. Siger, without looking to see whence the rock came, picked the stone up, and it was so hot that he was compelled to drop it. Upon looking up he beheld the unearthly monster not over fifty yards from him, and the air became quickly filled and inoculated with brimstone. Not wishing to be thought a coward, he did not mention

anything of this at the house; but upon walking home with his wife the same night he told her of what happened at the stable, and instantly she became alarmed, and was carried home in a state of apparent insensibility.

"The neighborhood is in a terrible state of excitement, and steps have been taken to investigate this frightening matter."

Source:

K., 1868. "A Ghost—or Something—in Prince William County, Va." *Petersburg* [Virginia] *Daily Index* (December 18). Reprinted from *Alexandria* [Virginia] *Gazette*.

1869: Poltergeist Goes on Trial

IN MAY LAST THE house of Mr. Childs (master machinist at the Chesapeake and Ohio Railroad shops) was rendered almost uninhabitable by ghostly visitations. The windows were broken, rocks thrown in every room, furniture turned topsy-turvy, groans, sobs and cries heard, bells rung by unseen hands, and such other further and frightened doings as only ghosts, goblins and witches could do. The strange affair attracted the town's [Richmond's] attention and everybody went to investigate it—including two detectives. These last, after viewing the premises and having no faith in the supernatural, arrested Mrs. Baggett, a widow lady living next door. The case was heard by the Mayor, who sent the accused on to the Grand Jury, where she was indicted and came before the Hastings Court for trial yesterday, with A. Moise as counsel.

"The decisive evidence in the case was given by a young priest, son of Mrs. Baggett, the accused. He testified that often, while standing in his yard and porch, with his mother and sister, he had seen large rocks flying through the air, had heard the tinkling bells and unearthly sounds. During an hour, he was listened to by the jury and spectators with breathless attention—so earnest, so honest did he seem that all were convinced of his mother's innocence. He plainly avowed his belief in the supernatural and when asked by Col. Jenkins how he could entertain such a belief, being a Catholic, he replied that he didn't know—but that seeing all these things transpire before his eyes—knowing no human being to be near, he was compelled to believe them supernatural.

"A few other witnesses were examined—and the jury acquitted Mrs. Baggett without leaving the box. Then the old lady bared her right arm and showed it to be withered and Dr. Beale testified that she couldn't throw an egg five feet and so the muddle is worse than ever and nobody to lay the blame on."

Source:
"Our Richmond Letter," 1869. *Petersburg* [Virginia] *Daily Index* (September 17).

1870: Vision of a "Horrid Beast"

JOHN W— RESIDES ON the Hill, in the eastern part of the city [Dayton]. He had for years proclaimed himself a materialist; and he took pride in saying that he believed in nothing that could not be recognized with the aid of the senses. He did not believe in 'supernatural manifestations,' and he utterly scouted the idea of an existence beyond the grave. His Sundays were mainly spent in hunting through the fields and forest contiguous to the city, with his dogs and gun. . . .

"Some time ago his wife was taken seriously ill; and in order that she should have the most careful nursing and patient attention it were possible to secure, he assisted in the duties of the sick room, and spared neither rest nor sleep, ministering to the wants of his sick wife, both day and night. Despite the best medical attention, and the most loving and devoted nursing, the wife grew worse, until all hope of her recovery was given up by the attending physicians. . . .

"The evening before the dissolution of Mrs. W., her husband, who was almost overcome with anxiety, watching, and fatigue, laid [*sic*] down on a lounge in the sick-room to obtain a little rest—not to sleep. A queer sort of drowsiness came over him. He did not sleep. He was

conscious of everything transpiring in the room. He was, for the moment, incapable of motion, he had no volition of will. Suddenly there appeared before him a monstrous, fierce looking bull, eyes blazing fire, his mouth open, smoke issuing from his nose, and he was lashing his sides with his tail! It was a fearful sight—it was too real for an apparition, and W. was appalled! What could he do? Making a desperate effort, he sprang from the lounge to the floor, and the same instant the terrible brute vanished!

"Observing that his sick wife was watching his strange movements, he inquired of her if she had seen the fearful sight. She had been awake all the while, but she had seen or heard nothing of the apparition which had so startled her husband. She told him it was all his own imagination; that he had worn himself down by constant watching, and that he needed rest and sleep; that she did not require so much attention, and that if he did not obtain the rest he badly needed, he would also be sick. At length W. also came to the conclusion that the thing was the fancy of an over-wrought mind. But he could not banish the incident, and he was much annoyed about it. He was perfectly wide awake

Suddenly there appeared before him a monstrous, fierce looking bull, eyes blazing fire, his mouth open, smoke issuing from his nose, and he was lashing his sides with his tail!

all the while, and how could the thing have merely been in his fancy?

"W. again laid [sic] down on the lounge, and was in that position only a few minutes when the same horrid brute again made his appearance, looking much more ferocious than before, and threatening to tear him with hoofs and horns! The huge brute approached W. [and] glared in his eyes, while the hot breath of the animal fairly stifled him! With a yell of terror W. sprang from the lounge, and the same instant the horrid brute disappeared! At the same time the stove in the room was enveloped in a sheet of flame, which shot up against the ceiling! And W. feared that the house would be set on fire. But the flames vanished as suddenly as they appeared, and there was not the smell of fire in the room, or the sign of smoke on the ceiling.

"W., who never knew fear, and had never prayed, now fairly trembled with terror, and prayed to the Great Being for protection for him and his. From that moment W. became a firm believer in a hereafter, and a system of rewards and punishments. He has not seen the wonderful apparition since."

Source:
"A Scared Sinner," 1870. *Petersburg* [Virginia] *Daily Index* (May 19). Reprinted from the *Dayton* [Virginia] *Journal.*

1871: Buchanan Poltergeist

THE REV. MR. JONES, pastor of the Baptist Church in Lexington, supplies . . . the substance of a letter received by him from the Rev. Mr. Thrasher, written on the preceding Monday [February 6], recounting some of the latest doings of the so-called 'ghost' which has recently made its appearance at Mr. Thrasher's residence in Buchanan, Botetourt county.

"He says that for five days during the week previous the manifestations were frequent, varied, and violent. Brickbats, old bones, chips, billets of wood, ears of corn, stones, &c., were thrown about the house in the most mysterious and unaccountable manner, and again and again everything would be turned topsy turvy in the

parlor, and chambers, without their being able to detect the agent.

"One day two young ladies being at the house, they determined to use every effort to ferret out the mystery. Accordingly they arranged the parlor; locked all the doors, sent Anna Pring to the kitchen with Mr. Thrasher's little boy to watch her and carried all of the keys to Mrs. Thrasher's room. They waited but a few minutes and returned to find that the doors had been opened, the books from the center table scattered over the floor, the lamps from the mantel-piece on the floor and things disarranged generally. And to increase the mystery they found a strange key that would neither unlock [n]or lock any door in the house, sticking in the key-hole of the parlor door. . . .

"One day Mr. Thrasher himself left the dining room, carefully locked the door and went up stairs to his wife's chamber. Just as he was about to enter he heard a noise down stairs and returned immediately, not having been absent from the room for more than *three minutes.* He found the door open, the furniture disarranged, and all of the dishes from the press scattered over the floor. One day the clock was taken from the mantel piece and put on the floor.

"Major Paxton says that he fully satisfied himself that the little girl, Anna Pring, could not have anything to do with it, and saw enough to convince him that there is some unexplained mystery connected with the affair. He says that one night while he was there a number of young men were on guard, and that the knocking at the door being very violent and frequent, they resorted to every stratagem and made every effort to detect it in vain. He also saw chips flying about the house in the most inexplicable manner. Mr. Thrasher says that they had not been disturbed at all from Friday evening up to the time he wrote on Monday."

Source:
"The Buchanan 'Ghost' Again," 1871. *Petersburg* [Virginia] *Index* (February 15).

1882: Virginia Midland Phantom

FIREMEN ON THE Virginia Midland railroad tell wonderful stories of the nightly appearance of a ghost on the track of that road, near Otto river, where a tramp was killed some time ago. His ghostship first appeared on two white horses, but, becoming more the form of a man, has dispensed with the steeds, and has several times, unattended, taken a position on the track in the attitude of a mad bull, and defied the iron horse. One night last week the fireman of an engine discovered what was supposed to be a man on the track. The engine, which was going at a high rate of speed, struck the man and apparently killed him. The train was stopped, and several hands were sent back to see what damage had been done. The body was seen a short distance down the road, but upon the men reaching it[,] it disappeared. At other times the ghost has appeared in the cabs of the engines, and, after surveying things generally, just stepped out into space."

Source:

"About Railroads," 1882. *Fort Wayne* [Indiana] *Daily Gazette* (October 12).

1882: Strange Meteors

MANY PEOPLE WERE yesterday [October 15] attracted by the strange phenomenon of what appeared to be small stars falling early in the evening. The stars, according to those who witnessed the sight, appeared to burst and cast off small particles which seemed to resemble snow-flakes and spider webs. Another informant mentioned having seen a splendid meteor also in the day-time, which after traversing the heavens for a while burst like a bomb and scattered its glittering particles in every direction. These disturbances are attributed to the presence of the intruding comet at this unlooked-for season. It is remembered that upon the appearance of the great comet of 1858 a section of the horizon seemed to be covered with a beautiful damask red, which rose and fell like a curtain veiling the sky late in the evening."

Source:

"Meteors at Richmond," 1882. *New York Times* (October 22). Reprinted from the *Richmond* [Virginia] *State,* October 16.

1887: Civil War Phantom

THE ALLEGED APPEARANCE of Stonewall Jackson's ghost to a sentinel of the Virginia military . . . a few nights ago, is the theme of much discussion here, and is held, by believers in spirit manifestations, to portend war."

Source:

"Stonewall Jackson's Ghost," 1887. *Elyria* [Ohio] *Daily Telephone* (February 5).

1889: Rocks Fall

CULPEPPER, VA., IS perplexed on account of a rock-dropping mystery. Rocks fall around, apparently from the heavens, and the phenomenon excites much interest."

Source:

Untitled, 1889. *Statesville* [North Carolina] *Landmark* (September 26).

1895: Falling, Sparkling Stones

ON LAST SUNDAY morning [May 5] about 7 o'clock a number of citizens of Fincastle witnessed a most remarkable solar display. At the hour mentioned Mr. S. B. Smith, deputy postmaster, noticed at home, on Roanoke street, what appeared to be balls of different colors, about the size of a man's head, falling about his garden, on the housetops and in the trees. He noticed some bounding about his garden, coming from the direction of the sun, and some falling straight to the ground. They seemed to fall in clusters of about 15 and were of different colors, some of brilliant hue, others pale and black, and in a brief time would dissolve into smoke. Others would seem to sparkle brilliantly and disappear.

"This unusual sight continued 15 minutes. At times the cluster of balls was in the shape of pyramids and other shapes. This phenomenon was seen by a number of residents on Roanoke street, in this place, who are worthy of belief. The evening before just before dark the sun presented the appearance of deep red, almost pink."

Source:
"A Polite Editor," 1895. *Olean* [New York] *Democrat* (May 10). Reprinted from the *Richmond Dispatch*.

1904: Fall of Leaves

A CURIOUS PHENOMENON was witnessed at Winchester last Friday [April 22] when a heavy downfall of dry leaves from the sky occurred. . . . The leaves for several hours fell thick and fast, and were principally oak. It is thought that they were taken by whirlwinds into the upper air currents from the mountains west of the town and then floated down over the city."

Source:
"Stop the Ship Quick, We Are Sea Sick," 1904. *Bluefield* [West Virginia] *Daily Telegraph* (April 28).

1905: "An Ordinary Snake Attached to a Strange Looking Bird"

A MOST REMARKABLE and uncommon 'flying snake' that was captured and killed at 'Berry Plain,' the home of John S. Dickinson in this county, a few days ago, has attracted wide attention and excited no little interest. . . . The curious reptile was first noticed flying about in the air with several feet of its horrid snakeship dangling around, presenting the appearance of an ordinary snake attached to a strange looking bird.

"It was finally killed and measured and proved to be five feet long and about one inch in diameter of body. It had perfect wings of good size, and these were covered with feathers. Berry Plain, where the curious thing made its unceremonious advent and met its untimely end, is one of the finest plantations and homesteads in King George and, being situated on the banks of the Rappahannock, it is conjectured that the 'flying snake' may have come from an impenetrable marsh of the river or some neighbor's creek.

"But this theory is not accepted by many, for, as far as can be learned, nothing bearing the least similarity to this serpent or reptile, or whatever it is, was ever before seen or heard of anywhere in this section of the country."

Source:
"Flying Snake," 1905. *Newark* [Ohio] *Advocate* (June 30).

1909: "An Automobile without Wheels"

THE SPECTACLE OF A mysterious airship, occupied by a man and a woman, passing over Sewell's point, was witnessed by more than fifty persons in that vicinity. Those who saw the strange air craft stated that it resembled an automobile without wheels."

Source:
"Saw Mysterious Airship," 1909. *Edwardsville* [Illinois] *Intelligencer* (May 3).

Washington

❝1868: Night Noise

DURING THE PAST THREE or four months there have been many mysterious transactions at the farm of Hon. J. G. Sparks, about two miles northeast of Olympia, occupied at present by Mr. J. G. Grimm and family, and to ascertain the facts in the case, we sent an unprejudiced reporter, who furnishes the following account:

"'About the above time, various members of the family heard what they supposed to be some person enter the front floor, go up stairs and cross the chamber floor. This occurred several times, both during the day and night. Not being able to find any one, they came to the conclusion that the noise was caused by the settling of the house.

"'About two months since[,] a Catholic priest, a friend of the Sparks and Grimm families, called. [All remained] quiet until about two o'clock when the reverend gentleman found himself lying in the middle of the room, with most of the bedding on top of him. He came to the conclusion (not having heard of the former demonstrations, nor either being spiritualist) that he had fallen out of bed. After recovering from the fall, which stunned him so severely that he was unable to get up for several minutes, he gathered his blankets and went back to bed. On awaking in the morning, he was somewhat surprised to find the very heavy feather bed he lay on when first retiring, and the pillows[,] lying on the opposite side of the room. After being thrown from the bed, he made no light, and being quite badly hurt, failed to discover the absence of the bed until daylight.

"'Shortly after this, the family went to the spring, only a short distance, to wash, leaving all the doors ajar. On returning, they were unable to open any of them, and Mr. Grimm, using his utmost strength to do so, came to the conclusion that some friend had come during his absence and fastened them out for a joke. He slipped around to the window, which he raised, and entered. On approaching the doors, he found none of them fast, and each one open as usual.

"'For the next two or three weeks, at different times, there would be a loud noise in an adjoining room, as though a window sash had fallen on the floor, and during the evening, while the family were sitting around the fire, the books would be thrown from the book-case in the most violent manner.

"'Some six weeks since, while the family were out in the yard, about one hundred yards from the house, they heard the most terrific crash, as though the cupboard had fallen. The children started to run to see what was the matter, but on a close inspection not a thing was found misplaced.

"'About one month since, two of the children were occupying the same bed from which the priest was thrown. During the night, the elder, a boy about 12 years old, while laying awake, saw what he thought to be a woman dressed in white. She crossed the floor and descended the stairs four steps to the landing, where she remained a short time and then returned to the foot of his bed. He then called to his brother[,] who is a few years younger, to look and see that woman. The younger boy rose up and saw the figure, when he became frightened and called to his parents, who occupied a room below, and then covered up his head. The older boy, however, continued watching, and before their father came up, the figure disappeared behind the chimney, which stands in the middle of the room.

"'The children have been brought up at home, and have never heard a 'ghost story' in their lives, and know nothing of fear or a 'booger' in the dark, and have occupied the same bed up to this time. Mr. and Mrs. G. are as reliable citizens as any in the county.'"

Source:
"Strange Phenomenon," 1868. *Morning Oregonian* (November 9).

1871: Rain from the Clear Heavens

LAST NIGHT [MARCH 2], between 7 and 8 o'clock, a young moon shone brightly in the west, while countless stars, glittering like diamonds, adorned the deep blue vault of heaven. Not a cloud was visible; yet it rained, and for nearly an hour, with sufficient force to be plainly heard pattering upon the roof."

Source:
Untitled, 1871. *New York Times* (1871). Reprinted from the *Pacific Tribune,* March 3.

1892: Ghostly Headlight

LOCOMOTIVE ENGINEERS are as a class said to be superstitious, but J. M. Pinckney, an engineer known to almost every Brotherhood man, is an exception to the rule. He has never been able to believe the different stories told of apparitions suddenly appearing on the track, but he had an experience Sunday night [presumably February 29] on the Northern Pacific east bound overland that made his hair stand on end.

"By the courtesy of the engineer, also a Brotherhood man, Mr. Pinckney was riding on the engine. They were recounting experiences, and the fireman, who was a green hand, was getting very nervous as he listened to the tales of wrecks and disasters, the horrors of which were graphically described by the veteran engineers. The night was clear and the rays from the headlight flashed along the track, and although they were interested in spinning yarns, a sharp lookout was kept, for they were rapidly nearing Eagle gorge, in the Cascades, the scene of so many disasters and the place which is said to be the most dangerous on the 2,500 miles of road. The engineer was relating a story and was just coming to the climax, when he suddenly grasped the throttle and in a moment had 'thrown her over'—that is, reversed the engine.

"The air brakes were applied and the train brought to a standstill within a few feet of the place where Engineer Cypher met his death. By this time the passengers had become curious as to what was the matter, and all sorts of questions were asked the trainmen. The engineer made an excuse that some of the machinery was loose, and in a few moments the train was speeding on to her destination.

"'What made you stop back there?' asked Pinckney. 'I heard your excuse, but I have run too long on the road not to know that your excuse is not the truth.'

"His question was answered by the engineer pointing ahead and saying excitedly, 'There! Look there! Don't you see it?'

"'Looking out of the cab window,' said Mr. Pinckney, 'I saw about 800 yards ahead of us the headlight of a locomotive.

"'["]Stop the train, man!["]' I cried, reaching for the lever.

"'["]Oh, it's nothing. It's what I saw back at the gorge. It's Tom Cypher's engine, No. 33. There's no danger of a collision. The man who is running that ahead of us can run it faster backward than I can this one forward. Have I seen it before? Yes, twenty times. Every engineer on the road knows that engine, and he's always watching for it when he gets to the gorge.["]

"'[']The engine ahead of us was running silently, but the smoke was puffing from the stack and the headlight threw out rays of red, green and white light. It kept a short distance ahead of us for several miles and then for a moment we saw a figure on the pilot. Then the engine rounded a curve and we did not see it again. We ran by a little station, and at the next, when the operator warned us to keep well back from a wild engine that was ahead, the engineer said nothing. He was not afraid of a collision. Just to satisfy my own mind on the matter I sent a telegram to the engine wiper at Sprague asking him if No. 33 was in. I received a reply stating that No. 33 had not come in and that her coal was exhausted and boxes burned out. I suppose you'll be inclined to laugh at the story, but just ask any of the boys, although many of them won't talk about it. I would not myself if I was running on the road. It's unlucky to do so.' . . .

"It is commonly believed by Northern Pacific engineers that Thomas Cypher's spirit still hovers near Eagle gorge."

Source:

"Ghost of the Gorge," 1892. *Ogden* [Utah] *Standard* (March 6).

1893: "Marine Monster"

A SEA SERPENT IS reported at Gamble Bay, Wash. The foreman of the mills at Fort Gamble says 'the marine monster was fully 100 feet long and its head projected into the air fully twenty feet and then sank in the water.' Several other[s] claim to have seen the same thing."

Source:

Untitled, 1893. *Woodland* [California] *Daily Democrat* (January 14).

1893: "Hateful Object"

PUGET SOUND, WITH its many traditions, has never until yesterday [June 16] given to the world an authentic story of a sea serpent. Not long ago a steamer coming from Tacoma to this city ran into a whale or at least what was thought to be a marine monster bearing that name, and was thrown out of its course several points. On various occasions huge mud sharks

have been captured, and from time to time fishes and reptiles of such strange form and color have hung out in front of the different markets in this city that many have wondered what the waters of Puget Sound would next bring forth to astonish the curious. . . .

"The experience of Mate F. A. Woodman, of the Steamer Edith, yesterday morning near Marrowstone Point, with what he believes to have been a sea serpent, will long be discussed by seafaring men, and unless by accident or otherwise the monster should be captured, it is probable that the waters in that part of the sound will be looked upon with misgiving and fear. . . . In the waters of every ocean [Woodman] has plied his calling, but the sight he saw yesterday morning was the climax of his experiences, and as he told of the long-bodied, big-headed, glistening monster that was sporting in the water near the steamer, he shook his head and said he did not want to go through another such experience.

"'We left Port Townsend in the morning and were passing Marrowstone Point, headed for Port Ludlow, about 9 o'clock,' said he, launching into his wonderful tale, 'when the wind commenced to blow a stiff breeze and the waves ran pretty high. I was on deck near the pilot house, when my attention was attracted by an object in the water less than 200 feet away on the starboard side. At first glance I thought it was a floating log, but a moment later that idea was dispelled in an astonishing manner and my attention was riveted to what I saw before me.

"'I am positive that I had all my wits about me and am sure of what passed before my eyes. A great, big, monstrous head, poised on a long, slim neck came out of the water fully five feet, and as the hateful object shook the water from itself it seemed as if its body fairly shone with reflected light. The head staid [*sic*] in mid-air but a moment, then sank beneath the waves and disappeared. I could not get a good look at the body, but from appearances judged it to be about 50 feet long and slender throughout.

"'The monster crossed our bow to the starboard, and although I watched closely I did not get another look at it until some time later, when it appeared about a quarter of a mile away, acting in the same manner as on its first appearance. I never saw anything like it before, and I don't care to again. Whatever it was, I will let other people decide, but I will take my oath that I saw what I have described.'"

Source:

"A Sea Snake," 1893. Associated Press. *Arizona Republican* (June 18).

1893: It Came from Beneath the Sea

WE LEFT TACOMA ABOUT 4:30 p.m., Saturday, July 1st, and as the wind was from the southeast we shaped our course for Point Defiance, intending to anchor off that point and try our luck with rod and line. We cast anchor about six o'clock, the wind having died out, and had fair success fishing. The wind coming up and down again pretty strong Mr. McDonald suggested getting under way for Black Fish Bay, Henderson Island, as he knew of a fine trout stream running into the bay and also an excellent camping place near the fishing ground.

"So about eight o'clock we weighed anchor and shaped our course for Black Fish Bay, which place we reached about 9:30. We landed and made everything snug about the boat and made a nice camp on shore, and as it was by this time eleven o'clock we all turned in to get a little sleep as it was agreed upon that at the first streak of daylight we should all get up. About one hundred yards from our camp was the camp of a surveying party, but as it was so late we decided that we would not disturb them but that we would call upon them the following morning and would probably get some valuable pointers as to the best places to fish and hunt on the island. After a few jokes had been cracked, the boys laid [*sic*] down, and in a short time everything about camp became as still as death.

"It was, I guess, about midnight before I fell asleep, but exactly how long I slept I cannot say, for when I woke it was with such startling suddenness that it never entered my mind to look at my watch, and when after a while I did look at my watch, as well as every watch belonging to the party, it was stopped. I am afraid that you will fail to comprehend how suddenly that camp was awake.

"Since the creation of the world I doubt if sounds and sights more horrible were ever seen or heard by mortal man. I was in the midst of a pleasant dream, when in an instant a most horrible noise rang out in the clear morning air, and instantly the whole air was filled with a strong current of electricity that caused every nerve in the body to sting with pain, and a light as bright as that created by the concentration of many arc lights kept constantly flashing. At first I thought it was a thunderstorm, but as no rain accompanied it, and as both light and sound came from off the bay, I turned my head in that direction, and if it is possible for fright to turn one's hair white, then mine ought to be snow white, for right before my eyes was a most horrible-looking monster.

"By this time every man in our camp, as well as the men from the camp of the surveyors, was gathered on the bank of the stream; and as soon as we could gather our wits together we began to question if what we were looking at was not the creation of the mind. But we were soon disburdened of this idea, for the monster slowly drew in toward the shore, and as it approached, from its head poured out a stream of water that looked like blue fire. All the while the air seemed to be filled with electricity, and the sensation experienced was as if each man had on a suit of clothes formed of the fine points of needles.

"One of the men from the surveyors' camp incautiously took a few steps in the direction of the water that reached the man, and he instantly fell to the ground and lay as though dead.

"Mr. McDonald attempted to reach the man's body to pull it back into a place of safety, but he was struck with some of the water that the monster was throwing and fell senseless to the earth. By this time every man in both parties was panic-stricken, and we rushed to the woods for a place of safety, leaving the fallen men lying on the beach.

The bands nearest the head seemed to have the stronger electric force, and it was from the first six bands that the most brilliant lights were emitted. Near the center of its head were two large horn-like substances, though they could not have been horns for it was through them that the electrically charged water was thrown.

"As we reached the woods the 'demon of the deep' sent out flashes of light that illuminated the surrounding country for miles, and his roar—which sounded like the roar of thunder—became terrific. When we reached the woods we looked around and saw the monster making off in the direction of the sound, and in an instant it disappeared beneath the waters of the bay, but for some time we were able to trace its course by a bright luminous light that was on the surface of the water. As the fish disappeared, total darkness surrounded us, and it took us some time to find our way back to the beach where our comrades lay. We were unable to tell the time, as the powerful electric force had stopped our watches. We eventually found McDonald and the other man and were greatly relieved to find out that they were alive, though unconscious. So we sat down to await the coming of daylight. It came,

I should judge, in about half an hour, and by this time, by constant work on the two men, both were able to stand.

"This monster fish, or whatever you may call it, was fully 150 feet long, and at its thickest part I should judge about 30 feet in circumference. Its shape was somewhat out of the ordinary insofar that the body was neither round nor flat but oval, and from what we could see the upper part of the body was covered with a very coarse hair. The head was shaped very much like the head of a walrus, though, of course, very much larger. Its eyes, of which it apparently had six, were as large around as a dinner plate and were exceedingly dull, and it was about the only spot on the monster that at one time or another was not illuminated. At intervals of about every eight feet from its head to its tail a substance that had the appearance of a copper band encircled its body, and it was from these many bands that the powerful electric current appeared to come. The bands nearest the head seemed to have the stronger electric force, and it was from the first six bands that the most brilliant lights were emitted. Near the center of its head were two large horn-like substances, though they could not have been horns for it was through them that the electrically charged water was thrown.

"Its tail from what I could see of it was shaped like a propeller and seemed to revolve, and it may be possible that the strange monster pushes himself through the water by means of this propeller tail. At will this strange monstrosity seemed to be able to emit strong waves of electric current, giving off electro-magnetic forces, which causes any person coming within the radius of this force to receive an electric shock. . . .

"I hardly need to tell you that we were not long in getting under way for Tacoma, and I can assure you that I have no further desire to fish any more in the waters of this bay. There are too many peculiar inhabitants in them."

Source:
"An Electric Monster," 1893. *Tacoma* [Washington] *Daily Ledger* (July 3).

1897: "Countries of Spiritual Beings"

THE FOLLOWING COMMUNICATION was brought in to the Chronicle office this morning by one of the most prominent spiritualists of the city. It is a novel explanation of the air ship problem which will be of interest not only to believers in spiritualism but to many others. It is as follows:

"'Notwithstanding the many frauds and fakes in spiritual investigation in Spokane, many of the most intelligent citizens believe in spiritual phenomena. There are mediums in many families through whom inspirational utterances of a high order are given. One of these, an estimable lady, whose mediumistic gifts are known to but few outside her own family, communicates the

startling intelligence that air ships of spirit creation are now hovering in the earth's atmosphere. Her communication on the subject is as follows:

"'["]The newspaper accounts of airships having been seen in various parts of your earth are not all mythical. Above your atmosphere there are peopled regions, the homes and cities and countries of spiritual beings, whose environing conditions, to their consciousness, are as real and tangible as are natural objects to you, and there are interlinking currents filled with sentient beings throughout the intervening ocean. There are no gulfs or desert wastes separating the different abodes of individualized intelligences, but all parts of the universe are rife with forms of intelligent life. The nearest available approach to you for the denizens of the spirit realms above is in the far north, where the terrestrial electrical currents converge. This accounts for the hope which for centuries has animated your navigators, of reaching the north pole. Their thoughts are drawn into that channel by spiritual attraction.

"'["]Life on the earth or in its immediate atmosphere (filled with coarse material emanations from it) is as unnatural to the spiritual beings who live above it as life in the water would be to your land animals. But it is possible for spirits to devise protecting and counteracting apparatus[es] by which they can safely sink through the intervening air ocean to the surface of the earth, analogously to the marine diver armoring himself, or the use of chemicals to counteract the effects of fire. The great difficulty to be overcome is to prevent sublimation of the atoms when the creation comes within range of human vision, and to maintain the necessary specific gravity to anchor it.

"'["]Spirits have long been experimenting with machinery materialized to traverse the lower depths of the earth's atmosphere, and have at last brought it to such a stage of perfection that it has been made visible to a few of the earth's inhabitants. In the polar regions of your planet it is quite easy to land these semi-spiritual craft on the higher elevations, and when your navigators discover the ocean currents through which they can sail into certain polar harbors, well known to spirit explorers, open communication will be established between the worlds, bringing to the people of the earth benefits and blessings your optimistic poets never conceived of in their wildest flights of imagination.[”]'"

Source:
 "Real Air Ships," 1897. *Spokane* [Washington] *Daily Chronicle* (June 5).

1911: Face at the Window

IF YOU WANT TO FORM the acquaintance of [a] real live, or rather dead, healthy spook, your ambition may be fulfilled by a visit to Porter, a few miles north of Centralia. The specter is, by all accounts, being sighted by scores of passengers on the trains. In addition to the passengers, several trainmen have sighted the Thing, and the sight sends a wobbly shiver down their spines even though they are [at] a safe distance and the sight recedes with all the speed of steam. Among those locally known who are said to have seen and been mystified by the spectacle are two engineers who are well known in Centralia where they make their stoping[*sic*]-off place between duty shifts. The sight has aroused such interest that camera shots are now being taken of the ghost and some interesting pictures will shortly be available. . . .

"Local tradition tells of an eccentric old man who lived the life of a recluse in a lone cabin near Porter many years before the railroad line was laid in these parts. The old man . . . wore a long white beard, and alternated fits of benevolence with fits of malevolence—being loved and feared as the moods changed. With the faith of an optimistic pioneer, the old man prophecied [*sic*] the opening of a mighty inland empire in Washington and deplored his age that made it impossible for him to see railroad trains speeding past his cabin door. The prescience of the recluse came true, for today finds the cabin crumbling with age and trains running past it. It is said that he was wont to declare that his cabin home would be the habitat of his spirit which would stay to see the growth of the country after his ashes were laid in the grave.

"Eye witnesses describe a crouched form, with an aged face resting upon two bony hands. A long white beard frames his face and the eyes are transfixed with earnest gaze at every train that passes. When the train goes, the face disappears from the window. Investigation has shown that the apparition has been seen when nobody was near the house, and the only man that has lived there for years is an old man who wears no beard and who knows nothing of any patriarchal visitant looking out of his window."

Source:
 "Have You Seen This Ghost Yet?," 1911. *Centralia* [Washington] *Chronicle* (April 12).

West Virginia

1863: Phantom Processions

BY THE SUMMER OF 1863, Lewisburg, in Greenbrier County, was officially no longer part of Virginia but was in the new state of West Virginia (admitted into the union on June 20). Most West Virginians had no sympathy for the Confederacy, and moreover, they had long chafed under east Virginia's domination of their affairs. Still, Lewisburg, no more than a few miles from the Virginia border, considered itself a Virginia town and backed the effort to break away from the union; thus, the following curious event was covered in two Confederate newspapers, the *Richmond Whig* and the *Staunton Spectator*.

At 3 p.m. on September 1, six persons—four adults and two teenagers—in a rural area a few miles west of Lewisburg were standing or sitting near Mrs. Pearcy's house when they saw a strange procession moving through the cloudless sky on a windless afternoon. The shapes came into view as they passed over and through the tops of trees covering hills to the south. They looked like cotton rolls or bundles with greenish fringes, and they were about the size of doors. An "immense number" of these peculiar phenomena sailed by until they were lost to view an hour later. The strange show was not over, however.

In their wake many thousands of figures appeared in an open field, several hundred yards in circumference, in the valley under the hills. They were marching in the same direction—north to northwest—that the rolls had taken. In lines thirty to forty long, separated from the other lines by a few feet, they were moving in strict military discipline at double-quick time,

ascending steep hills and stooping over like ordinary men as they did so. Yet they were not soldiers. They were dressed in loose white shirts and white pants, and they wore hats. They carried no guns, swords, or packs. They were of various sizes, from "very large" to "quite small." After an hour or so they had passed by and vanished over a mountain.

An unidentified Confederate officer conducted inquiries into the incident. He told the *Whig*, "I put myself to some trouble to ascertain the facts, and questioned the witnesses separately. They are above suspicion. I have given all the material facts, except that the so called men were marching north or northwest, right thro' the mountains. They were . . . as much like men as if they had been real flesh and blood."

A correspondent to the *Spectator* reported, in the September 22 issue, that on September 14, exactly two weeks after the first incident, "the same scene almost identical" was played out 4 miles east of Mrs. Pearcy's farm. The witnesses this time were eight to ten soldiers on picket duty as well as many local civilians. He wrote that it took about an hour for the men to pass by.

The *Spectator* subsequently (September 29) published a long letter from an anonymous theorist suggesting that a mirage had been responsible, though it is unclear exactly what earthly event was being reflected in the ghostly spectacle. The fire-breathing, pro-Confederate newspaper editor, however, preferred another explanation:

"The white bundles represent the cotton bales which the weak-kneed secessionists of Mississippi are now sending North, and the fringe of green—the emblem of that color being 'forsaken'—indicates how green these planters are for having forsaken their country, and the figures of men marching North at a rapid pace in the scanty garb described, represent that these traitorous cotton sellers should be reduced in their dress to a cotton shirt fastened on their 'tight hides' by a plaster of tar, and be made to march North at a 'double quick.' The absence of arms and equipments represent [*sic*] that they belong to that numerous class of 'men of war' who, before the war, were so willing to spill their 'last drop of blood,' but since the war have shown an unconquerable dread of spilling 'the first drop.'"

Sources:

"For the *Spectator*," 1863. *Staunton Spectator* (September 29).

Moore, Frank, ed., 1882. *Anecdotes, Poetry, and Incidents of the War: North and South, 1860–1865*. New York: Arundel.

"Remarkable Phenomenon—Interpretation Suggested," 1863. *Staunton Spectator* (September 22).

"A Strange Phenomenon," 1863. *Staunton Spectator* (September 22).

He called to the men, about 100 yards off, and told them to look up, and tell him what they saw. They declared they saw a white horse swimming in the sky, and were badly frightened.

1878: "White Horse Swimming"

EARLY IN THE SUMMER of 1878, the *Cincinnati Commercial* received a telegram from a Parkersburg correspondent who related a strange story of an "optical illusion or mirage." It was witnessed, the correspondent claimed, by farmers a few days earlier:

"The facts are these: A gentleman, while plowing in the field with several others, about 7 p.m., happened to glance toward the sky, which was cloudless, and saw, apparently about half a mile off in a westerly direction, an opaque substance, resembling a white horse, with head, neck, limbs and tail clearly defined, swimming, moving its head from side to side, always ascending at an angle of about 45 degrees. He rubbed his eyes to convince himself that he was not dreaming, and looked again, but there it still was, still apparently swimming and ascending in the ether. He called to the men, about 100 yards off, and told them to look up, and tell him what they saw. They declared they saw a white horse swimming in the sky, and were badly frightened. Our informant, neither superstitious [n]or nervous, sat down and watched the phantasm . . . until it disappeared in space, always going in the same direction, and moving in the same manner. No one can account for the mirage, or illusion, except upon the uneven state of the atmosphere. Illusions of different appearance have been seen at different times, in the same vicinity, frightening the superstitious, and laughed at by the skeptical."

Source:

"What a West Virginia Farmer Saw," 1878. *New York Times* (July 8). Reprinted from the *Cincinnati Commercial.*

1884: A Murdered Man Tells His Story

THERE IS GREAT EXCITEMENT among the mountaineers living in the eastern part of Wayne County over the alleged appearance of the ghost of Harvey Fairman, a farmer, who mysteriously disappeared in 1879. Fairman was supposed to have been murdered, but as no trace of crime was ever discovered, the matter was gradually forgotten.

"Several days ago Alexander Moore, a well-known and intelligent young man, was out hunting, and had a curious experience. When at the head of a dark ravine, he says, he was suddenly confronted by what seemed to be a goose or other large white bird, which attracted his attention and then ran into some bushes. Following it, he was startled by seeing before him the ghost of Harvey Fairman, his clothes torn and muddy, and his throat cut from ear to ear. In the conventional hollow tones the spirit told Moore he had been murdered and the body hid beneath the floor of his house for two days, and then brought to the ravine and concealed in a hollow tree. Moore returned home with every appearance of one suffering from extreme terror, and says he will make affidavit to the facts if necessary."

Source:

"Face to Face with a Ghost," 1884. *New York Times* (July 28).

1893: Monster on the Ohio

A FEW WEEKS AGO dispatches from below [Parkersburg] stated that a sea serpent was alarming people down the Ohio River. Boating parties here during the past week have been greatly frightened by the appearance of the monster with a head as big as a barrel. The freak has been variously described as 8 to 15 feet long, appearing to be floating on the surface of the river, often near Negle's Island. When approached the monster would dive with great commotion of the waters, making heavy waves that were dangerous to open boats.

"Last night a party of a dozen prominent young people, who are entirely reliable, were out on a large boat when the monster suddenly appeared, crossing the river in front of them. They were panic stricken and made frantic efforts to get away from the locality without making an investigation. Parties living along the river state that the monster can be seen daily, showing most conspicuously when heading up the river, and apparently changing the abode frequently. All descriptions of the strange monster agree, and the truth of the story is not doubted in this locality."

Source:

Untitled, 1893. *Pittsburgh Post* (July 8).

1895: Peculiar Light

THE FEW PEOPLE WHO reside at Middleport Junction [Ohio] complain of a strange occurrence in that region which savors of the supernatural. On dark nights a light can be seen to appear suddenly at the edge of the Ohio river on the West Virginia shore, pause there for a short time and then begin slowly to ascend the hills on that side of the river until it reaches the summit, when it suddenly disappears. In the course of half an hour it will appear at the point where first seen and go over the same ground as the first. On a dark night it may be seen to do this for nearly a dozen times. The light can only be seen from the Ohio side."

Source:

"Ghost Story," 1895. *Marion* [Ohio] *Daily Star* (August 4).

1896: Mystery Monster Bird

HUNTING DEER NEAR THE mouth of Vanatters Creek, Elias Midkoff and W. W. Adkins saw a very big bird circling high in the sky. As they watched, it rapidly descended and landed in the water. Adkins fired on it, crippling its wing, and waded into the creek to try to capture it alive. It fought him so furiously, however, that he was forced to kill it. It took five bullets to end its life.

A Baltimore newspaper, based on a description provided by Midkoff, stated, "The bird is 7 feet 4 inches from tip to tip, 4 feet from tip of bill to tip of tail, flat bill 4 inches long and 8 inches wide, somewhat similar to that of a duck; web feet, neck 19 inches long, and about 1 1/2

inches through below the feathers; plumage dark brown, relieved on the wings and breast by light-blue shading."

Midkoff had traveled to Charleston to urge the State Historical and Antiquarian Society to dispatch a taxidermist to Hamlin in order to preserve the body of the mysterious bird. Apparently nothing came of the request.

Source:

"Gigantic Feathered Creature," 1896. *Fort Wayne* [Indiana] *News* (February 1). Reprinted from the *Baltimore American*.

Early 1900s: Birdman

THOUGH MOTHMAN REPORTS since the 1960s have been rare to nonexistent, the creature, or something like it, may have visited West Virginia—and Point Pleasant—earlier in the twentieth century. According to folklorist James Gay Jones, who provides no further details, "In the early 1900's at Pt. Pleasant, a large bird with the head of a man and wingspan of at least twelve feet was seen. It appeared just prior to or immediately after the occurrence of a tragic event" (Jones 1979). It may be worth noting that late on the afternoon of December 15, 1967, the Silver Bridge spanning the Ohio River and connecting Point Pleasant with eastern Ohio collapsed, killing forty-six. Though the disaster is unrelated to the Mothman outbreak, even today some people in the Point Pleasant area associate the events, seeing the creature as a harbinger of tragedy. The bridge collapse is the climactic event of a movie starring Richard Gere, *The Mothman Prophecies,* released in 2002.

Jones writes that other sightings occurred around the time in various rural counties, including Mason, where Point Pleasant is located. "By World War I," he continues, "[the] birdman was observed flying over Looneyville, up Johnson Creek, down Gabe in Roane County thence down Elk Valley into the Kanawha. Its monstrous size and dark reddish feathers which glistened in the sunlight cast fear in all who saw it. Parents kept children indoors after sightings.

After World War II people said they were chased by a huge bird while traveling on the highways of Mason, Jackson, and Wood counties near the Ohio River."

Sources:

Coleman, Loren, 2002. *Mothman and Other Curious Encounters.* New York: Paraview.

Jones, James Gay, 1979. *Haunted Valley and Other Folk Tales of Appalachia.* Parsons, WV: McClain.

Keel, John A., 1970. *Strange Creatures from Time and Space.* Greenwich, CT: Fawcett Gold Medal.

———, 1975. *The Mothman Prophecies.* New York: Saturday Review Press/E. P. Dutton.

1908: Unexplained Brilliant Light

A STRANGE LIGHT OF great magnitude flashed over Fairmont on the eastern portion of the town at 9 o'clock last night [June 3] and remained with uniform brilliancy for more than two minutes. Hundreds of persons who were on the streets at the time, believing that a gigantic explosion had occurred, rushed to points at vantage to view the sight. The light, however, hovered far above the town. It was one of the strangest phenomena ever witnessed in this section and its nature is unexplained."

Source:

"Fairmont Sees Strange Light," 1908. *Indiana* [Pennsylvania] *Evening Gazette* (June 4).

Wisconsin

1835 and After: Underwater Pyramids

Rock Lake, in Lake Mills, between Madison and Milwaukee and not far north of the Illinois border, covers an archaeological mystery that has been known for a long time but only sporadically investigated.

The first white settlers came to the area in 1835. From local Winnebago Indians they learned that stone structures lay under Rock Lake's waters. During periods of drought, when the water receded, individuals boating on the lake had observed submerged rectangular artifacts of impressive size. Over the next decades a number of attempts were made to document the presence of the enigmatic artifacts, at least one of them described as a large pyramid. Though relatively small— less than 3 miles wide, and only 90 feet deep at its deepest point—and regularly fed by clear spring water, the lake has a bottom consisting of steep hills and trenches that can easily make even good-sized objects hard to detect. In addition, algae and silt darken the water and reduce visibility sometimes to as little as a yard.

A century after the white settlement, a series of dry summers reduced the water level sufficiently so that again, boaters could relatively easily glimpse a variety of geometrical forms, including triangles, cones, pyramids, and walls. A man named Max Knoll had the greatest number of sightings, made possible by his invention of a particular sort of diving gear that allowed him to stay under water for extended periods. Even so, many others, including professional archaeologists, scoffed.

There is, however, nothing inherently implausible about the notion that unusual structures could have existed there before the waters rose and submerged them. The immediate area houses somewhat comparable archaeological phenomena, including the remains of three flat-topped pyramid-shaped mounds. Three miles east of Rock Lake, Aztalan, a public park, displays two reconstructed earthen pyramids circled by a stockade, the remains of an old city apparently destroyed in the early fourteenth century by local Indians. An amateur archaeologist and Milwaukee judge, Nathaniel F. Hyer, dubbed the city "Aztalan" after a place named in an Aztec legend. Hyer thought the city looked like an Aztec ruin. Thus began a tradition, believed in popular legend though not in academic circles, that Aztecs fleeing the destruction of their empire in Mexico were the creators of the city. Though no direct evidence attests to that, archaeologists have determined that Aztalan once consisted of 500 residents and covered 21 acres. These archaeologists contend that the residents were related to the people who settled Cahokia, along the Mississippi River in southwestern Illinois.

One theory holds that the people of Aztalan were more advanced than their indigenous neighbors. Their superior weaponry enabled them to subdue the locals and to collect tribute from them. They may have come to the site as early as A.D. 1100. Beyond that, little is known about who they were.

In 1987, writer Francis Joseph began diving in the lake hoping to see the rumored artifacts. His efforts went unrewarded until the following year. Before then, however, Joseph flew to Loch Ness, Scotland, not to see its fabled monster but to find out what kind of equipment would best work in murky waters. Armed with this new knowledge and equipment, Joseph returned to Rock Lake in the company of like-minded seekers, including surveyors and geologists, on June 4, 1988. The sonar boat *Sea Search* detected a huge structure buried under 60 feet of water along the lake's northeastern shore. When he dived to examine it, Joseph observed a "gargantuan rectangle, its sides [running] in straight, parallel lines, the walls sloping steeply and uniformly inward. The apex rose almost to a point, but was flattened at the very top, running the full length of the structure" (Joseph 1989). Eighteen feet high and approximately 35 feet wide, the structure was an "ordered collection of round, black stones, selected for size, and piled up to form not so much a true pyramid, as a kind of elongated tent-like configuration" (ibid.). Joseph also saw a barbell-shaped stone, the remains of a plasterlike substance, and two L-shaped objects apparently made of copper.

In 1991 and 1992, Rock Lake's waters were unusually clear, allowing photographs of two structures to be taken. One of the structures was a circle consisting of stacked, fitted rocks. According to one school of speculation, the structures are not hundreds but thousands of years old, perhaps related to a prehistoric copper-trade culture circa 3000 B.C. Until more concentrated investigation is conducted, however, answers about who built the structures and why remain elusive.

Sources:

Corliss, William R., 1983. "The Rock Lake Pyramids." *Science Frontiers* 30: November/December, http://www.science-frontiers.com/sf030/sf030p01.htm.

Joseph, Francis, 1989. "Found—The Lost Pyramids of Rock Lake." *Fate* 42, 10 (October): 88–90, 92–95.

Smith, Susan Lampert, 1983. "Lake Mills' Lost Pyramids." *Wisconsin State Journal* (Madison, June 26).

1853: Unknown Water Animal

A SINGULAR REPTILE or fish was caught here a few days since, and is now in a glass jar before us. It has a skin like a catfish, a head and tail like an eel. The gills are on the outside of the neck, and it has four legs like a lizard, terminating in a miniature human hand. It is about fifteen inches long, and was taken with a hook. No one here has ever seen a creature like it, though Thompson in his book of Vermont, describes a similar one, caught at Colchester, near Burlington."

Source:

"A Strange Fish," 1853. *Alton* [Illinois] *Weekly Courier* (March 18). Reprinted from the *Janesville* [Wisconsin] *Free Press.*

1860: "Wonderful Freak of Nature"

WE LEARN FROM MR. Joseph S. Corey, who resides at East Lake, Polk county, Wisconsin, the following facts, which we give our readers, and, strange as they may appear, we doubt not their truth in any particular as they are well authenticated by many witnesses, upon whose veracity, as upon Mr. Corey's[,] we place implicit confidence.

"On Tuesday, June 26, while Mr. Corey and his two sons were at work in a field near his house, their attention was arrested by smoke, which appeared to rise from his stable. They hurried to the barn as quickly as possible, and discovered a pile of straw nearby on fire. This was immediately extinguished, and as they were returning to the field, the stable caught on fire in different places, which, by considerable exertion, was put out. Before, however, they had left the premises, another fire was discovered underneath the Granary, in a pile of boards. The bottom board was burned nearly through, but the others were not even scorched. After this was put out, Mr. Corey sent one of the boys into the house to ascertain whether all was safe there. He immediately came out and told his father that the house was on fire. Mr. Corey immediately ran up stairs, where he found some clothes that his wife had laid away the day before were burning. They were thrown out of the window, and from that time until late at midnight the fire broke out all over the house. First a paper would catch, then a mosquito bar, then a straw bed, &c., and it was only by the utmost exertions of Mr. Corey, aided by two gentlemen, Mr. Hale and Mr. Treadwell, that the building was saved. The fire continued at intervals until Sunday . . . and attracted many visitors. We shall not attempt to give any cause for this wonderful freak of nature."

Source:

"A New Mystery—A House That Will Burn Anyhow," 1860. *Western Journal of Commerce* (August 9). Reprinted from the *Taylor* [Wisconsin] *Reporter*.

1867: Man Beast

FARMERS IN THE OAK CREEK area grew alarmed and angry as some unknown animal broke into chicken houses and sometimes attacked animals as large as lambs. The victims were eaten on the spot, indicating that their killer was not a human being.

Determined to find out what was responsible and to see that it was stopped, one farmer hid near his hen roost under cover of darkness. One night around 11 p.m., an animal came out of the shadows and approached the roost with stealthy tread. Oddly, it alternately walked on two feet and on four. As it came into clearer view, the witness was astonished to see that its face and hands were humanlike but its body was hairy like a beast's. The farmer fired, and a piercing shriek—unnervingly like a man's—came from the creature. It fled into the woods and was lost to view.

Another witness observed the creature in the woods the next day. The article concluded, "Though wounded, it made its escape, and though subsequently seen again had not been captured at last accounts."

Source:

Untitled, 1867. *Morning Oregonian* [Portland] (October 9).

1881: "A Cloud of Smoke and Flame"

A PECULIAR CLOUD PHENOMENON was observed at Milwaukee about 3:30 o'clock last Wednesday morning [June 29]. . . . The atmosphere here had been extremely sultry. Suddenly a huge cloud of red and white light appeared in the north and traveled with inconceivable velocity toward the south, passing over the city. Following the passage of the cloud came a cold wave which lowered the temperature about 20 degrees almost instantly. The cloud looked as though it was composed of smoke and flame, and presented a very singular appearance. Its outlines were sharply defined, although it was very dark at the time."

Source:

Untitled, 1881. *Stevens Point* [Wisconsin] *Daily Journal* (July 2).

As it came into clearer view, the witness was astonished to see that its face and hands were humanlike but its body was hairy like a beast's.

1882: "A Big Thing Out There"

THE *LAKE MILLS SPIKE* observed on August 31, 1882, "Again has the lake monster been seen." The full story was recounted by Charles E. Brown in a monograph on the subject of sea serpents published in 1942.

The previous Monday evening (the 28th), Ed McKenzie and D. W. Seybert were on nearby Rock Lake participating in a rowing race. At one point they saw what they first took to be a floating log in front of them. One of the men pointed it out to the other, to be sure that they could take evasive action and not plow into it. He had no more spoken than the "log" reared 3 feet out of the water, revealing a head and jaws open a foot wide. Seybert shouted to McKenzie to hit the creature—as long as the boat itself, the color of a pickerel—with his oar, but the latter was so overcome with terror that he could only scream to watchers on shore to bring a gun.

From their vantage point the witnesses on land could see something thrashing about in the water. One thought it was a man, another had the impression that it was a dog. Another, John Lund, jumped into a boat and was at the scene of the sighting in short order. He detected a "sickening odor" in the air. McKenzie was white-faced, his teeth chattering.

If this story is in any way true, it may have arisen from the sighting of a large fish, most likely a pickerel. Be that as it may, the *Spike* had this to say (as quoted in Brown):

"This serpent has quite a history. Dr. R. Hassam saw it about fifteen years ago, in the rushes. At first he thought it was the limb of a tree. On close observation he saw it was a thing of life and struck it with a spear, but would no more hold it than an ox. Mr. Harbeck, now of Waterloo, who formerly resided across the lake, frequently saw the saurian while rowing back and forth. On one occasion it hissed at him just as he was entering the rushes. Fred Seaver, Esquire, has encountered his snakeship twice. On one occasion the monster seized his trolling hook and pulled his boat along over half a mile at a rushing speed before he let go."

The paper also reported that the previous week, the creature had "seized John Lund's troller." Lund attempted to pull it in, but "the line, before breaking, cut entirely through the skin of his finger."

Source:

Brown, Charles E., 1942. *Sea Serpents: Wisconsin Occurrences of These Weird Water Monsters in the Four Lakes, Rock, Red Cedar, Koshkonong, Geneva, Elkhart, Michigan and Other Lakes.* Madison: Wisconsin Folklore Society.

1883 and After: Lake Serpents

ACCORDING TO A BODY of lore and a handful of yellowing newspaper clippings, the Four Lakes surrounding Madison, the state capital and university town, once housed monsters. In 1942, Charles E. Brown of the Wisconsin Folklore Society collected reports and rumors and published them in a wryly composed monograph.

Brown remarked, without providing any source for the allegation, that one day in 1917, on the north shore of Picnic Point on Lake Mendota, a University of Wisconsin student found a thick, tough, very large something that appeared to be a fish scale. Unable to imagine how a fish could be so large, he showed it to a professor. "Being from New England and acquainted with the species," in Brown's words, the professor immediately identified it as a piece of a sea serpent. He did not explain what a reptile was doing with scales.

A few months later, however, a man fishing for perch off the point was unnerved when a huge snakelike head rose from the water 100 feet away from him. After watching it in disbelief and terror for a few minutes, he overcame his paralysis and fled the scene, abandoning his fishing pole. Nobody believed his story.

Not long afterward, two university students were sunbathing on the same beach. The couple were lying on their stomachs, looking landward, when something began tickling the soles of the young woman's feet. At first she thought her companion was doing it, but when she looked, he was resting with his eyes closed. So she turned around to see the head and neck of an immense serpent, its tongue extended. It was the serpent that had been tickling her feet, or so the story goes.

If this anecdote is surely no more than a joke, other witnesses, both students and regular Madison citizens, were serious in insisting that they had seen something out of the ordinary in the lake. Several fishermen swore that they had encountered the creature a few years earlier. Theories about the true nature of "Bozho," as the creature was nicknamed, abounded. One popular speculation had it that the animal was an especially large pickerel or maybe a gar fish. According to Brown, "Bozho was, on the whole, a rather good-natured animal, playing such pranks as overturning a few canoes with his body or tail, giving chase to sailboats and frightening bathers by appearing near beaches." The sightings subsided in due course, leading doubters to declare that the hysteria was over and proponents of Bozho's existence to theorize that it had left via a river linked to the lake.

The first known printed mention of this alleged creature was in the June 28, 1883, issue of Madison's *Wisconsin State Journal*. Allegedly, around 11 the previous morning, Billy Dunn, "perhaps the most famous of Madison's fishermen," was fishing in a boat anchored near Mendota's Livesey's Bluff when a black object began moving toward him. In no time he recognized it as a big, light-green, white-spotted snake, its head raised several feet above the water, its forked tongue darting. When the serpent attacked, Dunn clobbered it with an oar. Though stunned, the reptile wrapped the oar in its coils. Then it bit through the oar with its fangs. As it struggled to free it from its teeth, Dunn grabbed a hatchet and rained blows down on its head until the creature sank beneath the water, yielding the oar, which the fisherman reclaimed and soon sold to a traveling salesman from Chicago.

On June 12, 1897, the *State Journal* noted that a "sea serpent" had appeared in Monona, another lake in the chain, showing up "about two months earlier than usual in this season."

Residents of the eastern part of the city said they had seen a 20-foot-long creature, looking like an overturned boat but twice as long, swimming eastward along the water's surface. A man standing on the shore fired two shots at it. Their apparent effect was to cause the creature to turn around, then disappear under the water. "It is probably the same animal which is credited with having swallowed a dog which was swimming in the lake a few days ago," the *Journal* stated. It added that in recent years a similar animal had been observed from the east side of the lake by a number of individuals.

Another immense creature was reported to live in Lake Waubesa, a few miles south of Madison. Among the first to see it was an Illinois fisherman sitting in a boat one hot summer afternoon. When he spotted a disturbance in the water a few hundred feet away, he was unable to figure out what was causing it until a big body, followed by an equally large serpent head, rose up to the surface. It was 60 to 70 feet long, he estimated, and a dark green color. He escaped the scene, to tell his fantastic story to a mostly skeptical audience. Some were tactless enough to remark on the open bottle he had in the boat with him.

Soon, though, a couple who lived in a cabin by the lake claimed to have seen a gigantic serpent. They were swimming when it appeared uncomfortably close to them, moving in their direction. They moved in the opposite direction to the shore and dashed to their house.

Brown wrote that the sightings, or in any case rumors of sightings, continued into the following summer. The monster, if it existed in the first place, has not been reported since.

Sources:

Brown, Charles E., 1942. *Sea Serpents: Wisconsin Occurrences of These Weird Water Monsters in the Four Lakes, Rock, Red Cedar, Koshkonong, Geneva, Elkhart, Michigan and Other Lakes.* Madison: Wisconsin Folklore Society.

Mangiacopra, Gary S., 1979. "Water Monsters of the Midwestern Lakes." *Pursuit* 12, 2 (Spring): 50–56.

1886: Dark Day

AT THREE O'CLOCK ON an otherwise normal cloudy afternoon in La Crosse, on March 19, 1886, darkness fell abruptly. Within five minutes blackness as deep as midnight blanketed the city, generating mass confusion and panic. The air was perfectly still, and nothing in the sky—such as drifting smoke from a large fire—seemed to account for the sudden diminishing of all light.

After eight to ten minutes the darkness passed as if moving from west to east. (Later it was learned that towns to the west had experienced the same phenomenon earlier in the afternoon.) "Nothing could be seen to indicate any air currents overhead," a correspondent reported to the *Monthly Weather Review.* "It seemed to be a wave of total darkness passing along without wind."

Source:

"Atmospheric Phenomenon," 1886. *Monthly Weather Review* 14: 79.

1888: The Supernatural on Trial

WILLIAM ROBERTS IS a well-to-do farmer of Princeton, a small village near here [Dartford]. He cultivates something over 100 acres of land and has a wife and four children. Across the road from his farm lives a wrinkled old German woman named Albright. Her home is a little old cottage, and though a resident of the place for many years she is unable to speak or understand a word of English. Since last Christmas Mr. Roberts asserts that his family and his house have been bewitched, and he swears to any number of astonishing occurrences, which cannot be well accounted for by ordinary means. His entire family corroborates him in a series of most surprising statements.

"Gradually the man came to a belief that the witch who was exerting such occult influences was his little old German neighbor. He gave her on two separate occasions due warning that she must cease casting her evil eye over him. The manifestations did not stop, and so the other day he took his gun and started out to kill her.

He announced his intention to some of his neighbors, and they induced him to postpone the slaughter. He insisted that her death alone could dispel the evil charm, and vowed before an awe-stricken throng that he would become her executioner, saying that even were it his own father who had so tormented his family he would kill him. It became evident that Mr. Roberts should be placed in custody lest he murder the supposed witch. He was therefore arrested and placed on trial, and thus the witchcraft disclosures came about.

"A book on the Occult lay at Squire McConnell's elbow yesterday, and it was evident that the squire had been preparing himself for the novel trial of [*sic*] consulting the authorities on witchcraft. When called for the defense, Mr. Roberts went to the witness stand. He swore that as long ago as last Christmas things at his home began to act very queerly. The first he noticed was a spot of blood on the sheets of his bed. He slept alone, and it frightened him. The sheets were changed continually, but in a short time the bloody spot would again appear. One morning about 11:30 he lay down on the outside of the bed to await the call for dinner. When he arose there was a large pool of blood beneath him. He was terribly frightened, and took off his coat and shirt to see if there was any blood on his back or garment, but there was none.

"Mrs. Robertson [*sic*] said that many a time she has put the meals on the table and stepped to the front door to call the rest of the family to dinner, and when she turned again she found the table nearly stripped of its contents. The bread and meat had disappeared, and plates, knives, forks and saltcellars had absolutely vanished as if by magic. Sometimes they would be found in an out of the way place under the wagon house, in the corn crib, out in the garden, and often they were never discovered. These queer demonstrations and many others were sworn to [by] Mr. Roberts and his wife in the most honest and solemn manner, and no amount of cross-questioning could divert them from their straightforward story. The wife testified that many times when the dishes disappeared there was positively no one else in the house. Roberts averred that he was kept busy most of the time searching for missing articles which had disappeared almost under his eyes.

"He was using a hammer on one occasion and laid it down at his feet. He turned his eyes away for a moment and then reached down for it, when lo! it was gone. His jackknife, hoe, shovel, and innumerable other things all acted in the same peculiar way.

"Both the children, Anna and John, were placed on the stand, and they told in a frank and artless manner the same and many other equally astonishing occurrences. A ring had suddenly disappeared from Anna's finger. She had seen dishes come sailing out of the cupboard when no one was near, and settle softly down on the floor without breaking. Both she and the boy had seen the clothing thrown off the beds, and coats and dresses pulled from the walls and hurled across the room. A married brother named Edward Albright and his wife, who were here on a few weeks' visit, were called to testify. They both had similar unaccountable tales to relate. On the witness stand the family told how they had summoned the parish priest and asked him to dissolve the charm, and he had confessed to them that it seemed as if supernatural agencies were at work, and he would have to consult the bishop in order to obtain the interposition of the church in the afflicted family's behalf.

"The evidence was concluded on Friday [July 20], and, thoroughly at a loss what to do, the justice continued the case for three weeks. It is the all-absorbing topic for miles around, not only among the country folk, but among the guests in the summer hotel. During the three weeks intervening a diligent effort will be made by physicians, county officers, and a number of prominent people, who have become interested in the case, to arrive at some intelligent conclusion."

Source:
"Wished to Kill Her," 1888. *Atlanta Constitution* (July 25). Reprinted from the *New York Sun*.

1889: Lake's Sea Lion

A DISPATCH FROM Menasha, Wis., says during the past two years people have at various times claimed to have seen a sea serpent in lake Winnebago. Last Saturday afternoon [July 20] while two boys were spearing frogs near the lake they saw the monster in a shallow pool. Assistance was summoned and a sea lion eleven feet long was captured. It escaped from a circus four years ago."

Source:
Untitled, 1889. *Statesville* [North Carolina] *Landmark* (July 25).

1890 and Before: "Ten Feet High and Six or Seven Inches in Diameter"

FOR SEVERAL YEARS THE existence of a ghost has been reported here [Cochrane]. Last Friday a party was organized to investigate. Arriving at the ghost's walk, we saw an object about ten feet high and six or seven inches in diameter. It slowly advanced to within a rod of the most daring. Some of the party fled, but those who remained saw it suddenly disappear."

Source:
"A Real Spook," 1890, *Bismarck* [North Dakota] *Daily Tribune* (January 12).

1895: Nocturnal Enigma

MONDAY EVENING [December 16] a large number of Green Bay citizens observed, some of them with not a little awe, a peculiar phenomenon high in the eastern sky. It was seen first at 7 o'clock in the evening and was visible for about ten minutes, when it disappeared. Those who saw it unite in describing its appearance as a ball of fire. It was perfectly round, about ten feet in diameter, and of a dark red color. The entire heavens elsewhere were dark."

Source:
Untitled, 1895. *Stevens Point* [Wisconsin] *Daily Journal* (December 17).

1896: Kite or Airship?

A HUGE KITE HOVERING over Milwaukee on Sunday [December 6] was seen by hundreds of people and for a time many insisted it was an airship, even asserting they could see a man moving in the thing."

Source:

> Untitled, 1896. *Oshkosh* [Wisconsin] *Daily Northwestern* (December 7).

1897: "Terrific Aerial Monster"

TWO OSHKOSH PEOPLE affirm that they have seen the Kansas airship. The thing, be it an airship or terrific aerial monster, as some would have it, answered all the specifications and descriptions of the monster that has disturbed the citizens of Kansas, Nebraska and Iowa for the last six weeks. At least John C. Thompson and C. D. Cleveland, Jr., say it did. It carried a red light to port and a green light to starboard, which is eminently proper in aerial circles, and the only course of action which a respectable and well regulated airship would think of pursuing. It wobbled and wiggled . . . and seemed to oscillate as if breasting the bellowing swells of the blue cerulean. Likewise it took the form of a cross and flamed at rare intervals.

"This course of action is to be particularly noted as being entirely original. . . . Both gentlemen solemnly asseverated that they do not drink nor smoke the 'black smoke.' Their story is that while walking on Washington street toward Main at 8:30 p.m. last evening, they sighted an object moving swiftly across the western heavens; apparently it was two or three miles outside of the city limits and a mile from the surface of the ground. According to Mr. Thompson the outlines of the ship could be plainly seen, the forward portion being cigar-shaped and the rear part square or box-shaped. The center portion of the ship was sunk below the other two portions and probably the well or the pit in which the occupants were seated.

"When last seen the boat was heading in the direction of Lake Poygun. . . . Messrs. Donelly and Harper, two well-known Ashland gentlemen who were in the city last evening, also claimed to have seen the ship."

Source:

> "Seen in Oshkosh," 1897. *Oshkosh* [Wisconsin] *Daily Northwestern* (April 10).

1897: Airship over Milwaukee

PROF. [G. W.] HOUGH and his theory that the supposed airship is the star 'Alpha Orionis' don't stand as high in the minds of thousands of citizens of this city to-night [April 12]. They have seen the wonder and are convinced that it is a machine which navigates the air, and not a star which has been wandering around for 10,000,000 or more years through space without an object or a destination. The strange machine made its first appearance in the wee small hours yesterday morning. It was not seen by many at that time, but the few who did see it are convinced that it is a machine. The lights which appeared on it seemed to move backwards and forwards toward each other, as if signaling to the earth.

"It was first seen on the northern horizon, and about the only persons who were up at the time and were not seeing things double, were a few newspaper men, police officers, and a guard at the house of correction. All of these are willing to make oath they saw an airship come from the north a little before the break of daylight and that it disappeared again, reversing itself and fading from view in the north.

"Last night the stranger made its appearance again in the heavens about 9 o'clock. It came from the northeast from out over the lake [Michigan]. There was no possibility of a mistake this time. Thousands of people saw it, and in a few minutes they were following the machine as it floated over the city. It traveled towards the southwest until it reached a point directly over the City hall, where it stopped for a quarter of an hour. Then the excitement in the down-town districts became intense. It was reported that attempts were being made to anchor the machine.

"A Mr. Mayor, a traveling man, had a field glass ranged on the machine and said he distinctly saw four men in it. Stationkeeper Harry Moore of the Central police station saw it distinctly and was one of the few who did not at the same time lose his head. He says: 'The machine, or whatever it was, anchored or stopped directly over the city hall. The light which I saw was suspended from a large, dark oval-shaped object, the shadow of which could be distinctly seen. In fact it could be seen so plainly that I could discern the wheels working. I did not see any one in it, but any one who claims that the thing I saw floating over the city hall is a star simply don't [*sic*] know what he is talking about. I saw it too distinctly to be fooled. It was, I should judge, about 1,000 feet above the city hall.'

"After hovering for about fifteen minutes, it went back and disappeared in the northeast."

Source:

"Airship Is Plainly Seen," 1897. *Burlington* [Iowa] *Hawk-Eye* (April 13).

Wyoming

1888: Ghost Rider of the Rails

CONDUCTOR [W. H.] SMITH'S train, a freight [on the Denver Pacific Railroad], left Cheyenne in the evening about two weeks ago [early April]. The sky was clear and the moon was shining. About fifteen miles this side of Cheyenne, at the foot of the big hill, is the point known as Big Springs. The conductor was in his caboose, seated in the lookout. The train was running somewhat more rapidly than schedule time, and Mr. Smith gave the signal to the engineer for a whistle for brakes. The whistle was given, and just at that moment the conductor saw, about three car-lengths ahead, seated on the brakewheel, a man whom he mistook for one of his brakemen. The figure did not move at the whistle and the conductor climbed down to ascertain the reason of [sic] such apparent neglect of duty.

"As he stepped out of the caboose door he discovered his rear brakeman busily twisting the wheel, and farther up the train, beyond the figure, was the other brakeman also at work. The conductor called the attention of the rear brakeman to the figure, and together they started toward it. When they were within a car and a half's length the man deliberately rose to a standing posture and stepped off the side of the car. There was no movement as if the man had lost his balance and fallen off, but he apparently walked off just as a man would step from a curbstone into the street. The watching trainmen were horrified, because they could realize no other effect from such an apparently suicidal movement than the death of the man dashed against the ground many feet below. There was, however, no such result; nor, in fact, could the man be seen at all. He had vanished.

"Several trainmen, on Mr. Smith['s] relating his experience, stated that they had heard something like the same experience occurring to others at the same time on the road. One brakeman, named Malone, was found who said that not more than a week previous to Mr. Smith's experience he had turned around as he was going forward on his train, setting the brakes, and beheld a man sitting on a wheel three cars behind him. He was not startled until he went ahead three cars more and, looking back, saw the fellow sitting on the brakewheel at the place where he had stood himself when he first saw him. The spectre or man, or whatever it was, kept just three car-lengths behind him all the way until he reached the engine. He called the attention of the fireman to the apparition, and they turned back to see what it was. The figure, as had been the case in Mr. Smith's instance, rose from the wheel and deliberately walked off the end of the car. No crushed or mangled remains of an unfortunate could be found nor could any one be seen. The man had vanished in thin air.

"Neither of the railroad men who tell this peculiar story is a bit superstitious, and both were in their soberest senses when this peculiar apparition appeared to them. No accident has occurred at Big Springs within four years. About that time a freight-train went to pieces on account of a broken rail and a brakeman was killed. There is something funny about the affair. If it isn't one, what is it?"

Source:
"A Spectre Riding on the Brakes," 1888. *Indiana [Pennsylvania] Democrat* (April 26).

1890 and After: Whispers

A RARE BUT STRANGE PHENOMENON periodically occurs at Shoshone and Yellowstone, two nearby lakes in Yellowstone Park. The effect is sometimes called the "Yellowstone whisper." Presumably, it has a natural explanation, perhaps related to seismic activity, but its curious qualities make it difficult to classify.

One experience of it took place in 1890 and was reported by Professor S. A. Forbes. *Science* quoted him as relating that while on Shoshone one bright, quiet morning, he and his companions heard what reminded him of the "vibrating clang of a harp lightly and rapidly touched high up above the tree tops, or the sound of many telegraph wires swinging regularly and rapidly in the wind, or, more rarely, of faintly heard voices answering each other overhead" (quoted in Corliss 1983).

The sound had begun in the far distance, then seemed to approach quickly, growing louder all the while, then disappeared at the opposite horizon. It manifested more than once, though not exactly the same as it had the first time. "We heard it repeatedly and very distinctly here and at Yellowstone Lake, mostly frequently at the latter place," Forbes said. Usually it happened on still mornings just after sunrise, though Forbes said, "I heard it clearly, though faintly, once at noon when a stiff breeze was blowing."

In a 1915 book on Yellowstone, H. M. Chittenden wrote that the phenomenon was noted by the earliest white explorers of the region. The sounds "have a general motion through the air, the general direction . . . being from north to south. They resemble the ringing of telegraph wires or the humming of a swarm of bees"

(ibid.). Nine years later, Robert E. Sellers was in a small boat on a still, early morning when he heard a low roar from the west. It grew in pitch, "then gradually faded away as the pitch lowered again, while the sound seemed to soar rapidly to the southward as it faded into silence. Then from another direction a similar sound was heard, and again from still another direction, the whole phenomenon lasting only half a minute" (ibid.).

In two succeeding days in the summer of 1929, tourists and others heard the sounds. They compared them variously to musical notes, moans, or the beating of numerous birds' wings. Among those who attested to the occurrences were Chief Ranger George Baggley and Assistant Chief Ranger Edward E. Ogston.

Source:

Corliss, William R., ed., 1983. *Earthquakes, Tides, Unidentified Sounds, and Related Phenomena: A Catalog of Geophysical Anomalies.* Glen Arm, MD: Sourcebook Project, 176–177.

1896: Raining Clay

A SINGULAR PHENOMENON was a shower in this city [Lander]. First red and then yellow clay fell, followed a little later by almost black mud. People who were out in the storm looked as though they had been churned in a tub full of mud."

Source:

"Rain All Kinds of Mud," 1896. *Bismarck* [North Dakota] *Daily Tribune* (April 14).

Index

Snakes, flying, 46–47, 109, 178–179, 180–181, 242, 280, 295, 308, 321, 332
  pink eyes and yellow wings, 179
  train-attacking, 17–18
Snakes, giant, 1, 16, 17, 72, 94, 128–129, 134, 173, 252, 255, 259, 266
  alleged capture, 103–104
  circus escapee, 287
  heart-shaped scales, 284
  horned monster, 112
  human bones within snake skeleton, 257–258
  hydra-headed monster, 143
  killed by boy, 199
  killed on a reaper, 81–83
  large coachwhip, 45
  mountain monster, 320
  "multidimensional reptile" apparitions, 282–283
  odiferous rattler, 115–116
  patent medicine salesman versus, 179–180
  rattler, 44, 87
  See also Lake and inland water serpents or monsters; Sea serpents or monsters
Snow, 39
  black and dirty, 159–160
  black bugs on, 325
  caterpillars on, 273
  "cold fire," 31
  fishes in, 304
  greasy flakes, 241
  multiple colors, 72
  strange precipitation, 88–89
  worm-infested, 223–224, 229
Snyderstown, Pennsylvania, 277
Soap-bubble sky, 227
Solomon, Kansas, 113
Sonora, California, 16, 312–313
Sounds and noises:
  drumming ghost, 19
  ghostly fifing, 94
  musical thunder, 37
  piano-playing haunter, 264
  screaming sky, 29–30
  "silent battle," 326
  underground rumblings, 33–34
  Yellowstone Park acoustic phenomenon, 358
  See also Ghosts and apparitions
South Butler, New York, 231–232
South Canaan, Connecticut, 34–35

South Carolina, 293–295
South Dakota, 297
South Iredell, North Carolina, 242
South Orange, New Jersey, 204
Sparta, Ohio, 255
Sparta, Tennessee, 299
Spiritualists and UFOs, 96, 338–339
Spokane, Washington, 338
Spontaneous combustion. See Fire
Springfield, Illinois, 71
Springfield, Michigan, 152
Springfield, Ohio, 266
Springheeled Jack, 119, 203
Squid, fallen in hail storm, 140
Starfish-like flying apparition, 184–185
"Star jelly," 139, 142, 216
Steamships in the sky, 246
Stephenville, Texas, 124
Sterling, Colorado, 32
Stone-throwing ghosts, 56–57, 80, 252
Stratford, Indiana, 97–98
Sulfur showers, 304
Sulfur snow, 276
Summerset, Iowa, 103
Swamp creature, 50
Swamp fire, 152
Sword from the sky, 224–225
Sword shape in the sky, 317
Symmes, John Cleves, 247–250
Symzonia, 249

Tallahassee, Florida, 43–44
Tavern, haunted, 250
Taylor, Georgia, 51
Taylorville, Illinois, 64–65
Telephone-answering ghost, 287
Tennessee, 299–306
Texas, 307–321
Tiffin, Ohio, 254, 257, 262
Tiskaville, Iowa, 105
Titusville, Florida, 46
Tornadoes and whirlwinds. See Whirlwinds and cyclones
Train, attacked by flying serpent, 17–18
Trains, phantom, 50, 69, 75–76, 110, 141, 142–143, 255, 279, 325–326, 334–335
  See also Railroad-associated ghosts or apparitions
Tree, largest Sequoia, 20–21
Trenton, New Jersey, 204, 208
Troy, New York, 216, 224
Turlock, California, 19–20
Turtlelike sea monster, 129–130

Twain, Mark, 191

UFOs. See Flying objects
Unidentified flying objects. See Flying objects
Upper Sandusky, Ohio, 256
Utah, 317–321

Van Wert, Ohio, 250
Vanishings and abductions:
  David Lang of Tennessee, 303–304
  fiery ghost carries off family, 104–105
  Oliver Lerch, 86, 304
Venice, California, 30
Vermont, 323–324
Vincennes, Indiana, 90, 96
Virgin Mary, visions of, 224–225, 276–277
Virginia, 325–332
Virginia City, Nevada, 191
Volcanolike swamp phenomenon, 43–44
Voliva, Wilbur Glenn, 73–75

Wabash, Indiana, 85
Wabash River monster, 91
Wabash River mysterious light, 99
Wahhoos, 184, 192–194
Wakulla volcano, 43–44
Waldoboro Maine, 127–128
Waltham, Massachusetts, 145–146
Warren, Pennsylvania, 286
Warrenton, Virginia, 327
Warsaw, Indiana, 95
Washington, 333–339
Watertown, Massachusetts, 143
Watseka, Illinois, 75–76
Waukegan, Illinois, 232
Weather and atmospheric phenomena:
  beautiful effects of light and electricity, 327–328
  black bugs on snow, 325
  black cloud and pillars of fire, 297
  black cloud and precipitation, 241
  black rainbow, 141
  black snow, 159–160
  boiling hot rain, 24
  cloud of smoke and flame, 349
  colored snow, 72
  continuous mist, 260
  falling meat, 117–118
  fiery whirlwind, 50
  finger-pointing cloud, 261

fireball parts lake waters,
36–37
localized continuous rain,
1–2, 55, 59, 110–111
luminous electric snow,
31–32
mountain lights, 226–227
mud shower, 111
muddy snow, 88–89
musical thunder, 37
pebble shower, 109
phosphoric magnetizing
cloud, 41
queer little cloudburst, 310
rain from clear sky, 334
rainbow cloud, 109
raining chalk, 151
raining clay, 358
raining meat, 15
reflected fire, 143
salt shower, 61
shower of flesh and blood,
237, 239–240, 299–300
slimy rain, 250, 280
small fireball, 20
snow and brilliant
illumination, 254, 261
storm of shot, 229
sudden fog, 229
sulfur snow, 276
sulfurous luminous incendiary
cloud, 273
thunder on cloudless day,
217–218

whirlwind a pillar of fire, 301
yellow precipitation, 134, 304
See also Animal precipitation;
Ball-lightning and related
phenomena; Falling
objects; Lightning
Weather phenomena, raining
animals. *See* Animal
precipitation
Wells:
boiling water in, 53, 240
snake in, 51
West Atchison, Kansas, 110
West Huron, Ohio, 263
West Philadelphia, New Jersey,
209
West Virginia, 341–345
Westfield, Massachusetts, 143
Westfield, New York, 216
Whirlwinds and cyclones, 52,
132
fiery with sulfur smell, 50
grasshopper cloud, 194
pillar of fire, 301
White whale, 249–250
Wild child, 80
Wild hairy anthropoids. *See*
Hairy bipeds
Williams Creek, Ohio, 252
Williamsport, Pennsylvania, 274
Willoughby Lake, New
Hampshire, 199
Wilmington, Massachusetts, 146
Winchester, Virginia, 332

Winemas, Indiana, 122
Winthrop, Maine, 130
Wisconsin, 159, 347–355
Witchcraft suspect, 352–353
Wolf Creek, Iowa, 106
Women in white, 85–86, 95,
333
airborne, 52–53
country club haunter,
209–210
giant, 54–55
haunted Maine town, 130
hundreds of witnesses,
252–253
on railway, 55, 254
Woodland, California, 24
Wooly mammoth, 7–8
Worcester, Massachusetts,
143–144
Worlds within the world theory,
247–250
Worms on snow, 39, 223–224,
229
Wyoming, 357–358

Xeni, Ohio, 251

Yahoo, 235–236
Yazoo City, 173
Yellow rain, 134, 304
Yellowstone Park acoustic
phenomenon, 358

Zion, 73–75

About the Author

JEROME CLARK IS THE author of the multivolume *UFO Encyclopedia* (Omnigraphics, 1990–1998) and other books, including *Extraordinary Encounters* (ABC-CLIO, 2000). Besides having a lifelong interest in anomalous phenomena, he is a serious devotee of folk and roots music as well as a songwriter whose works have been recorded by Emmylou Harris, Mary Chapin Carpenter, Tom T. Hall, and others. His other enthusiasms include history, politics, folklore, literary fiction, and good beer. He lives in southwestern Minnesota with his wife, editor and writer Helene Henderson.